D0205011

Plant Diseases:

Their Biology
and
Social Impact

Gail L. Schumann

APS PRESS
The American Phytopathological Society
St. Paul, Minnesota, USA

Library of Congress Catalog Card Number: 91-70960
International Standard Book Number: 0-89054-116-7

The poem on p. 270 is from *The Poetry of Robert Frost*, edited by Edward Connery Lathem. Copyright 1936 by Robert Frost, ©1964 by Lesley Frost Ballantine, ©1969 by Holt, Rinehart and Winston. Reprinted by permission of Henry Holt and Company, Inc. Reprinted for distribution outside of the United States and Canada by permission of the Estate of Robert Frost and the publisher, Jonathan Cape.

Printed in the United States of America

The American Phytopathological Society
3340 Pilot Knob Road
St. Paul, Minnesota 55121, USA

To Michael and Esther

Preface

The history of plant diseases is an important component of the history of human culture and civilization. In many cases, plant diseases have changed the course of human history, and yet few people know very much about plants or their pathogens. Most of the people who will read this book have lived in a time of abundant food—indeed, food surpluses and low prices for farmers—but they know that for much of the world, sufficient food remains a daily challenge and that famines still exist. It is important to understand the vulnerability of the world's food supply, the narrowing of the world's gene pool, and the continuing battle to stay one step ahead of the new strains and races of pathogens that may appear in any season to destroy our crops and forests.

Plant pathology is an important biological science. A plant disease can be understood only by studying the biology of both the vulnerable plant and the pathogen that threatens it. Thus, this science introduces a student to botany as well as to the various sciences associated with the pathogens: virology, bacteriology, mycology, and nematology. Beginning as the study of diseased crops, plant pathology now includes studies of epidemics and the ecology of agriculture. Genetic studies of plants and their parasites have led to disease-resistant cultivars. Cell biology and plant physiology have resulted in new industries to rapidly propagate pathogen-free crops from microscopic pieces of plant meristems and even single cells. At the molecular level, plant pathologists are now able to manipulate single genes using genetic engineering to create new kinds of plants.

The goal of this textbook is to explore plant pathology as a biological science with important ties to human welfare. All citizens should be concerned about legislation governing environmental protection, pesticide use, release of genetically engineered organisms, land preservation for agricultural use, and genetic reserves. The following chapters provide an introductory course in the biological principles necessary to understand plant health in a world dominated by the human species.

The book was written to provide a text for a plant pathology course taught at the University of Massachusetts as part of the general education curriculum. Although rarely considered in discussions of core curricula for college students, a basic understanding of agriculture and food production is a valuable component of higher education. Many agricultural scientists are concerned about the inability of the general public, as voting citizens, to make rational decisions about the controversial issues of agriculture. A final purpose of the course is to introduce a broader cross section of students to the study of plant diseases and encourage them to consider further study.

Students with no previous study of biology will be able to use this text. Its primary goal is to introduce biological principles. However, there might also be students who have had some college-level introductory biology but for whom the traditional course in general plant pathology is not appropriate. For such students, more detailed discussions are provided on certain topics.

I am indebted to the many colleagues who have encouraged me to complete this book and contributed to it with invaluable stories, ideas, and photographs that make a general textbook attractive to students. Special gratitude is extended to H. D. Thurston, who was my graduate advisor and who expanded my view of plant pathology to a global level. He also contributed many of the photographs. All contributors of figures are acknowledged in a section at the end of the text. Artist Nancy Haver produced the line drawings with good humor, imagination, and interest, converting my scribbles into useful illustrations. She also redrew or adapted a number of figures. Several people read the entire manuscript with helpful suggestions: M. S. Mount, G. Safir, D. N. Schumann, M. C. Shurtleff, M. S. Switzenbaum, and my APS editor, M. C. Heath. Special thanks is offered to M. C. Shurtleff for his detailed editing and for a number of the glossary definitions. Several people generously reviewed specific chapters: S. V. Beer, C. L. Schoulties, and T. S. Schubert (parts of Chapter 4), J. J. Jenkins (Chapter 7), B. M. Zuckerman (Chapter 8), R. W. Stack (Chapter 11), F. E. Gildow (Chapter 12), and G. C. Smith (Chapter 13). L. Karis and F. SanMartin made it possible to include Figure 9-19. As always, any errors remain my responsibility. I would like to particularly acknowledge the important role of my 1988 Genesis Award from the American Phytopathological Society Foundation in the completion of this project. With gratitude, I also thank my husband Michael, my daughter Esther, and my mother, who support me through all my projects, little and big, and make the whole process worthwhile.

Gail L. Schumann

Contents

Plant Diseases:

Their Biology
and
Social Impact

The Irish Potato Famine and the Birth of Plant Pathology

In the early summer of 1845, the days were sunny and the potato crop was growing well. There was no warning of the disaster that would strike, causing misery, suffering, and death. Then, the weather turned overcast and rainy for weeks, and the potato plants rotted as the Irish peasants watched helplessly. The horrors of the Irish potato famine are still remembered—one more wedge between the English and Irish, contributing to political conflicts that continue to this day.

The story of this disaster is an important one. It introduces many of the ideas to be presented in this text, including the political aspects of food supply, the risks of genetic uniformity and dependence on only a few food crops, and the distribution of new crops from their origins throughout the world. But these same concepts could be derived from nearly any agricultural failure. The Irish potato famine is of specific importance because the debate surrounding the study of the stinking mass of rotted potatoes gave birth to the science of plant pathology. Infectious microorganisms were finally to be accepted as causes of disease rather than its result, predating even Pasteur's work with bacteria. The **theory of spontaneous generation** of microorganisms from decaying tissues was soon to be replaced with the **germ theory**.

This idea is so commonly accepted today that we assume that a disease is caused by "germs" unless proven otherwise. Today's mysterious problem diseases are those that do not appear to be caused by bacteria, fungi, viruses, and other less familiar **pathogens**. In the mid-1800s, however, most people viewed disease, whether in plants, animals, or people, as the result of bad weather, or punishment from God, or perhaps just bad luck.

The Arrival of the Potato in Europe

In the 19th century, the peasant farmers of Ireland were so dependent on the potato that its common name, the Irish potato, still reflects its connection with these poor people. Its origin was not the Emerald Isle, however, but the highlands of South America, specifically the **Lake Titicaca** region between Peru and Bolivia. The Spanish Conquistadors discovered this plant in South America while searching for gold in the Andes in the 16th century. The starchy tubers were, and continue to be, an important food crop in South America, particularly at higher elevations where corn does not grow well. Religious and agricultural records of the ancient Incas contain many references to potatoes, and there is archeological evidence of potatoes as a South American food crop dating to 400 B.C.

Some potatoes were brought back to Europe aboard ships, although it is likely that many failed to survive the long voyage because of poor storage conditions. The early history of the crop is confused because of the similarity of the word *la batata*, which refers to the sweet potato, an unrelated plant, and *la papa*, the Spanish name for what we now call the "white" or "Irish" potato. The first potatoes to reach Europe probably arrived in Spain about 1570. Each language now has its own common name for this valuable plant. The French use *pommes de terre* and the Dutch call them *aardappeln*, both of which translate to "apples of the earth," certainly a poetic description.

Acceptance of the new food crop was reluctant. Many religious advisors discouraged production of a food that grew in soil, a food more appropriate for animals than humans. Besides, there was no mention of the potato in the Bible, which suggested that perhaps it was evil or might instill sinful desires in its eaters. Slowly, however, the crop spread throughout Europe

Fig. 1-1. Map of South America. The potato originated in the region around Lake Titicaca between Peru and Bolivia.

Fig. 1-2. An ancient pottery vessel from South America depicts the pattern of a potato tuber.

Fig. 1-3. Indian women sell many different potato cultivars in a Colombian market.

as hungry people discovered the many advantages of a highly nutritious crop that could be grown beneath the ground, safe from the trampling feet of invading armies.

The exact time of arrival of the potato in Ireland is unclear, but it was a well-established food crop by 1800. Harvests were low in some years, but they generally produced large amounts of the nutritious food. The Irish population grew from about 4.5 million in 1800 to more than 8 million by 1845. When the harvest was good, Irish peasants often ate 8–14 pounds (3–6 kilograms) a day, with little else, unless some milk was available from a cow. Although decidedly boring, this daily intake provided substantial nutrition, including protein, carbohydrates, and many vitamins and minerals, particularly vitamin C.

The grain crops, which grew poorly in Ireland anyway, were needed to pay the rent to the landowners, most of whom lived in England. Most peasants lived in one-room, windowless huts with little furniture or other possessions, but potatoes thrived in the cool, moist climate of Ireland, which was similar to their place of origin in the highlands of South America. A family could grow enough potatoes to feed themselves on half the land required to produce the same amount of calories in grains.

Fig. 1-4. Woman sowing potatoes in the Montaro Valley of the Andes.

Fig. 1-5. Logogram of the International Potato Center (Centro International de la Papa [CIP]) in Lima, Peru. The diagram is from fourth century Nazca culture and depicts harvested potato plants, one (in the figure's right hand) a healthy plant and the other diseased. The figure is one of many found in art forms and indicates the importance and high regard that ancient Peruvian cultures gave to the potato.

Fig. 1-6. Harvesting potatoes in Maine in the early 1930s.

The Potato Plant

A field of growing potato plants is a beautiful sight, especially in midsummer when the plants begin to bloom. The **leaves** are usually composed of a number of dark green **leaflets**. Sometimes the lowest leaves are composed of a single blade that connects to the main stem by a leaf stem called the **petiole**. In the angle between the stem and the petiole is a tiny **bud** that can grow into a side branch. There is also an **apical bud** at the highest growing point of the plant. The hormones produced at the apex inhibit the growth of the lateral buds, a phenomenon known as **apical dominance**. If the top of a plant is pinched off, apical dominance is eliminated, and the **lateral buds** begin to grow. This is a common practice to produce bushier house plants or other ornamental plants.

Most of the leaves of a potato plant are divided into leaflike sections or leaflets that all connect to the petiole. Such leaves are **compound** and, in the potato plant, consist of an odd number of leaflets. When leaves are not divided into leaflets but whole, like those of a maple tree, they are called **simple**. Buds are present where the petiole of a leaf meets the stem, but not where leaflets connect to the petiole. There should never be any confusion about whether a leaf is simple or compound once the location of the bud is determined.

The presence of flowers on potato plants may not be familiar to those who have not seen potato fields. They are similar to those of the tomato, a close relative, but they vary in color depending on the cultivar and can range from white to deep purple. Potato flowers consist of a single **pistil** surrounded by five pollen-producing **stamens** within a set of five fused **petals**. After blooming, fertile plants produce green **berries** similar to a small tomato and filled with numerous tiny seeds.

The potato is in the botanical family **Solanaceae**, which includes the tomato as well as eggplant, green and red pepper, tobacco, petunia, and some poisonous members such as deadly nightshade. The similarities in

Table 1-1. Top Food Crops of Developing Market Economies[a]

Energy Production		Protein Production	
Crop	Megajoules of Edible Energy	Crop	Kilograms Produced per Hectare per Day
Potatoes	216	Cabbages	2.0
Yams	182	Dry broad beans	1.6
Carrots	162	Potatoes	1.4
Maize	159	Dry peas	1.4
Cabbages	156	Eggplants	1.4
Sweet potatoes	152	Wheat	1.3
Rice	151	Lentils	1.3
Wheat	135	Tomatoes	1.2
Cassava	121	Chickpeas	1.1
Eggplants	120	Carrots	1.0

[a] Source: *Potato Atlas*. 1985. International Potato Center (CIP), Lima, Peru. Used by permission of the CIP.

flower structure make the relation between these plants clear even though they are quite different in other characteristics. The name is derived from the Latin *solamen* which means "comforting," reflecting the sedative effects of some of the **alkaloids** produced by some members of the family. Deadly members of this family slowed the acceptance of the edible members as food plants, but eventually they were discovered to be safe. Related alkaloids are present in the leaves of potato plants, causing digestive upsets to those who eat them. If potato tubers are left in the light, they too will develop green color and alkaloids and should not be eaten. The Solanaceae family is worldwide in distribution, but some members, such as tomato, tobacco, and potato, are native to South America and were unknown to Europeans before they began their explorations of the New World.

The familiar starchy "potato" is actually a **tuber** as reflected in the plant's Latin name, *Solanum tuberosum*. All organisms are given a similar **Latin binomial** (two-word name) based on a system created in the 1700s by the famous Swedish naturalist, **Carolus Linnaeus**. The first word is the **genus**, a name shared by closely related organisms. Other plants in the genus *Solanum* are closely related to the potato but sufficiently different to be considered different species. The second word is the **specific epithet**, a descriptive word that separates this species from all others in that genus. Because the words are Latin, they should be italicized or underlined. Once the genus has been written out fully, it can then be abbreviated, when repeated, by the uppercase first letter followed by a period and the specific epithet, as in *S. tuberosum*. A **species** was once defined as organisms that would produce fertile offspring if mated, but our expanding knowledge of the variety of life has made this definition difficult to apply in the case of many microorganisms. In such cases, species names change as our understanding of the relationship between organisms improves.

At about the same time as potato plants begin to flower, small swellings start to develop at the ends of underground stems called **stolons**. These swellings are the new tubers, which store the excess food the plant produces during photosynthesis. The starchy tubers may be harvested and eaten at any time. The small, freshly harvested "new potatoes" are tender and delicious, but, to maximize yield, most potato growers wait until later in the season to harvest the potatoes. As the mature vines die above ground, the tubers develop an outer cork layer for survival in the soil during the winter. This layer also protects them from **desiccation** (drying out) and wounding when they are harvested and put into storage. Stored potatoes must be kept cool to prevent rotting by bacteria and fungi present on the tubers. They also need to be kept in the dark to prevent "greening" of tuber tissues.

Because tubers grow underground, they might appear to be roots, as carrots are. Closer examination reveals that the **eyes** are really buds in the **axils** of tiny scalelike leaves. Roots do not produce buds and leaves, so tubers are actually underground stems adapted to storage of nutrients. After several months in storage, the buds begin to sprout. Apical dominance, as described earlier, can be observed on tubers as well as on aboveground stems. Because the buds at the apex of a tuber produce hormones that inhibit growth of lower buds, the sprouts are most developed at the apex of the tuber.

Most commercial potato growers and home gardeners do not plant the tiny seeds from the green berries produced after flowering. Potatoes are grown by **vegetative propagation**; that is, small tubers or pieces of tubers are planted. To maximize their planting stock, farmers may cut the tubers into several pieces. Each piece can grow into a new plant as long as an "eye" is present. Tuber pieces that do not have an eye are called "blind" and will not grow into a new plant. The cutting of tubers also releases the lateral buds from apical dominance, so that each eye may produce a plant.

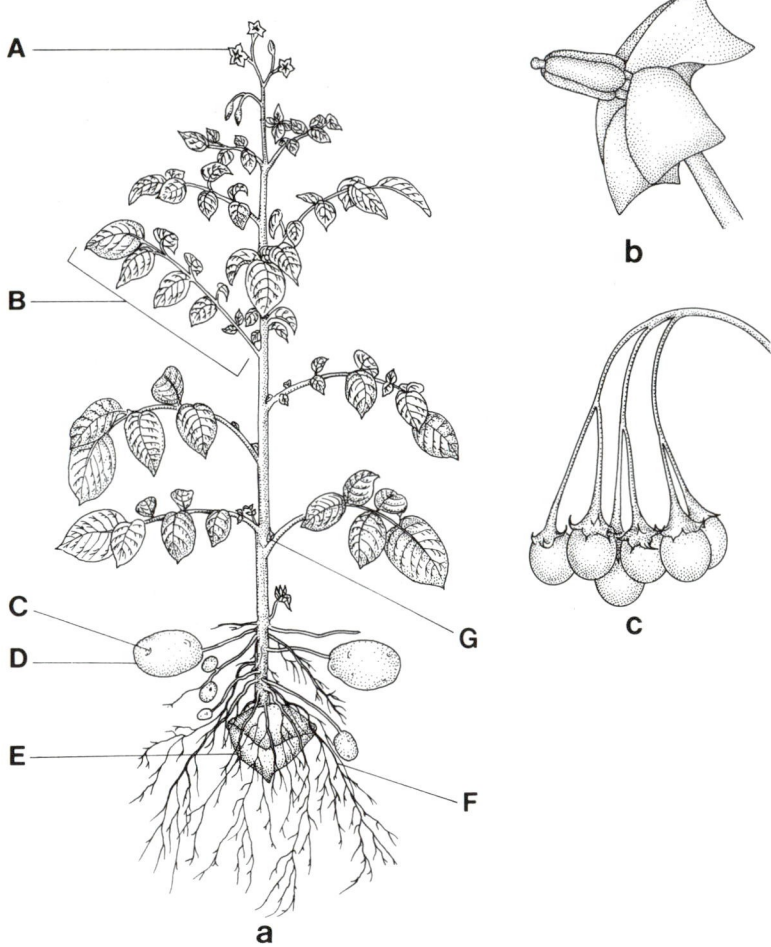

Fig. 1-7. The potato, *Solanum tuberosum*. **a,** Diagramatic illustration of an entire plant: A, flower; B, compound leaf; C, "eye"; D, tuber; E, tuber piece used for planting; F, roots. **b,** Flower. **c,** Fruits (berries).

Vegetative propagation has several advantages. The tuber pieces, known as "seed" to farmers, contain substantial food reserves, so that a vigorous, green shoot pushes up through the soil more quickly than does the shoot from a tiny seed. In addition, each tuber piece grows into a plant that is genetically identical to the parent plant in color, taste, maturity, and other important characteristics. True botanical seed, from the fruit of a plant, is the product of sexual reproduction, which results in genetic variation. Each seed produces a plant that is slightly different from all of its siblings. **Genetic uniformity** is a great advantage when uniformity has an economic benefit, as in flowers, fruits, and ornamental plants, so vegetative propagation has become a common practice in modern agriculture.

There are, however, some disadvantages to vegetative reproduction. The advantages of uniform characteristics may be outweighed by the disadvantage of uniform susceptibility to pests and pathogens. If one plant in a field can be destroyed by disease, then so can all of its neighbors. Genetic uniformity increases the risk of loss. A second important risk lies in the large pieces of plant tissue that are planted during vegetative reproduction. Rather than a tiny seed, cuttings, roots, tubers, bulbs, and other relatively large plant pieces may be planted. These large plant pieces often carry pathogens, especially viruses and other systemic parasites, into the planting area at the very beginning of the season. The yield and quality of the crop may be greatly reduced due to their presence.

Of course, many kinds of parasites were not discovered or well understood until relatively recently, but farmers of the past observed the ravages of the diseases they caused. Farmers suspected that reduced yields were related to the use of vegetative reproduction which, despite its convenience, was considered "unnatural." Periodically, they allowed the plants to reproduce sexually, "the natural way," and harvested the seeds to begin new selections for acceptable plants. They believed that sexual reproduction restored the vigor of their weakened plants, when, in fact, the seeds had simply escaped infection by the pathogens present in the parent plant.

Only small collections of potatoes were brought from the New World in the early days of European potato production, and, from these tubers, further selections were made in an attempt to find potato **cultivars** suitable for consumption. As a result, the genetic variation among the potato cultivars was very limited. Crop losses occurred every year from various causes, and occasional food shortages were not unusual. When crops were good, tubers were plentiful in the months after the harvest, but as the winter months passed, supplies diminished and the tubers began sprouting, ready to plant for the next crop. The summer months were a hungry time, usually requiring the peasants to spend what few coins they had for grain to feed themselves until the next potato harvest.

The Components of the Epidemic

With this background, the stage was set for the impending disaster. What were the important components that led to the tragedy? The human component should always be considered, and, in this case, a large population of human beings was dependent on a single food crop. The Irish population had grown rapidly following the introduction of the potato, and no significant alternative foods existed when the crop failed. The grains were needed to pay the rent to the English landlords. If the rent was not paid, the people were thrown off their land to certain starvation. Perhaps this dependence on one crop sounds foolish or ignorant today, but the current human population, now past the five billion mark, relies on essentially three species of plants to feed most of its people (wheat, corn, and rice), an issue we will return to in Chapter 14.

In addition to the human component, we should consider the important biological components that combined to cause the blight of the potato crop. All plant pathologists are familiar with the **disease triangle**, a memory aid describing the three factors necessary for disease: a susceptible plant, a pathogen capable of causing disease in the plant, and environmental conditions favorable for disease development. The potato crop was derived from a small supply of tubers that survived the lengthy trip across the sea to Europe. Thus, every plant in every field was nearly genetically identical, a desirable situation for the agricultural characteristics but a potentially dangerous situation for disease.

In the early weeks of summer in 1845, records show hot and dry weather overall. The weather then changed, and overcast weather continued for six weeks, with temperatures 1.5–7 degrees below the average for the previous 19 years. Records throughout Europe indicate a particularly cool and rainy period. In just a few weeks, the vigorous green potato vines became a

Fig. 1-8. "An Eviction." Tenants are thrown out of their homes in Ireland for not paying rent. Printed in the *Illustrated London News*, December 16, 1848.

blighted mass of decaying vegetation. When the tubers were dug from the ground, some were rotted, but many appeared to be sound. Later, however, these potatoes, too, rotted away in the storage bins. Throughout Europe, the potato crops failed, but the disaster was worst in Ireland because of the nearly complete dependence of Irish peasants on the potato for their food. Note that even though the rainy weather affected all crop yields to some degree, the blight epidemic was specific to the potato. Something new and frightening was occurring in the potato fields, and the answer to the mystery lay in the third component necessary for disease—a virulent pathogen capable of infecting the potato crop.

It is interesting to consider, at this point, the state of science in the mid-1800s in Europe. Only wealthy people generally had the leisure time and education to look about and consider how the world operated, so most scientists were men trained in medicine or religion. The microscope had been invented nearly 200 years previously by **Anton van Leeuwenhoek** in The Netherlands. **Robert Hooke**, a physicist in England, first described the cells in plant tissues, although the concept that cells are the basic unit of plant and animal tissues was not proposed until 1838, nearly 150 years later. Hooke also first observed fungi through his microscope in 1667.

Fig. 1-9. "After the Eviction." Homeless families had to find shelter in the countryside. Printed in the *Illustrated London News*, December 16, 1848.

Spontaneous generation was the prominent theory to explain the presence of teaming populations of microbes in diseased or dead tissues. When, in 1845, a white fungus was found on the blighted potato vines, it was considered to be the result, rather than the cause, of decay. One common explanation for the rotted crop was that the plants took up too much water in the rainy weather. Many observations of diseased plants had been made over the centuries, however, and many thoughtful people had contributed clues to the solution of the puzzle of plant disease. The potato blight attracted the attention of a number of biologists and brought many of the clues together. Rainy seasons had occurred before without such losses. This new phenomenon required a new explanation.

A pathogen is something capable of causing disease. In the case of the potato blight, the pathogen is a fungus that parasitizes the potato plant. A **parasite** derives its food from another living organism and lives in intimate contact with its **host**. In nature, parasites are associated with nearly all living organisms. We are all familiar with the fleas, ticks, and intestinal worms that annoy most pets. A healthy animal can maintain reasonable populations of such parasites without necessarily being diseased. Similarly, with wild plants and animals, parasites that do not cause noticeable disease should not be called pathogens.

The blight fungus parasitizes potato plants and closely related plants

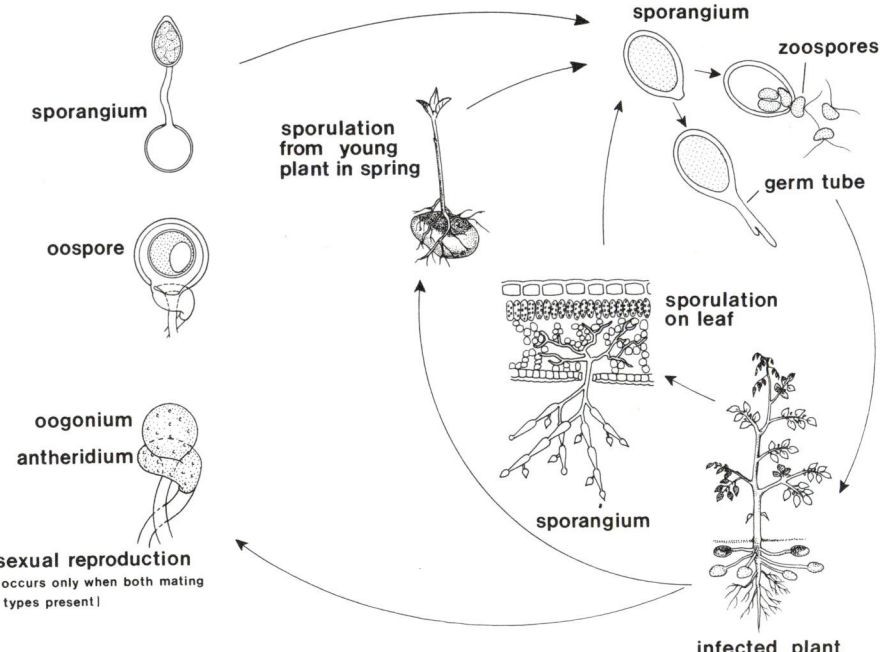

Fig. 1-10. Life cycle of *Phytophthora infestans*, causal agent of late blight of potato and tomato.

in South America. Farmers there grow an enormous number of potato cultivars that vary in their susceptibility to the parasite, so that even though some loss to the parasite occurs, devastating epidemics are avoided. The technology of modern agriculture, and of vegetatively propagated plants in particular, leaves crops genetically uniform and thus uniformly susceptible to a parasite. This situation can result in a severe epidemic in which the parasite not only derives food from its host but increases so rapidly that it becomes a pathogen, causing disease and loss.

The blight fungus arrived aboard a ship from South America just as the potatoes had. It is likely that blighted tubers carrying the fungus had been taken aboard many times previously, but that the tubers were eaten or had rotted before arriving. Perhaps faster crossings allowed some of the tubers carrying the fungus to survive and be planted in Europe. Perhaps, also, the environmental conditions were so favorable for infection by the fungus in 1845 that the parasite was able to establish itself rapidly upon arrival. Certainly, the narrow genetic base of the existing potato plants left them vulnerable to the invading fungus.

The fungus itself is barely visible. It appears as a whitish mildew on the surface of infected potato leaves and stems when it emerges through the **stomata** (air exchange pores) in humid weather. Microscopic treelike **hyphae** grow away from the plant tissue, producing lemon-shaped **sporangia** at their tips. A **sporangium** (plural: sporangia) is a structure that contains **spores**, the reproductive units of a fungus. The sporangia are usually dispersed by air currents to neighboring plants and can travel easily to nearby fields when the air is moist. They dry out and die at high temperatures or after traveling long distances. At cool temperatures such as those that prevailed in 1845, a change occurs in sporangia when they land on a wet potato leaf or stem. After a few hours, the cellular material inside the

Fig. 1-11. Sporulation of *Phytophthora infestans* on potato leaves. Sporangia are produced on hyphae that emerge through stomata in the leaves.

sporangium is converted into about eight wiggling **zoospores**, each with two **flagella** to aid in movement. The zoospores swim out of the sporangium and attach themselves to (encyst on) the leaf surface. A small germination tube infects the plant. From the tiny infection sites, an extensive network of threadlike hyphae penetrates between the cells, absorbing nutrients to feed the growing fungus. Brown lesions of dead plant cells and newly produced sporangia begin to appear as soon as three to five days after infection, increasing the epidemic. The fungus can produce many generations of sporangia in a short time and rapidly colonize all available potato tissue. A whole potato plant can be left a slimy mass in less than three weeks.

Germination and infection always require water on the leaf surface. At higher temperatures (over 68° F or 20° C), the sporangia germinate by producing a single germination tube rather than zoospores. Thus, the number of potential infections from a single sporangium is greatly reduced in warmer environments, although disease is not eliminated.

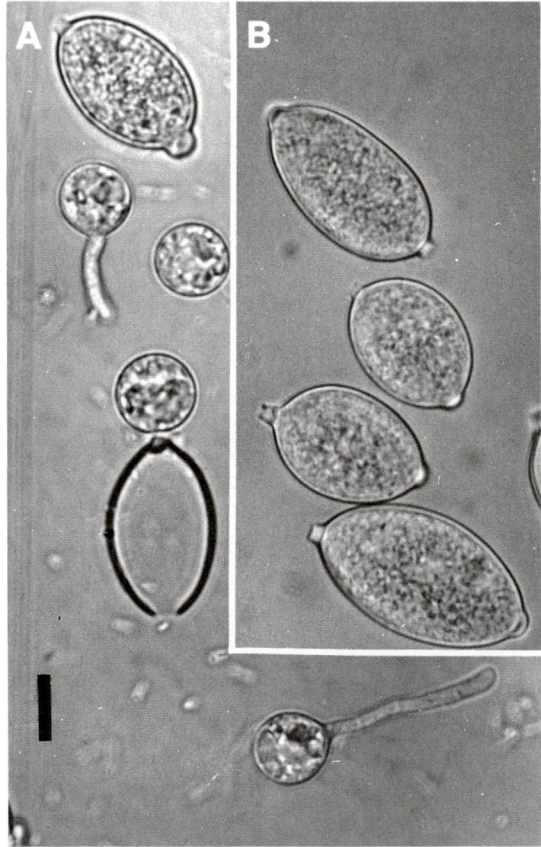

Fig. 1-12. *Phytophthora infestans.* **A,** zoospores produced within the lemon-shaped sporangia (**B**).

The favorable conditions for the blight, extended periods of cool and wet weather, are explained by the biology of the parasitizing fungus. Sporangia best survive dispersal by air when the air is humid, and water is necessary for infection by the fungus. In addition, zoospores are produced at cooler temperatures, which maximizes the number of potential infections possible from a single sporangium. The cool, wet environment of Ireland was particularly favorable for this new disease, now known as **late blight**.

A final explanation is still needed. How do the tubers become infected by a fungus that invades the leaves and stems? It is now known that sporangia from the leaves are washed by water into the soil, where they can infect the tubers. Potato growers have learned to "hill up" the soil around the stems of young potato plants to increase the depth of soil that sporangia have to traverse to attack the tubers. In addition, potato vines are usually killed before harvest with herbicides, mechanical cutting, or by frost, to protect newly dug tubers from exposure to sporangia on the leaves and stems. The fungus is incapable of surviving temperate winters in the soil,

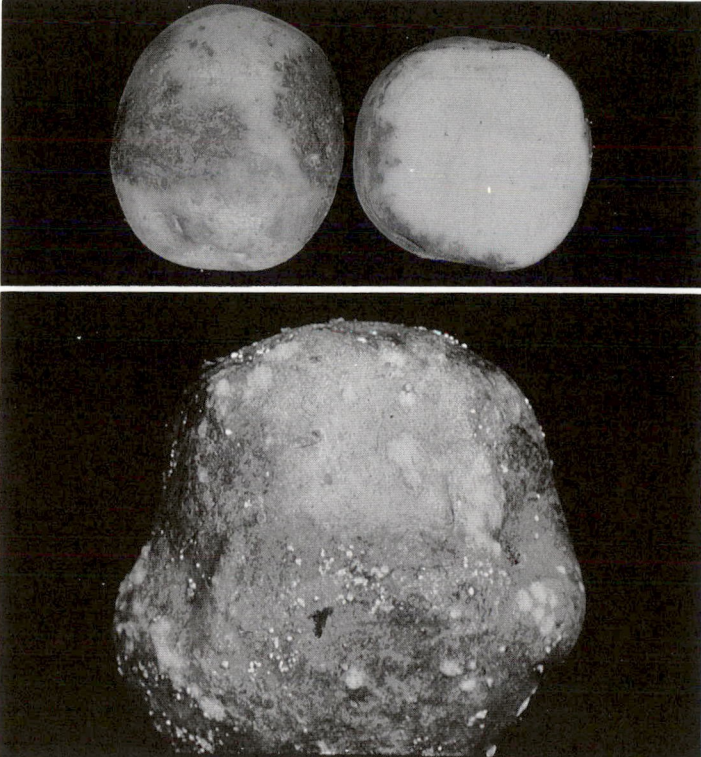

Fig. 1-13. Potato tubers infected by *Phytophthora infestans*. **Top,** tuber in the early stages of infection; **bottom,** tuber showing sporulation of the fungus after storage under moist conditions.

but it spends the winter, well supplied with food, in infected tubers in storage bins and discard piles at field edges. The cool environment of storage bins is very conducive to fungus growth in infected tubers, causing further losses. If conditions are moist in storage, the fungus may even spread to previously healthy tubers. Any infected tubers that survive the winter without being completely destroyed by the fungus will serve as a source of new sporangia on the developing plants. So begins a new blight epidemic.

The Birth of Plant Pathology

The history of the scientific study of the blight is well chronicled in E. C. Large's popular book *The Advance of the Fungi*. Some of the important characters in this drama should be mentioned. The fungus found on the potatoes was first described and named by Dr. C. Montagne, a French physician in Napoleon's army. He shared his observations with Rev. M. J. Berkeley, in England, who recognized that this new fungus was probably connected with the blighting in some way. Berkeley's rival was Dr. John Lindley, a botany professor at University College in London, who did not believe that the fungus was the cause of the blight. Their arguments, published in a periodical called *The Gardener's Chronicles*, reflected some of the most intense philosophical arguments among scientists throughout Europe. Were the rusts, smuts, and other fungi a product of diseased tissue or could they be the cause of the disease itself?

Fig. 1-14. A political cartoon printed in the United States at the time of the Irish potato famine.

While those of better means argued and philosophized, the Irish were starving. Food was in short supply the first winter, leaving fewer seed potatoes to be planted the next season. The summer of 1846 brought new hope as the plants began green and healthy, but the blight came again and destroyed any hope of a better winter. Help from England was slow to come. It was difficult to convince landowners that this famine was any worse than those that periodically left the poor hungrier than usual. Grain, and other food collected for rent, was exported from Ireland throughout the blight years. These cruelties and their political consequences are well described in *The Great Hunger* by Cecil Woodham-Smith.

An important character in the political drama was **Sir Robert Peel**, the English prime minister, who used the blight as an excuse to repeal protectionist trade laws by requesting the import of American maize (corn) to feed the hungry Irish. The corn mash was cooked in open kitchens, often miles from the villages. Those too weak or sick to walk to the food had to share the portions of the stronger ones. The cooked mush also

Fig. 1-15. Sir Robert Peel, Prime Minister of England during the Irish potato famine.

prevented the sale and distribution of the grain on a black market. The rations were dubbed "Peel's Brimstone" by the recipients, who were considered ungrateful since food was not only provided but cooked for them as well. The rainy weather and resulting blight continued throughout what came to be called "The Hungry Forties." Poor people throughout Europe found themselves hungry because of the severe losses in the potato crop and increased prices of other foods. It was a time of economic depression and political unrest. The *Communist Manifesto* was published in 1848. Nowhere was the suffering as grim as in Ireland, however, and in a 15-year period, the Irish population of 8 million decreased by at least 1 million due to disease and starvation and an additional 1.5 million due to emigrants who found a way to leave the island, many traveling to the United States and Canada.

The scientific philosophers were unable to save the potato crops or prevent the blight in succeeding years, but from this disaster came an important new understanding of plant disease. **Anton deBary**, a German botanist,

Fig. 1-16. Anton deBary.

performed the experiments that proved the role of the fungus in the blight. He subjected potato plants to the cool, wet environmental conditions that favored blight. To some plants he applied sporangia from blighted plants, but others he kept as "control" plants and applied no fungus. Even though both sets of plants were exposed to the same favorable environment, only the plants inoculated with the fungus became blighted. It was clear that the plants did not rot away because they absorbed too much water. In the absence of the pathogen, no disease occurred. It is important to appreciate the intellectual significance of deBary's experiments. They triggered major advances in the study not only of plant diseases but also of human and animal diseases, due to a new understanding of the importance of contaminated food and water and unsterilized medical instruments in infectious and contagious diseases. For 200 years, people had been observing many disease-causing organisms, but these had been seen as the result, rather than the cause, of disease. Scientists were now prepared to abandon the theory of spontaneous generation for Louis Pasteur's germ theory of disease, which he proposed in 1863.

The blight fungus was recognized as new to Europe. Like all other organisms, the fungus has a Latin binomial. It was first named *Botrytis infestans* by Montagne, but the name was changed to *Phytophthora infestans* by deBary. In a formal designation of a Latin name, initials follow the name to allow one to find the description of the organism that is officially recorded at the time of the naming. Many plant and animal names are followed by "L." because Linnaeus first created the system and named many of the commonly known plants and animals. Most of those names were well accepted because the organisms were already familiar and well studied. Names of less familiar organisms are often changed as their biology and relation to other organisms become better understood. This is particularly true of microorganisms because of the rapid improvements in the technology for their culture and study. In such cases, the specific epithet usually remains, but the genus is changed. The name of the original describer is then put in parentheses, followed by the name of the new describer. For example, the potato blight fungus is now called *Phytophthora infestans* (Mont.) deBary. Montagne's specific epithet, *infestans*, was retained, but deBary realized that the fungus was not really similar to other species in the genus *Botrytis*, so he created the genus name *Phytophthora* to describe the blight fungus.

Linnaeus used Latin to describe organisms because that was the international language of educated Europeans. Many of the names can be translated from their Greek and Latin roots to reveal the ideas of the describers. *Phytophthora* comes from the Greek (*phyto* = plant and *phthora* = destroyer), and the species name, *infestans*, suggests the devastating infestation. People use common names for their plants and animals in their various languages, but scientists communicate internationally using the Latin names. Most microorganisms have no common names because usually no one but a scientist observes or talks about these species.

DeBary's work led many scientists to observe diseased plants and describe the parasitic fungi associated with them. Of course, the presence of a fungus in diseased tissue is not proof that it is the primary causal agent in the disease. In fact, most fungi are incapable of parasitizing healthy plant tissue and can invade only dead plant tissue. Some means was necessary to determine whether a fungus present on a diseased plant had caused the disease or was simply an opportunistic **saprophyte**, that is, an organism that obtains its nourishment from nonliving organic matter.

The **proof of pathogenicity** was derived from work by a German microbiologist named **Robert Koch**, who worked with anthrax disease of sheep. By chance, the bacteria that cause anthrax are particularly large and were visible even with 19th century microscopes. Koch created a process to convince himself and others that the bacteria in the blood of the cattle actually caused the anthrax disease. **Koch's postulates**, which can be applied to diseases of plants as well as animals, include the following steps:

1) The symptoms and any evidence of the pathogen in the diseased host are carefully described.

2) The suspected pathogen is isolated from the diseased host and from all other contaminating microorganisms, usually on a nutrient medium that will keep the organism alive. A description is made of the suspected pathogen.

3) A healthy host is inoculated with the suspected pathogen. It is later observed for symptoms, which must be identical to those described in Step 1.

4) The pathogen is reisolated from the inoculated host and must be identical to the organism described in Step 2.

When all four steps are completed, the proof of pathogenicity is established. These rules continue to guide our studies of new diseases, although modifications must be made for pathogens that cannot be cultured outside a living host. New disease reports published today still frequently indicate that Koch's postulates have been fulfilled in the described experiments.

Following deBary's work with *P. infestans*, other scientists began to report numerous fungi capable of infecting plants and causing disease. In the early years, bacteria seemed to be associated mostly with animal and human diseases, and an inaccurate separation of pathogens was created by scientists too quick to look for differences between plants and animals. Near the end of the 1800s, however, bacteria were also shown to be plant pathogens— but only after intense argument. Plant pathology, the study of plant diseases, had become an important part of scientific study.

Protecting Potatoes from the Blight

Scientists had come to a consensus about the cause of the blight, but they still had no solution to the problem of saving the potato crop from invasion by the fungus. Protective copper fungicides still lay in the future, so that the poorest Europeans went hungry in the blight years that occurred periodically. According to Carefoot and Sprott in their interesting book *Famine on the Wind*, the last major famine due to *P. infestans* resulted in 1916 in the deaths of 700,000 German civilians, who were unable to protect their potato crop because the copper was needed for bullets rather than fungicides. Even today, more than 150 years after the Irish epidemic, frequent applications of fungicides are necessary to grow potatoes in moist climates, and losses occur even in Israel and the drier western United States. Potatoes remain a chemically intensive crop despite years of study of *P. infestans* and the disease it causes.

The search for potato cultivars resistant to infection by *P. infestans* remains an important goal. For years, farmers have selected potato plants that survived blight better than others, and more potatoes were collected in South America in an attempt to find blight-resistant plants. Although some cultivars were more resistant than others, no potatoes could survive cool, rainy blight years without fungicides.

In the first half of the 20th century, plant breeders were able to transfer resistance genes from a closely related plant, *Solanum demissum*, that had been collected in Mexico. They cross-pollinated the plants, harvested the seeds, and selected the most blight-resistant plants. Eventually, new cultivars were created that remained blight-free even when deliberately inoculated with the fungus. Some plant pathologists believed that late blight had been eliminated. Unfortunately, the success did not last. Within a few years, new races of the fungus appeared that were capable of infecting the resistant plants. Farmers reported that the resistance had "broken down," but, in reality, the fungus had "overcome" the resistance through genetic change. For some years, *S. demissum* remained an important source of resistance genes, but each time a cultivar with a new resistance gene was planted, it was only a matter of time before the fungus would develop a new race and cause a disastrous epidemic. This **boom-and-bust cycle** in the use of resistance genes has been observed in many important crops. Plants with resistance to nearly every important disease have been produced in breeding programs, but the production of plants with durable resistance remains a difficult challenge, which will be considered more fully in Chapter 5.

Fig. 1-17. Potatoes with resistance to *Phytophthora infestans* survive an epidemic better than the blighted susceptible cultivar in the foreground.

Lessons from the Potato Famine

From its beginnings in the Irish potato famine, plant pathology has faced a clear set of important problems. Human beings wish to produce food with the greatest efficiency and the least effort. Even in the earliest agriculture, this has meant growing large numbers of individual plants close together in the same area. In addition, the growers prefer that the crop be fairly uniform, meeting standards of taste, color, yield, and even maturation, so that harvesting is simpler. In many cases, genetic uniformity, via vegetative propagation, is highly desirable for the production of particularly good apples or an unusual tulip color. Unfortunately, these horticultural demands create an environment at risk for disease epidemics. An important goal of a plant pathologist is to determine ways to balance the needs of agricultural production with the need for plant protection in a world filled with highly adaptable parasites.

Human beings have probably caused all of the major plant disease epidemics in history, mostly by moving plants and pathogens from their points of origin to every corner of the earth. Plants are most vulnerable to attack by a parasite to which they have never been exposed. Plants and parasites in their place of origin have had millions of years in which to coevolve. The chestnut blight and Dutch elm disease are examples of the devastation caused by introduced pathogens in Europe and North America. Similar problems have arisen when foreign insect pests, such as the Japanese beetle and the gypsy moth, have been introduced. Many Native Americans tragically died of European "childhood" diseases, such as measles and chicken pox, when they were first exposed to those viruses. More recently, astronauts were kept in isolation following the early space explorations to ensure that they had not brought back some pathogen capable of destroying the earth's human population. A similar situation can exist when a plant is taken to a new environment where it is exposed

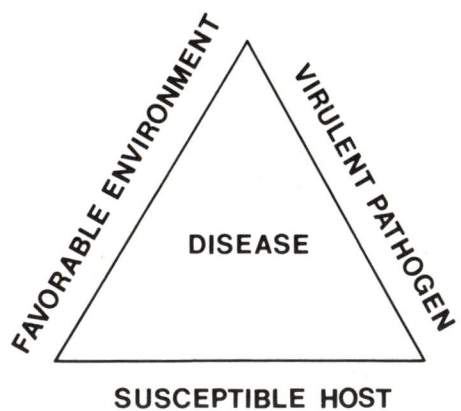

Fig. 1-18. The disease triangle.

to new parasites, as demonstrated by the severe pear losses from fire blight in North America. Plants and their associated pathogens were carried all over the world without restriction in the mid-1800s. Plant quarantine laws, first passed in the United States in 1912, could not be created until the basic principles of plant pathology were understood. Tremendous plant losses could have been prevented with earlier restrictions. Even today, most people who travel do not understand the continuing threat of foreign pests and pathogens.

The disease triangle demonstrates that three components must be considered in the study of any plant disease—the plant, the pathogen, and the environment. It also warns that significant change in any one of the components should be investigated. The unusual weather conditions of the 1840s certainly contributed to the devastation, but only because *P. infestans* had arrived from South America. Cool and rainy weather were not such a threat before that.

From the tragedy, however, grew a new perspective of the natural world. Disease was no longer a dark and magical force, but a biological activity involving tiny parasites that invaded the plants and caused their decay. The new science of **plant pathology** involved the study of plants and their pathogens. Crop protection could move from pagan rituals to a scientifically based understanding of the means by which farmers and growers can favor crop growth and discourage infection by parasites. People could consider ways to prevent the fungus from infecting the plant, using chemicals, resistant plants, or environmental management.

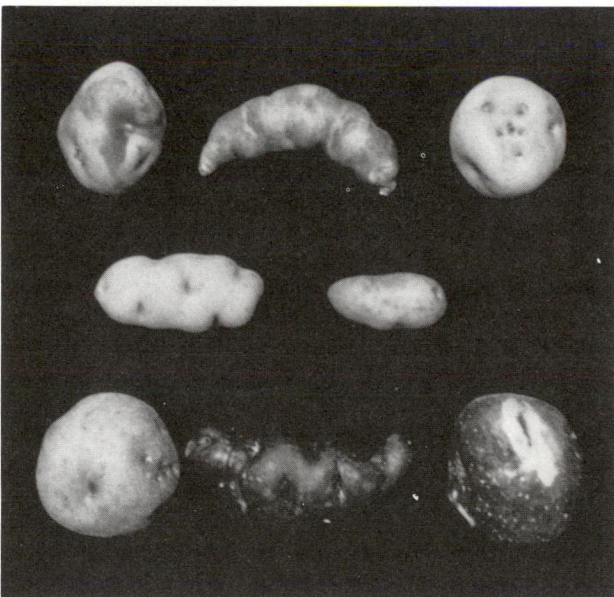

Fig. 1-19. Examples of the variety of potato tubers available in the center of origin of the species in the highlands of the Andes.

It soon became clear that every crop is subject to many diseases, each of which is unique. Most of the 100,000–250,000 species of known fungi are incapable of infecting living plants, and even parasitic fungi are usually able to infect only a few closely related species of plants. What is the basis of this specificity, and how are plants able to defend themselves from the multitude of microorganisms in their environment? These questions have not yet been fully answered, but current research continues to work on these fascinating basic questions as well as on the applied problems of crop protection.

The human component of plant pathology was probably never so clear as in Ireland in the Hungry Forties. Yet even today, we remain totally dependent on plants for our food, and, as the human population increases, food production needs increase. At the same time, the amount of arable land decreases each year, as does the number of living species of plants. The **centers of origin** of the major food crops, which are an important source of the genetic diversity needed for continued plant protection, lie mostly in the areas of the world with the greatest population pressure. An important principle of plant pathology, learned since the potato famine, is that we will never eliminate the parasites of plants, so we must learn to coexist with them and still produce enough food for an ever-growing human population.

The study of the epidemic caused by *P. infestans* also introduces the important points of a **disease cycle**. To understand any plant disease, one must know how the pathogen is dispersed to other plants, how it survives unfavorable conditions (such as winter or hot, dry weather), and the effects of environmental conditions at various stages of the life cycle of the parasite. The study of plant diseases requires some knowledge of the biology of the pathogens. Since fungi cause many important plant diseases, the next chapter introduces the fungi and their role as plant pathogens, beginning with *P. infestans* and some closely related pathogenic fungi.

Selected Readings

Agrios, G. N. 1988. Plant Pathology, 3rd ed. Academic Press, San Diego.

Carefoot, G. L., and Sprott, E. R. 1967. Famine on the Wind. Rand McNally and Co., Chicago.

Large, E. C. 1940. The Advance of the Fungi. Dover Publications, New York.

Roberts, D. A., and Boothroyd, C. W. 1984. Fundamentals of Plant Pathology, 2nd ed. W. H. Freeman and Co., New York.

Woodham-Smith, C. 1962. The Great Hunger. Harper and Row, New York.

Introduction to the Fungi
and Their Life Cycles

CHAPTER 2

Human beings are always trying to categorize. By relating new objects to those already familiar, we can more easily discern their similarities and differences. Early biologists tried to divide the living organisms of the world into plants and animals. At first this presented few problems because most large organisms fall easily into one of these two groups. Most larger plants are green, sedentary, and their cells possess a cell wall in addition to the cell membrane that encloses animal cells. The mushrooms and other fungi, so commonly seen in fields and forests on decaying vegetation, presented something of a problem. They do not move, and their cells have cell walls, but they do not possess the green pigments for photosynthesis so characteristic of plants. In fact, their whole biology presented an array of mysteries, such as the sudden appearance of mushrooms after rain and the growth of fungi in dark recesses where plants cannot normally live. However, fungi are certainly not animals, so botanists accepted this confusing group, and the study of fungi has generally remained within the realm of botanical scientists, frequently ignored by zoologists and even many microbiologists. Modern **systematics**, the study of the relationships between organisms, now separates the fungi from true plants, and, as will be seen shortly, current evidence suggests that fungi themselves are not even closely related to each other.

Vegetative Structure of Fungi

Most **fungi** consist of a threadlike mass of **hyphae**. A single thread is a **hypha**, the term originating from the Greek word for *web*. Many scientific terms are written according to the linguistic rules of Latin, including the means of making words plural, so we see a single *hypha*, but several *hyphae*. The word *fungus* (plural: *fungi*) itself is from Latin. The mass of hyphae that comprises the vegetative, or nonreproductive, part of a **fungus** is called the **mycelium**, using the Greek root *mykes*, which refers to fungi. Similarly, the study of fungi is called **mycology**. The words *hyphae* and *mycelium* are often used synonymously.

The fragile hyphae are very vulnerable to destruction. They are easily damaged by physical pressure, temperature extremes, and desiccation. The hyphae themselves are also not suitable for dispersal in the environment, except perhaps by water or plant parts such as bulbs, roots, or tubers that would prevent desiccation. If a nutrient source is not available, the mycelium usually dies quickly. Yet, despite their simple structure, fungi are highly evolved organisms. Their cellular organization is similar to that of all higher organisms, with such features as a membrane-enclosed nucleus containing organized **chromosomes**, and **organelles**, called **mitochondria**, for respiration. Such organisms are called **eukaryotes** (from the Greek *eu* for *good* or *true* and *karyon* for *nucleus*) because they have a true nucleus,

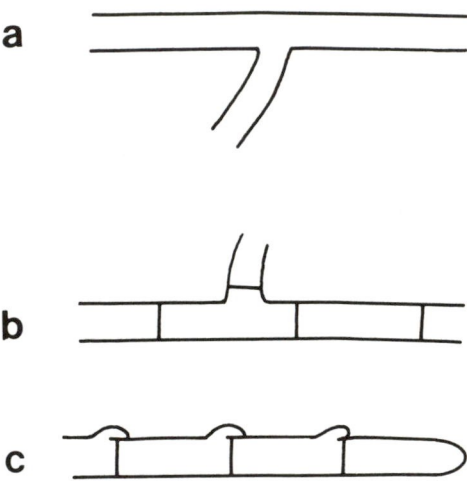

Fig. 2-1. Various forms of fungal hyphae: **a,** without septa (crosswalls); **b** and **c,** with septa.

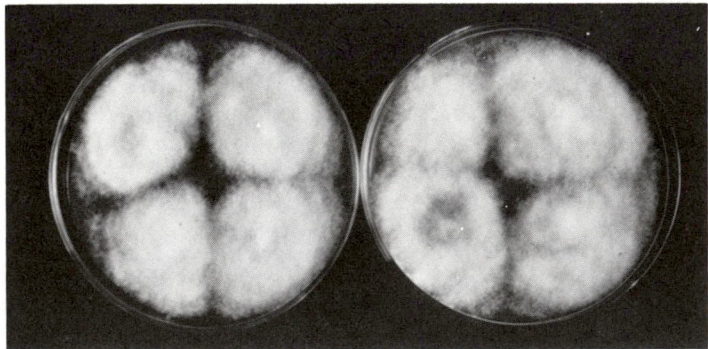

Fig. 2-2. *Phytophthora infestans* growing on a nutrient medium in petri plates. Each plate contains four individual colonies. Abundant white mycelium is visible.

unlike the **prokaryotes**, such as bacteria, that lack a nucleus and other membrane-bound organelles. All plants, animals, and fungi are eukaryotes.

Mitosis of the Eukaryotic Cell

The nucleus in a typical eukaryotic cell contains the genetic information necessary for cellular control. This genetic information is stored in the chromosomes, which are composed of **deoxyribonucleic acid** (DNA). The structure of DNA and the **genetic code** are common to all living organisms and are discussed in Chapter 5. Cells must have processes to allow this genetic information to be passed on to new cells during growth and to succeeding generations during reproduction. Whether in a human being or a fungus, these processes are the same. At the cellular and molecular level, we can see the biological links that connect all of life despite vast differences in appearance.

The study of plant pathology allows us to especially appreciate these connections because the relationship between a parasite and a host is a dynamic genetic interaction. Genetic changes in the host plant correspond to genetic changes in the parasite as an ever-changing balance is pursued.

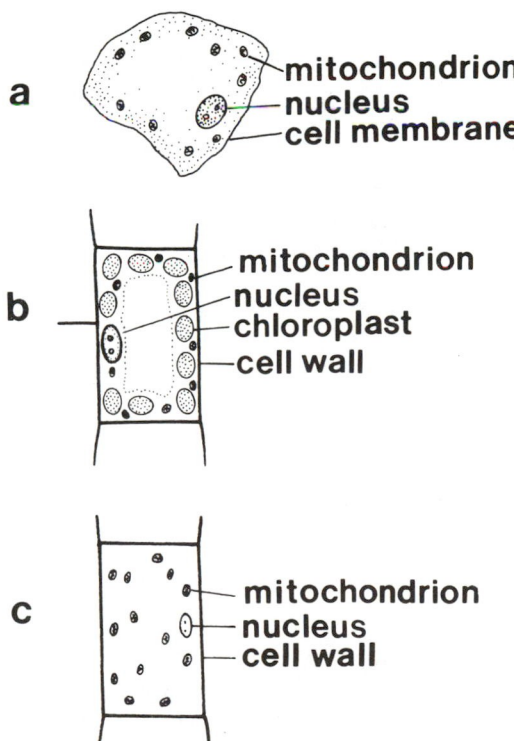

Fig. 2-3. Cell structure of various organisms: **a,** animal; **b,** plant; **c,** fungus.

Effective agricultural practices require that the interactions of the genetics and life cycles of hosts and parasites be well understood. Such studies have demonstrated how similar these functions are in all life forms.

Each time a cell divides, the DNA in the nucleus must be replicated and divided between the resulting daughter cells so that both cells receive an identical set of the genetic blueprints necessary for cellular control. The carefully controlled replication and subsequent division of the nucleus of a cell is called **mitosis** (Fig. 2-4). When cells prepare to divide, the chromosomes double and become visible in **prophase**. During prophase, each chromosome is seen as a duplicated pair of **chromatids**. During **metaphase**, the chromatid pairs become arranged along an equatorial plane. In **anaphase**,

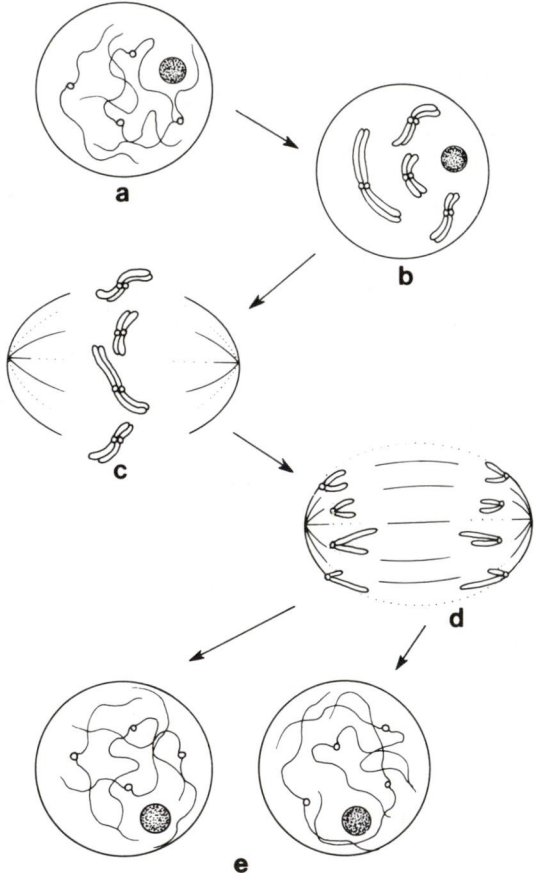

Fig. 2-4. Mitosis. **a,** Interphase: chromosomes are present but not visible in the nucleus; **b,** prophase: chromosomes visible as duplicated pairs of chromatids; **c,** metaphase: chromosomes line up along an equatorial plane; **d,** anaphase: pairs of chromatids are drawn apart; **e,** telophase: nuclear envelope surrounds each new set of chromosomes.

the pairs of chromatids are drawn apart so that two identical sets of chromosomes collect at each end of the dividing cell. New nuclear envelopes surround each set of chromosomes in **telophase**, and mitosis is complete. The period during which the chromosomes are invisible between the end of telophase and the beginning of the next prophase is called **interphase**. Mitosis occurs with each cell division and ensures that both resulting daughter cells are genetically identical. The nuclei of all eukaryotes, including fungi, undergo mitosis during cell division.

Genetic Recombination Through Sexual Reproduction

Despite the tremendous variety of life forms on earth, eukaryotes also share the processes by which genetic information is passed from one generation to the next. Maintenance of a species requires genetic stability between generations, but enough variability to allow the species to survive in a changeable environment is also desirable. **Sexual reproduction** serves both of these functions. Genetic information is contributed from two different nuclei, and the genes are recombined to produce new, genetically

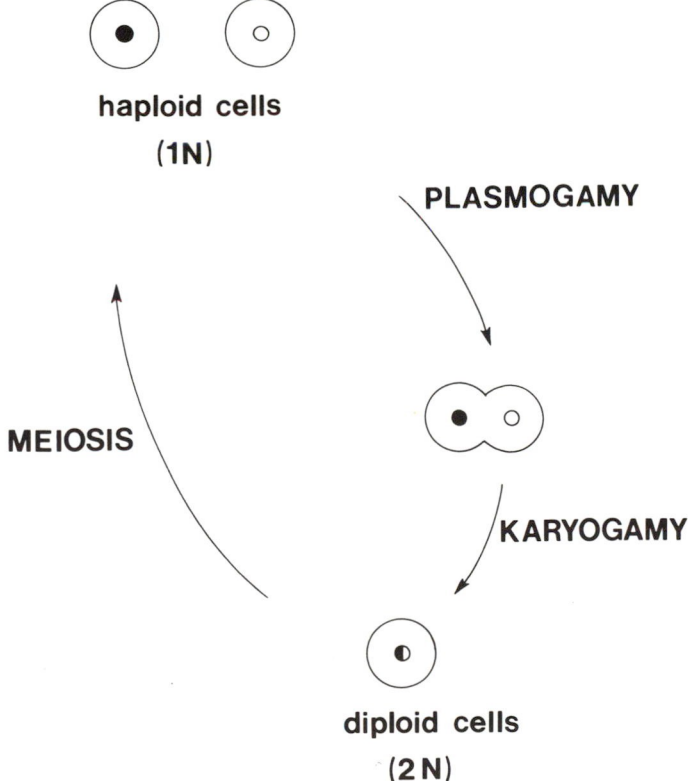

Fig. 2-5. Sexual reproduction.

diverse nuclei. The resulting individuals are very similar, but not genetically identical, to the parent organisms.

How does this **genetic recombination** occur? In sexual reproduction, each "parent" nucleus contributes its set of chromosomes to the "offspring" nucleus. There are actually two steps to this process (Fig. 2-5). In the first step, the two nuclei join together in the same cell (**plasmogamy**). In the second step, the two nuclei fuse (**karyogamy**). The resulting nucleus consists of pairs of chromosomes and is therefore termed **diploid**. (The parent nuclei that contain only one of each kind of chromosome are **haploid**.) If this process were to continue, the number of chromosomes would double with each succeeding generation. To avoid this problem, a **reductive division** occurs in the diploid nucleus that reduces the nuclei back to the haploid state. In this way, each parent contributes part of its genetic makeup to the offspring, but the final chromosome number remains stable for the species.

This reductive division is called **meiosis** (Fig. 2-6) and is a process similar to mitosis. The diploid nucleus contains a pair of each type of chromosome. All chromosomes double and divide as in mitosis, but the process continues to a second division, so that four haploid nuclei are produced from the original diploid nucleus, each containing only one chromosome of each pair. During meiosis, there is a random division of chromosome pairs, so each haploid nucleus contains a different assortment of each type of chromosome. Each of the four haploid nuclei is genetically different but still contains a complete set of genetic blueprints for that species. For example, human beings receive a complete set of genetic information from each parent, but the specific genes, such as those for eye color, may be different for each individual.

In animals, the products of meiosis are the haploid, single-celled **gametes**, the sperm and eggs. During fertilization, each sperm and egg contributes one set of chromosomes to the new nucleus. After these gametes fuse (plasmogamy), the nuclei also fuse (karyogamy), and the diploid state is restored. The new diploid nucleus contains one set of chromosomes from each parent. The resulting fertilized egg divides and grows, creating a diploid nucleus for each new cell by mitosis. All cells in an animal's body are diploid except the haploid eggs and sperm. Thus, sexual reproduction maintains a stable chromosome number in the species but allows a reshuffling of the genetic material to produce variability in the next generation. Each new animal produced by sexual reproduction is genetically similar, but not identical, to its parents.

The processes of meiosis and karyogamy are characteristic of sexual reproduction in all eukaryotic organisms, but the predominance of the diploid body is not. In plants and fungi, the life cycles show great variation. For instance, plants produce two different vegetative bodies, one composed of haploid cells and another composed of diploid cells. This **alternation of generations** is connected by karyogamy and meiosis to accomplish genetic recombination.

In many lower plants, such as mosses, the haploid body predominates. Following meiosis, haploid spores are produced that germinate and grow into the small green plants commonly found in moist and shady places. Each cell of the plant is haploid. At this stage, moss is a **gametophyte**, which means gamete-producing plant, because later, by mitosis, it produces

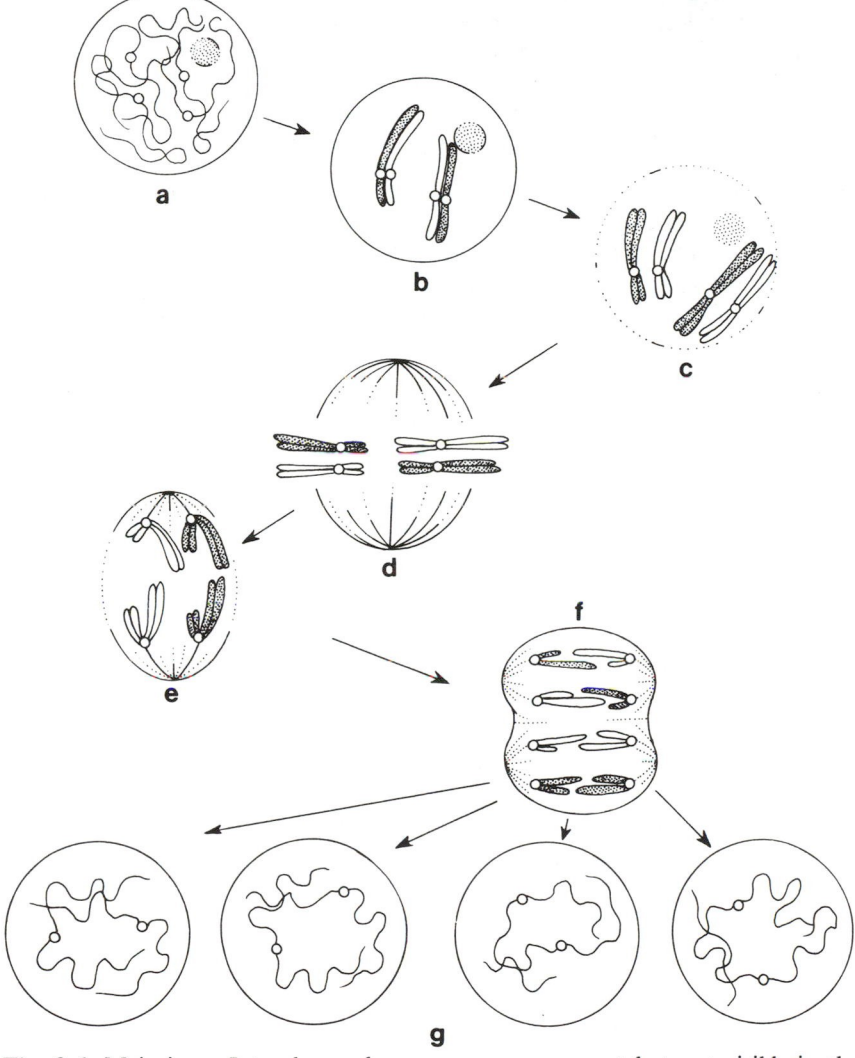

Fig. 2-6. Meiosis. **a,** Interphase: chromosomes are present but not visible in the nucleus; **b,** prophase I: chromosomes become visible and come together in homologous pairs; **c,** late prophase I: chromosomes visible as duplicated pairs of chromatids; **d,** metaphase I: paired chromosomes move into position on either side of an equatorial plane; **e,** anaphase I: paired chromosomes separate and move to opposite poles. At this point, a second division occurs similar to mitosis. Only anaphase II is illustrated (**f**). Each of the four resulting haploid nuclei receives one chromosome of each type (**g**).

sperm and eggs. When the sperm and egg fuse, karyogamy restores the diploid state. From the diploid nucleus, a small diploid plant grows from the tip of the haploid plant and is nutritionally dependent on it. This diploid plant, called a **sporophyte**, produces a spore capsule. Inside the capsule, some cells undergo meiosis to produce haploid spores. This completes the alternation of generations. The production of a gametophyte composed of haploid cells represents a separation between meiosis and karyogamy that does not occur in animals.

In contrast to the primitive moss plants, in which the haploid gametophyte is dominant, the diploid sporophyte predominates in flowering plants, and the haploid gametophyte is greatly reduced. In the anthers of flowers, meiosis is followed by mitosis to form a haploid pollen grain, usually consisting only of two cells, that travels to the stigma of a flower. The **pollen tube** grows down from the stigma to reach the haploid embryo sac in the **ovule** of the flower. The number of cells in the embryo sac of a flower varies

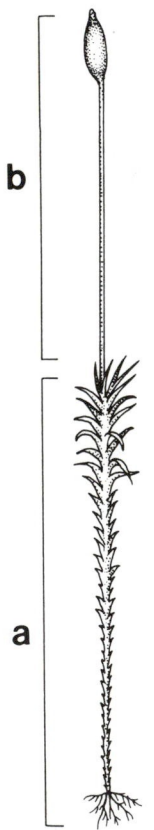

Fig. 2-7. Moss plant. **a,** Haploid gametophyte; **b,** diploid sporophyte with spore capsule.

depending on the species, but only one cell functions as the egg. A generative cell in the pollen grain produces sperm by mitosis. During karyogamy, the egg and sperm nuclei fuse to produce a diploid cell that is protected and nourished in the seed until germination. All cells of the new plant contain diploid nuclei. However, some plants have more than two sets of chromosomes and are known as triploids, tetraploids, or polyploids. These complexities are of agricultural importance and are discussed in Chapter 5.

Life Cycles of Fungi

Most fungi also undergo sexual reproduction. To understand a life cycle, it is necessary to make detailed studies of chromosomes in cells at various stages to determine chromosome numbers. Then one can determine which cells are diploid and which are haploid and, thus, the point at which meiosis occurs. Such studies may be difficult because fungal chromosomes are often very small and difficult to count accurately. Some important fungi grow only in host tissue rather than on nutrient media in a laboratory, where they are more easily studied. In addition, many fungi require different mating types, special nutritional requirements, or precise environmental conditions to reproduce sexually. It is known, however, that the nuclei in the vegetative mycelium of fungi may be diploid, haploid, or **dikaryotic** (containing two different haploid nuclei), depending on the fungal group and the stage of development.

In some fungi, such as *Phytophthora infestans*, the hyphae are diploid, and meiosis occurs in the reproductive structures just before fertilization, similar to the situation with animals. This is typical of the **Oomycetes**,

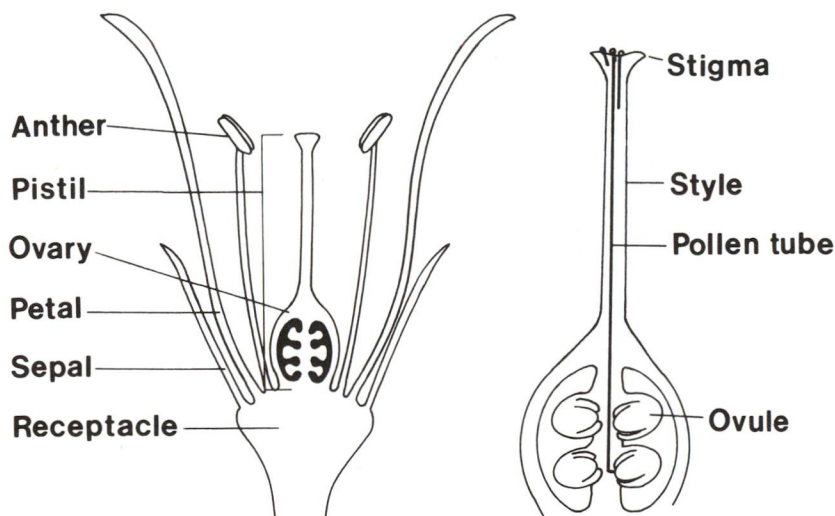

Fig. 2-8. Generalized flower anatomy.

the fungal group to which *P. infestans* belongs. Fungi in this group have similar life cycles but have adapted to their hosts and various environments as described later.

In the **Ascomycetes**, another fungal group, the vegetative hyphae are haploid. In the early stages of sexual reproduction, one nucleus joins another in the same cell (plasmogamy), but karyogamy is delayed, so that a limited dikaryotic mycelium develops. In the dikaryotic mycelium, the two haploid nuclei exist in the same cell and divide by mitosis, in tandem, as the hyphae grow. Eventually, karyogamy occurs but the resulting diploid nucleus undergoes meiosis almost immediately to create haploid spores that will germinate into the new haploid vegetative mycelium. Thus, in the Ascomycetes, the haploid state predominates. Many important plant pathogens are Ascomycetes, including fungi that cause powdery mildews, leaf spot, canker, wilt, and root rot diseases.

In a third major fungal group, the **Basidiomycetes**, haploid spores produce a limited haploid mycelium, which must fuse with the haploid mycelium of a genetically different strain in order for an extensive mycelium to develop. Plasmogamy brings the nuclei into the same cell, but karyogamy is usually delayed, so the predominant vegetative mycelium remains dikaryotic. As in the Ascomycetes, karyogamy does not occur until just before meiosis, providing haploid nuclei for the developing spores. Thus, diploid nuclei have a very limited existence in both the Ascomycetes and the Basidiomycetes. Rusts, smuts, and many wood decay fungi are important plant-pathogenic Basidiomycetes. The familiar mushrooms found in fields and forests are the spore-producing structures of Basidiomycetes.

It is probably more clear now that the fungi are a diverse group of organisms with complex life cycles. Despite all these variations, however, fungi show a consistent pattern of alternation between the haploid and diploid state through karyogamy and meiosis. Sexual reproduction, in all its forms, provides a stable process by which genetically different nuclei can contribute genetic information to the next generation with the subsequent recombination of that information to produce unique new individuals.

Why is knowledge of life cycles and genetic recombination so important? As an example, recall the interaction of *P. infestans* and the potato. The genetic uniformity of the potato crop resulted in the disastrous epidemics in Ireland, and the potato remains vulnerable to disease today. Early attempts to create resistant cultivars using single resistance genes failed because of the ability of the fungus to quickly develop new genetic races. If we wish to protect crops with genetic resistance, we must understand the genetic interaction between parasites and their hosts. Many factors determine the success or failure of genetic resistance in a plant. Important factors for the plant include whether it is a perennial or an annual (replanted each year), whether or not it is vegetatively propagated, and the genetic diversity available for breeding selections. Factors in the life cycle of the parasite include the number of generations produced during a growing season, how the parasite is dispersed to other plants, and the inherent genetic variability of the organism. Many fungi do not commonly appear to reproduce sexually but still find ways to overcome resistance in plants. Others require different **mating types** or strains for sexual reproduction to occur. Since sexual reproduction is an important means of genetic variation, we need to understand the role of sexual reproduction in the parasite's life cycle if we intend to find ways to manage the disease it causes.

A second important reason to study fungal life cycles is that spores produced during sexual reproduction are often survival structures. They may have a thickened wall or be produced in a special fruiting body that protects the fungus from adverse conditions. Thus, sexual spores often initiate epidemics after a dry or cold period or after the prolonged absence of a susceptible crop.

The fungi are a rather diverse and perhaps heterogeneous group of organisms. One reason that their evolutionary relationships have been difficult to study is that the hyphae decay easily and have not been well preserved in the fossil record. Some of the best fungal fossils are of parasitic species preserved in or on their host plants. The origin of the major fungal groups and their relation to each other is not yet clear, but the great variation in life cycles, as well as in other important aspects of their biology, have convinced scientists that the fungi are not a closely related group of organisms. However, it is convenient to use the general term *fungi* for this diverse group in discussions of plant diseases in which pathogen dispersal and epidemic development are similar. Because it is difficult to generalize further about life cycles in the fungi, specific life cycles will be discussed as the important pathogens in the major fungal groups are introduced in later chapters.

The ability of fungi to be plant pathogens is governed by their biological characteristics. Within each of the major fungal groups can be found an array of adaptations to host plant species and plant tissues. Thus, root-rotting species can be found in each major fungal group, but the conditions under which various species predominate are governed by their basic biology. As an example of the variety of diseases found within a single fungal group, the following section considers the Oomycetes. One species, *P. infestans*, is already familiar from the previous chapter. Although the diseases caused by this group of fungi vary, they are governed by the biological characteristics of the group. Similar variations within the other fungal groups are discussed in later chapters.

Plant-Pathogenic Oomycetes

P. infestans and some other very important plant pathogens belong to the Oomycetes, a fungal group that seems to be particularly distinct from other fungi. Oomycetes produce a filamentous mycelium that is multinucleate. An individual hypha is essentially an undivided tube of **cytoplasm** containing a number of nuclei. The mycelia of Ascomycetes and Basidiomycetes, in contrast, are divided into sections by **septa**, or crosswalls.

A cell wall helps maintain the shape of the hypha and consists mostly of cellulose and various cellulose polymers. Chitinous compounds, rather than cellulose, predominate in the cell walls of other fungal groups, suggesting that they are probably not closely related to the Oomycetes. A hypha grows from the tip. All fungi are **heterotrophic**, which means that they require an external source of organic compounds for food. This is in contrast to **autotrophs**, such as green plants, that make their own food by photosynthesis or chemosynthesis. Simple compounds can be absorbed through the fungal cell wall, but most potential food sources consist of more complex molecules too large to be absorbed. Fungi, therefore, secrete various **enzymes** that degrade, or break down, large molecules into smaller ones that can be absorbed. As long as a nutrient source is available, the hyphae continue to grow.

The fragile mycelium is subject to desiccation and starvation if environmental conditions become unsuitable, so the fungus must be prepared to survive inhospitable circumstances or be able to disperse to a more suitable place. Fungi produce various reproductive structures that function as dispersal and survival devices. Many fungi commonly produce various types of **asexual spores**. These spores involve no genetic changes and are simply containers for one or more nuclei and sufficient cytoplasm for renewed growth of the fungus. The dusty powder found on mildewed bread or clothing is a mass of asexual spores produced by the mycelium of a fungus. Such spores can serve as a means of dispersal and/or survival. Those that function for survival are usually protected by a thickened wall or fruiting body. In some fungi, a section of the cell wall of a hypha thickens to protect the cytoplasm and one or more nuclei. The resulting survival package is called a **chlamydospore** and is commonly produced by many species of Oomycetes as well as other types of fungi. They are strictly asexual because they simply protect part of a hypha containing at least one nucleus and involve no sexual reproduction or genetic changes.

P. infestans is able to disperse by either water or air once the lemon-shaped sporangia have been produced by the mycelium. They are well adapted for air dispersal due to their production on treelike hyphae that protrude through the stomata away from the leaf surface. Several hundred thousand sporangia can be produced from a single leaf or stem lesion. Rapid changes in relative humidity at the leaf surface during air movement cause the hyphae to twist suddenly and eject the sporangia. Even a small propulsion helps remove the sporangia from the quiescent boundary layer of air on the leaf surface and send them into the swirling atmosphere that will carry them to other parts of the same potato plant or to different ones. Although the sporangia are not well adapted for long-distance travel because they are relatively susceptible to desiccation, they are able to travel many hundreds of meters when relative humidity is high. After dispersal to a new field, new sporangia can be produced in a few days, so even distant fields may become diseased during a growing season.

As discussed previously, zoospores are produced from sporangia at cool temperatures (optimum 55–60° F, 12–15° C). At higher temperatures, disease can still occur because sporangia are able to germinate and infect directly, eliminating the zoospore stage, although the infection potential is reduced. Germination and infection by sporangia or zoospores require the presence of water. Water must be absorbed by the sporangium for germination, and zoospores require water for movement and infection of the host plant. Because of the importance of water in the life cycles of Oomycetes, the common name for this group is the "water molds."

Clearly, reproduction and dispersal of sporangia is rapid and effective

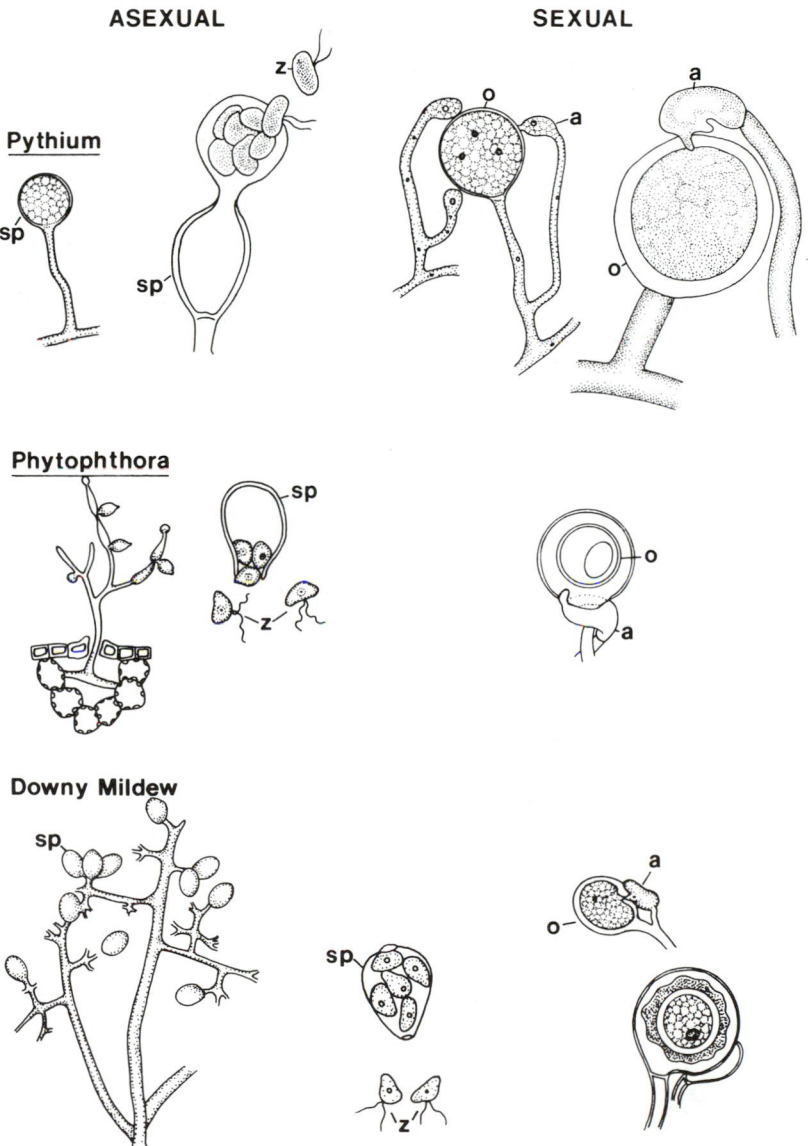

Fig. 2-9. Some important plant-pathogenic Oomycetes. Asexual structures: sporangium (sp), zoospore (z); sexual structures: oogonium (o) (female) antheridium (a) (male).

in a late blight epidemic, but this means of reproduction has some limitations. When sporangia are produced from the mycelium of *P. infestans*, reproduction is asexual, that is, meiosis and karyogamy do not occur. As with chlamydospores, the sporangia simply enclose a nucleus or nuclei created during normal nuclear mitosis. The nuclei of the vegetative mycelium, the sporangia, and the zoospores are all diploid. There is no genetic recombination through sexual reproduction, and no thick-walled survival structures are produced.

Is there no sexual reproduction in *P. infestans*? This same problem puzzled Anton deBary in the 1800s, and the answer to the question was discovered on potato plants in the Toluca valley of Mexico, where the sexual spores were first found. *P. infestans* produces mycelium of two different mating types. The mycelium of each mating type produces both male and female structures, termed **antheridia** and **oogonia**, respectively. The structures were named by botanists who observed that their functions were analogous to those of the anthers and ovules of flowers. Sexual reproduction occurs only when both mating types are present. Some *Phytophthora* species are self-fertile and do not exist as different mating types. Chromosome studies show that the nuclei of the mycelium are diploid and that meiosis occurs in the antheridia and oogonia. A haploid nucleus then travels from the

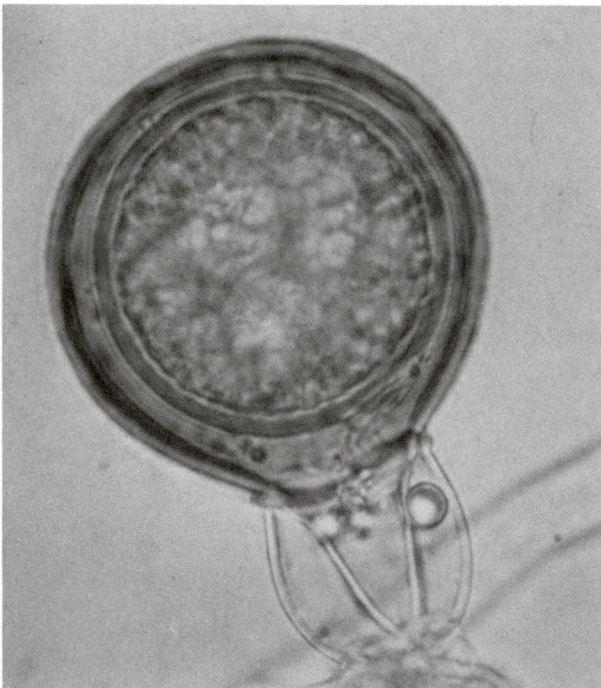

Fig. 2-10. Sexual structures of *Phytophthora cinnamomi*. The collarlike antheridium (male) is below the round oogonium (female), which contains the thick-walled sexual oospore.

antheridium into the oogonium, which also contains a haploid nucleus, and karyogamy restores the diploid state. The resulting diploid oospore has a thick wall that resists adverse environmental conditions. Its name, **oospore**, reflects the egglike appearance of the spore. Sexual reproduction is similar in all members of the Oomycetes.

Oospores of *P. infestans* are common in potato-growing areas of Mexico and other parts of Central and South America wherever both mating types of the fungus exist. Apparently, the original samples of *P. infestans* that were accidentally transported to Europe, and eventually to North America by European colonists, were of only one mating type, so oospores could not be produced.

Without oospores, the fungus cannot survive in the soil in temperate winters. Farmers take advantage of this weakness in the life cycle of *P. infestans* by careful sanitary practices in their fields. Infected tubers are destroyed; fields are carefully cleaned of volunteer plants from tubers missed in the previous harvest; and tuber pieces for planting are carefully inspected to prevent introduction of the fungus to the field along with the "seed." Even with these precautions, some of the fungus escapes detection and threatens crops each year.

Recently, the second mating type of *P. infestans* has been discovered in several European countries. In the Netherlands, the presence of the second mating type was discovered repeatedly in small garden plots. This suggests that the second strain may have arrived illegally in foreign potato tubers brought in by adventurous but uninformed gardeners. The survival of the strain has probably been enhanced by the less thorough sanitary practices of amateur gardeners. At the time of this writing, the second mating type has not yet been detected in North America.

While the full implications of its arrival in Europe are not yet known, the following possibilities exist. First, the sexual spores may allow the fungus to survive in the soil outside of tuber tissue. Thus, sanitary practices to reduce the fungus population may become less effective. The fungus may be able to survive in the soil for years as thick-walled oospores waiting to attack the next potato or tomato crop. In addition, the genetic recombination that occurs during sexual reproduction may increase the genetic diversity of the fungus. One important result may be changes in the ability of the fungus to infect potatoes. Another concern is that the fungus may be able to develop strains resistant to the new systemic fungicides even more quickly than it does now, an important problem discussed in Chapter 7.

Even without sexual reproduction, *P. infestans* has exhibited considerable genetic variation. How is this possible? Sexual reproduction is not the only source of genetic recombination and variability. Simply put and without going into detail, the hyphae of some fungi contain genetically different nuclei that may undergo genetic recombination by means other than the usual sexual process.

In addition, mistakes occur during the replication of DNA that change the genetic information in the chromosomes. These mistakes, called **mutations**, are relatively rare, estimated to occur in any particular gene at a rate of once in every 200,000 to perhaps one in a million cell divisions. Despite its low frequency, mutation is an important and basic source of genetic change in the evolution of all organisms. In an organism that produces many millions of sporangia every few days during an epidemic, mutation can be a very significant source of genetic variation. The tremendous reproductive capacity of microscopic parasites greatly increases their ability to change genetically relative to the much slower rate of reproduction in their host plants. Thus, even without sexual reproduction, *P. infestans* has been able to overcome the specific resistance genes of potato plants after only a few growing seasons.

P. infestans exhibits both sexual and asexual reproduction, a phenomenon common in fungi. A very broad generalization is that sexual spores have two functions: 1) as a survival structure for adverse conditions, and 2) as an important source of genetic variation. Thus, sexual spores often initiate epidemics. Asexual spores contain only mitotic nuclei but often serve the important function of rapid multiplication during an epidemic. In temperate climates, they are sometimes called "summer spores." In *P. infestans*, the sporangia produce zoospores, which are asexual, and rapidly increase disease during the growing season. The thick-walled oospores are the sexual spores, but they are produced only in areas where both mating types exist. Where no oospores exist, the mycelium of the fungus survives adverse conditions only in tuber tissue.

The Oomycetes consist of over 400 species. These water molds can be easily found in almost any surface water. To observe them, simply float a popped kernel of popcorn in a cup of pond water. In a few days, a growth of white mycelium will become visible, and examination with a

Fig. 2-11. Phytophthora blight of tomatoes. Tomato fruits in direct contact with excessively wet soil are highly vulnerable to infection.

microscope will reveal the hyphae, sporangia, and perhaps even sexual structures. This fungal group also includes many important plant pathogens.

Other species of the genus *Phytophthora* have life cycles nearly identical to that of *P. infestans*, but *P. infestans* is unique in its adaptation to air dispersal. Other species of the genus are strictly soilborne, relying on the movement of moist soil and water for dispersal of their sporangia. Oospores of all species allow the fungus to survive adverse conditions such as dry soil or the absence of a host. Infection by zoospores occurs on underground plant parts, such as roots, or aboveground plant parts within the **splash zone**, the area in which sporangia might be carried by drops of water.

A particularly important species of this genus is *P. cinnamomi*. It has a wide host range, causing important losses of ornamental trees and shrubs due to root rots in poorly drained soils that inhibit root growth and favor zoospore production. The fungus infects roots and plant parts near the soil line, particularly through wounds, killing the feeder roots. Rhododendrons are frequently subject to this decay. Dead plants must be replaced with care because new plants will be exposed to the oospores and mycelium left in the contaminated site, especially if soil drainage is not improved to foster good root growth and reduce the movement of zoospores.

P. cinnamomi has also caused extensive losses in avocado groves, especially in California, where 60–75% of the groves are affected. If the fungus is introduced to the soil in a grove of susceptible trees, very little can be done to stop its spread. Soilborne pathogens are among the most difficult to control because soil application of fungicides is difficult and expensive, may be environmentally hazardous, and is often ineffective. It is very difficult to find durable genetic resistance in trees that must remain

Fig. 2-12. Phytophthora root rot in a Christmas tree plantation. When infected young trees are transplanted to a poorly drained site, many of them will die.

in soil for many years continually exposed to new generations of a genetically variable pathogen.

P. cinnamomi was introduced into Western Australia and has spread rapidly throughout the country in the past 40 years, including the state of Victoria in the last 10 years, probably on contaminated road gravel. Zoospore movement is fairly limited, but downhill patterns of water movement have allowed the fungus to move 400 meters (over 400 yards) per year. The jarrah tree (*Eucalyptus marginata*) is particularly susceptible. In Victoria, entire plant ecosystems are being destroyed by this root-invading fungus.

Many other *Phytophthora* species are important plant pathogens. Some have narrow host ranges while others are pathogenic to many plant species, but all are difficult to control because of our current inability to adequately protect underground plant parts.

Another important genus in the Oomycetes is *Pythium*. The life cycles and morphology of species in this genus are so similar to those of *Phytophthora* species that some taxonomists would like to see the genera combined. One species, *Pythium debaryanum*, reflects the name of the scientist who took such an early interest in this group. Nearly every outdoor soil contains some species of the genus *Pythium*. Most species are well-adapted soil inhabitants that can survive equally well as saprophytes or oospores.

In cold, poorly drained soils, seeds germinate slowly. *Pythium* species can be an important cause of **damping-off**, a term used to describe the loss of seedlings at various stages of growth, typically toppling over at the soil line. Germinating seedlings exude nutrients that unfortunately attract the zoospores. Usually *Pythium* species infect only slowly growing young seedlings or plants that are wounded. Infections may occur in the splash zone when sporangia are carried to an infection site in a splashing drop of water.

Fig. 2-13. *Phytophthora parasitica* on peperomia. Note the blackened tissue near the soil line that has caused the plant to collapse.

Because *Pythium* species are ubiquitous, natural soils should be heat-treated to kill fungi before the soil is used to pot house plants or garden seedlings. In heavy soils and poorly drained gardens, raised beds improve drainage and make plants less vulnerable to attack by these water molds. Allowing soils to warm in the spring before planting seedlings will also hasten germination and reduce losses to *Pythium* species. Many seeds are coated with a fungicide to provide a short protection to germinating seedlings. Usually plants become less susceptible to attack as they mature. *Pythium* species can potentially cause damping-off of nearly every species of plant, so genetic resistance is less likely to be a suitable control solution than careful attention to environmental conditions at planting time.

One other important pathogen group in the Oomycetes is the **downy mildews**. The life cycle of most downy mildews, like those of other Oomycetes, involves sporangia that produce zoospores and sexual oospores for survival. Of the previously discussed Oomycetes, however, only *P. infestans* had evolved adaptations for air dispersal of sporangia. In contrast, the downy mildews, as a group, are pathogens of aboveground plant parts and well adapted to air dispersal. The common name for these fungi reflects the downy mass of hyphae and sporangia that appears on the underside of infected leaves in wet weather. The sporangia are produced on treelike branched hyphae, where they can easily become airborne on a passing breeze. While the sporangia of most downy mildew genera produce zoospores, some genera form asexual spores that never produce zoospores despite their appearing similar to the sporangia of other genera. This is considered an advantageous adaptation to a terrestrial environment because the fungus is no longer dependent on a zoospore stage, which requires water for locomotion and infection. Whether or not zoospores are produced, the

Fig. 2-14. *Pythium* species are a common cause of damping-off of greenhouse bedding plants (in this case, Verbena plants in the center container).

sporangia still require water for germination and are quite subject to desiccation during dispersal, so downy mildew diseases, like other diseases caused by Oomycetes, are most severe in wet weather.

The downy mildew fungi are very **host-specific**. Thus, one species causes disease only in grapes, another in tobacco, and another in spinach. Unlike *Phytophthora* and *Pythium* species, downy mildews are also **obligate**

Fig. 2-15. Downy mildew of grape (*Plasmopara viticola*) infects foliage and destroys fruits. The downy appearance of the hyphae and sporangia give the fungus its common name.

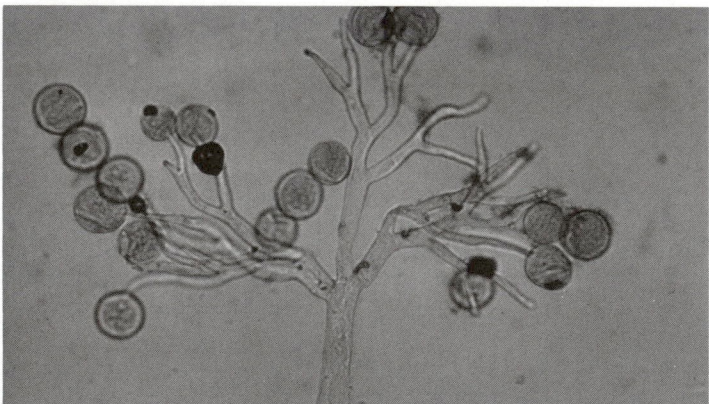

Fig. 2-16. A photomicrograph of downy mildew sporangia on tree-like hyphae. These hyphae emerge from leaf stomata and thus elevate the sporangia above the leaf surface, which aids in the air dispersal of the sporangia.

parasites, which means that they are unable to exist as saprophytes and rely totally on oospores for survival in the absence of available host tissue. As the host tissue dies, oospores form, so the fungus can survive in plant debris and soil until new host plant tissue becomes available. It is difficult or impossible to grow obligate parasites on nutrient media in the laboratory, so all studies of downy mildews must be accomplished with the fungus in living host tissue.

Downy mildews cause a number of economically important plant diseases. In wet seasons, they can rapidly reproduce and disperse via sporangia. Each growing season ends with the production of sexual oospores, ensuring genetic variation in the next generation of the fungus. New races of the fungi have been able to overcome resistance genes in host plants. Fungicide protection is required on crops that are grown in moist climates. In fact, the first discovery of an effective foliar fungicide was made in France in the 1880s during the downy mildew epidemic of the grape vineyards caused by *Plasmopara viticola*. Hop production in the northeastern United States was virtually eliminated by the hop downy mildew, *Pseudoperonospora humuli*, which appeared in 1928. U.S. hop production is now predominantly found in drier western states. In a more recent epidemic, Cuba lost most of its tobacco crop to blue mold, caused by the tobacco downy mildew, *Peronospora tabacini*, which resulted in economic hardships that may have contributed to the release of many Cubans in the "Freedom Flotilla" of 1980.

The previous examples of diseases caused by fungi belonging to the Oomycetes show great variety among these fungi despite the biological similarity of their life cycles. In species in which chromosome studies have been successful, the mycelium, sporangia, and zoospores all contain diploid nuclei. Meiosis occurs in the sexual reproductive structures, and the diploid state is restored with the production of the thick-walled oospore. All Oomycete diseases are associated with wet environmental conditions. Water is necessary to prevent desiccation of sporangia, for movement of zoospores, and for germination of sporangia and oospores. Some pathogenic Oomycetes are strictly soilborne, causing infections only on underground or splash-zone plant parts. Others, like the downy mildews, are airborne and infect aboveground plant parts. Some species have very wide host ranges, such as species of *Pythium* that attack vulnerable seedlings but generally exist as soil saprophytes. At the other end of the spectrum are the host-specific downy mildews, which are also obligate parasites incapable of saprophytic existence.

These examples demonstrate that the ability of organisms to be parasites is always conditioned by their basic biology. If a parasite is known to belong to a certain fungal group, some important facts about its life cycle are immediately known. It is then necessary to determine the specific adaptations that the parasite has made to its hosts and its environment to better understand the disease it causes. Life cycle patterns provide the basic framework of plant pathology. The individual life cycles of the many thousands of pathogenic fungi can be divided into just a few categories, allowing us to quickly discern the important aspects of each disease. The search for such patterns is an important component of the science of plant pathology.

Selected Readings

Alexopoulos, C. J., and Mims, C. W. 1979. Introductory Mycology, 3rd ed. John Wiley & Sons, New York.

Coffey, M. D. 1987. Phytophthora root rot of avocado. Plant Disease 71:1046-1052.

Lucas, G. B. 1980. The war against blue mold. Science 210:147-153.

Spencer, D. M., ed. 1981. The Downy Mildews. Academic Press. London.

Weste, G., and Marks, G. C. 1987. The biology of *Phytophthora cinnamomi* in Australasian forests. Annual Review of Phytopathology 25:207-29.

Pathogens and Quarantines

When the potato, *Solanum tuberosum*, was transported from its native home in South America to Europe, a new and important food crop began its journey to all parts of the world. Today the potato is the fourth most important world crop—behind rice, wheat, and maize (corn)—and the rate of increase in its production is greater than that of any other crop. It is grown in climates similar to those of the cool South American highlands, but some cultivars are adapted to warmer climates, thus extending its range to tropical lowlands as well. It is grown in each state in the United States and in hundreds of countries on many continents. About 200 years after its introduction to Europe, one of the important fungal parasites of the potato inadvertently crossed the ocean in infected tubers, leading to late blight epidemics and human starvation. Since no one understood the nature of plant disease at that time, there could be no safeguards to prevent such an introduction. *Phytophthora infestans* is now an established pathogen in all potato-growing areas of the world, along with various other fungi, bacteria, viruses, nematodes, and other pathogens that traveled along with the tubers of their host.

How to move useful and important plants around the world without transferring dangerous pathogens and pests is one of the great difficulties facing the human race. We now know that foreign pests and parasites represent a dangerous threat to native plants. If a pest is capable of attacking a plant, the initial exposure can be disastrous because the plant is not likely to have defenses against the attack. After years of selective pressure, plants evolve various means of resistance to their attackers, but at the first exposure, they have not had the opportunity to coevolve with the pest to create a genetic balance. Without well-developed resistance mechanisms, plant losses can be great.

Also vulnerable are the various introduced species that have been removed from the selective pressure of pests or parasites in their native land. This occurred with the potato in Europe and many other species that have been moved to new continents. During this "protected time," the plants usually become quite uniform genetically to meet the agricultural requirements of farmers, and, in the absence of the pest or parasite, they have no means of selectively maintaining resistance. When a pest or parasite is then accidentally introduced to the new land, the threat of disaster is great because of reduced resistance and reduced genetic diversity.

As human travel increased and the time required to cover long distances decreased, plants and, unfortunately, many of their parasites were collected and distributed to new lands. The primary production of certain crops has shifted to areas far from their place of origin to escape native parasites. With the birth of the science of plant pathology in the 1840s, people began to understand the role of introduced pathogens in some of the important epidemics of that time. They also began to see how some traditional agricultural practices make crops more vulnerable to plant diseases. But many of these lessons were accompanied by tremendous economic loss and important changes in the agricultural development of the areas in which these epidemics occurred. Major changes in agriculture always have political and economic consequences. This chapter describes some historical examples of pathogen movement and agricultural practices that led to epidemics and how quarantine legislation helps protect plants within political borders against invasion by dangerous pests and parasites.

Coffee Rust

Coffee is a crop of the tropics surpassed only by oil in its value as a world commodity. For centuries, it has been a significant import crop in Europe and economically important to the European countries that ruled tropical colonies. It remains an important crop to the independent nations created from those colonies.

Coffee became a popular drink in Europe in the 1600s when contaminated drinking water limited people to fermented beverages or those made with boiled water, such as tea or coffee. Coffee houses were major social centers in England in the 1650s. The Dutch were the first major European coffee importers, transporting coffee from their colonial plantations in Ceylon (now Sri Lanka), Java, and Sumatra. The small, nondeciduous tree, *Coffea arabica*, produces red berries that contain the seeds or beans that are roasted

Fig. 3-1. Coffee (*Coffea arabica*) with the berries that contain the seeds or "beans" that are harvested and roasted.

and then brewed into a potent caffeinated drink. The trees grow best in cool, humid climates but cannot survive frost and are thus limited to tropical highlands.

During Napoleon's time, much of the coffee-producing area was lost by the Dutch to the English. In 1825, the British began development of their property in Ceylon (now Sri Lanka), and every suitable piece of land was planted to coffee plantations. By 1870, Ceylon was the world's greatest producer of coffee. *Java* remains a slang term for coffee, reflecting the time when coffee production centered in that part of the world. Today, however, 90% of the world's coffee comes from the tropical Western Hemisphere. Sri Lanka is now known best for its tea production, and the cup of tea, rather than coffee, has become a familiar part of England's culture. As with the Irish potato famine, a fungus was responsible for these changes, but only because of the agricultural practices of human beings.

The fungal parasite probably arose in southern Ethiopia, the origin of the coffee plant itself. It is a Basidiomycete, a fungal group containing many important plant parasites, and belongs to a subgroup known as the rusts. The rusts are such important plant pathogens that they are discussed in detail in Chapter 10, and the complete biology of the coffee rust fungus is explained there. It is sufficient at this point to consider the problem of a rapidly reproducing fungus capable of infecting the foliage of the coffee tree, a nondeciduous, perennial plant that grows in a frost-free climate.

A single tiny rust pustule on a coffee tree leaf can produce 150,000 spores, and a single leaf can contain hundreds of pustules. When the coffee rust fungus, *Hemileia vastatrix*, reached Ceylon in 1875, nearly 400,000 acres (160,000 hectares) were covered with coffee trees. No effective chemical fungicides were available to protect the foliage, so the fungus was able to colonize the leaves until nearly all the trees were defoliated. The spores produced on the leaves are quite resistant to desiccation, unlike the sporangia of *P. infestans*, and are capable of long-distance movement in a viable

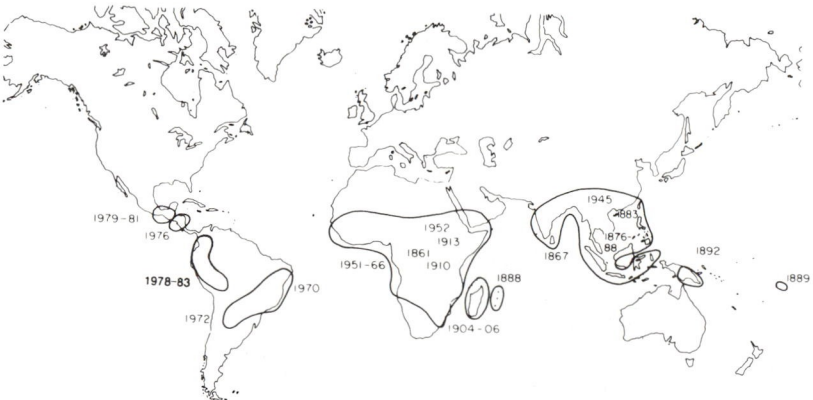

Fig. 3-2. World distribution of the coffee rust fungus, with the dates it was first discovered.

state. They easily moved through the acres of coffee trees, feasting on the banquet prepared by unsuspecting plantation owners. In 1870, Ceylon was exporting 100 million pounds (45 million kilograms) of coffee a year. By 1889, production was down to 5 million pounds (2.3 million kilograms). In less than 20 years, many coffee plantations were destroyed, and production had essentially ceased.

Fig. 3-3. Symptoms of coffee rust infection (*Hemileia vastatrix*).

Fig. 3-4. H. Marshall Ward, a student of Anton deBary, who was sent to Ceylon by the British government to save the coffee plantations from the rust epidemic.

H. Marshall Ward, a student of Anton deBary, was sent to Ceylon by the British government to save the coffee plantations. Even though he failed, he presented the infant science of plant pathology with two important concepts that are still fundamental to plant protection. His studies of the life cycle of the rust fungus convinced him that the germinating spores represented a vulnerable stage for attack. He recommended that, to effectively protect the plant from invasion, fungicides should be present as a protective coating on the leaves before the spores arrived. Once infection had occurred, the hyphae inside the leaf tissue were no longer vulnerable to the fungicide. Thus, it was important to anticipate the disease and not to wait for symptoms to appear before spraying was initiated. Unfortunately, the sulfur fungicides of that time were neither readily available nor very effective, and the rust epidemic was too well established to save the coffee trees.

Ward also warned about the dangers of **monoculture**. He observed that the continuous plantings of coffee trees over the island had created a perfect environment for a fungus epidemic. Rusts, like downy mildews, are obligate parasites and require living host tissue for their growth and reproduction. The rapid epidemic of the coffee rust was enhanced by the many acres

Fig. 3-5. Coffee plantations on a steep hillside in Colombia.

of the host plant. His warnings, unfortunately, were ignored, and most of the dead coffee trees were replaced with tea bushes. Luckily, no fungus immediately invaded the tea crop, and newly discovered fungicides were soon available to protect the tea from its fungal parasites.

In an attempt to escape the rust disease, coffee production moved to the Western Hemisphere. Coffee had been grown in the Caribbean Islands since the 1700s, but plantings quickly spread to the tropical highlands of Brazil, Colombia, and Central America. Today, Brazil, followed by Colombia, dominates the world coffee market. Coffee production centers in the tropical Americas because the coffee rust was successfully excluded by careful quarantines.

The quarantine was successful for over 100 years, but, in 1970, coffee rust was discovered in Brazil. It is not completely clear how the fungus arrived in Brazil, but intercontinental movement of the rust spores from coffee plantations in East Africa is a likely means. The dustlike spores could also have been easily carried on luggage, people, plants, or airplanes that continuously move between the continents. Eradication of infected trees has failed to eliminate the parasite, and the fungus has slowly spread throughout the coffee-growing areas, moving into Colombia and the countries of Central America. The spread was delayed by careful quarantines between many of the countries, but political unrest and human travel, along with natural dispersal of spores by wind, have allowed the fungus to circumvent the quarantines.

What will be the consequences of the importation of such a dangerous pathogen? Frequent fungicide applications will be necessary to protect the highly susceptible cultivars of *C. arabica*, which produce the best quality of coffee. Chemical inputs are particularly demanding for small growers, who now must purchase fungicides and spraying equipment. Trees must be grown at lower densities to allow complete fungicide coverage of the susceptible foliage. Wider spacing of trees also increases air movement between the trees. When the foliage dries more quickly, infections are reduced

Fig. 3-6. Billboard near the airport of Bogata, Colombia, warning about the danger of importing the coffee rust fungus, before its introduction to South America.

since, like almost all fungal spores, rust spores require water for germination. Chemical inputs and changes in planting practices increase the costs of production and hence the price to consumers.

Rust-resistant cultivars of *C. arabica* and other species such as *C. canephora* exist, but the crop is of poorer quality. Plant breeders must often struggle with the problem of combining desirable genetic traits for crop quality with genes for resistance in the same plant. Rust fungi are capable of producing many genetically different races, and 32 races of *H. vastatrix* have been detected. It is always particularly difficult to find durable resistance to a pathogen when the crop is a perennial growing in a frost-free environment. The pathogen population is not reduced by winter stresses, and replanting with new cultivars is expensive and infrequent. Resistance that is effective against all races of the parasite remains the long-term goal. In the meantime, fungicide applications are becoming part of the routine production practices on coffee plantations in the Western Hemisphere.

South American Leaf Blight of Rubber

While the British switched to tea production as a result of the coffee rust epidemic, many of the remaining Dutch holdings in southeast Asia became important rubber plantations. Like the potato, the rubber plant had its origins in South America. Until the year 1900, almost all rubber production was in Brazil and Peru. Once again, a fungal parasite is at the center of our story, greatly assisted by the agricultural practices of human beings.

The rubber plant, *Hevea brasiliensis*, produces natural rubber in a latex sap contained in a system of tubes throughout its trunk. Over 12,000 species of plants exude a milky latex, including such familiar species as milkweed,

Fig. 3-7. Spraying coffee with fungicides to prevent rust disease in the Western Hemisphere.

poppy, lettuce, poinsettia, and their relatives. The exact function of the latex is not known, but, in some cases, it contains compounds that protect against insect pests. The latex of many of these plants contains the polymer compounds known as rubber, and the latex of *H. brasiliensis* is a particularly efficient source, containing 40–50% rubber. In the days before the abundant variety of plastics and other polymers derived from oil, there were many industrial, medical, and household needs for a durable and flexible waterproof substance. In 1840, Charles Goodyear developed the process of vulcanization that made rubber tough and resistant to melting at high temperatures but not brittle in cold weather.

With the advent of motorized vehicles, the demand for rubber greatly increased for use in tires. Unfortunately, rubber trees grew in the jungles of South America at very low densities, only a few in a single acre. Rubber was collected by natives who cut down whole trees and collected the mass of latex that bled from the tree. This process became very inefficient as the collectors were forced to travel farther and farther into the jungle searching for rubber trees.

Seeds of these valuable plants were smuggled out of South America by an Englishman, H. A. Wickham, in the 1870s. Plantation production was begun in Asia after a tapping system was developed that allowed the latex to be repeatedly harvested from the trees without killing them. Plantation production followed in South America as jungle harvesting became more and more difficult. Following World War I, there was particular pressure to develop plantation rubber in South America to provide secure sources of rubber for the United States in times of war. Henry Ford developed 8,000 acres (3,200 hectares) in Brazil for rubber production in 1929, but once again a microscopic fungus was responsible for the failure. The trees were all killed by a disease called the South American leaf blight, resulting from infection by the fungus *Microcyclus ulei.*

This fungus is a member of the Ascomycetes, a major fungal group mentioned in Chapter 2. The mycelium of an Ascomycete can be easily distinguished from that of an Oomycete due to the presence of septa, or crosswalls, that separate a hypha into compartments. Such hyphae are described as **septate** to separate them from the nonseptate hyphae of Oomycetes. In Ascomycetes, the number of haploid nuclei in each compartment can vary because each septum contains a pore that allows nuclei and cytoplasm to move throughout the mycelium.

As with many other fungi, both sexual and asexual spores are produced. The asexual spores, called **conidia**, are important in the rapid reproduction of the fungus but represent no genetic variation except that produced by mutation. Conidia of *M. ulei* are produced on the surface of infected leaf tissue, where they are easily dispersed by air to other plants. Only a few days after infection, conidia begin to be produced in numbers too large to imagine, continually infecting the tender new tissue of young leaves.

M. ulei also produces a survival stage through sexual reproduction (Fig. 3-8). The vegetative mycelium is haploid, and, at the time of sexual reproduction, a nucleus from a male structure (often an **antheridium**) joins the nucleus in the **ascogonium** (analogous to the oogonium of the

Oomycetes). The nuclei divide in tandem to produce a limited dikaryotic mycelium. In the ascus mother cells, karyogamy results in a series of diploid nuclei, each of which immediately undergoes meiosis to produce four haploid nuclei. These nuclei undergo mitosis to produce eight nuclei, each of which becomes enclosed by a wall and some cytoplasm to form an **ascospore**. Each set of ascospores is produced in a sac, or **ascus**, which is characteristic of sexual reproduction of all members of the Ascomycetes. **Asci** (the plural of *ascus*) often contain four or eight ascospores because they are the products of meiosis (four ascospores) and frequently a secondary mitosis (eight ascospores).

The asci are produced in a vaselike **perithecium**, a protective fruiting body that usually consists of dark-colored hyphae that are resistant to desiccation. In addition to their ability to survive adverse conditions, ascospores also represent a genetically diverse new generation because they are a product of meiosis and genetic recombination. The ascospores are forcibly discharged through an opening at the top of the perithecium by a puffing mechanism caused by changes in humidity as drier air passes by the liquid-filled asci. This small propulsion is sufficient to send ascospores on a journey to other rubber plantations or to renew an epidemic in the home plantation.

The Ascomycete group includes many important plant pathogens and other well-known fungi as well. Probably the best known Ascomycetes are the yeasts, so important in the production of bread and alcoholic beverages. Rather than having the mycelium of typical fungi, the yeasts have a vegetative structure that is reduced to single cells that bud to produce new cells (Fig. 3-9). Sexual reproduction involves the fusion of the nuclei of two single cells, followed by meiosis, resulting in an ascus of ascospores.

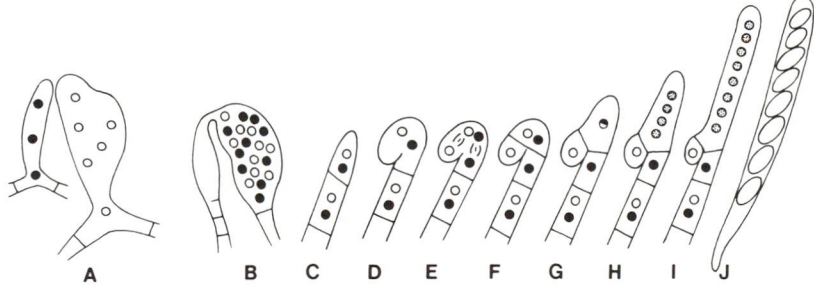

Fig. 3-8. Diagram of generalized ascus formation. **A,** Male structure (often an antheridium) and female structure (ascogonium); **B,** the male structure contributes nuclei to the female structure; **C,** a limited dikaryotic mycelium develops in which each cell contains two genetically different nuclei; **D-F,** through a complex mechanism, a single cell, the ascus mother cell, becomes delimited; **G,** karyogamy occurs and the two nuclei fuse to form a single diploid cell; **H,** meiosis produces four haploid nuclei; **I,** in many Ascomycetes, mitosis follows meiosis to produce eight haploid nuclei, and the sac, or ascus, containing the nuclei elongates; **J,** a spore wall forms around each nucleus. The ascus now contains eight ascospores.

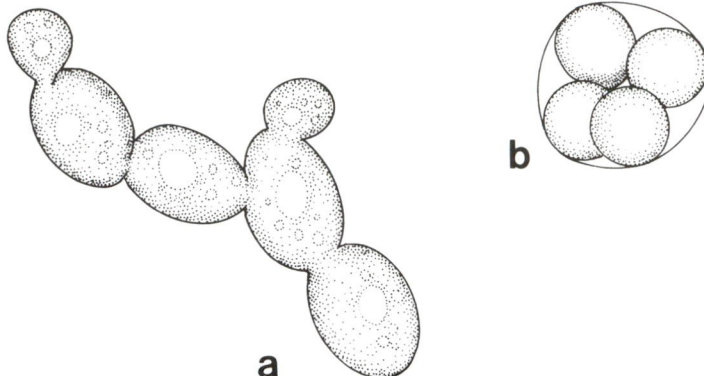

Fig. 3-9. Yeast: **a,** the greatly reduced vegetative cells bud to increase their numbers; **b,** in sexual reproduction, an ascus containing ascospores is produced.

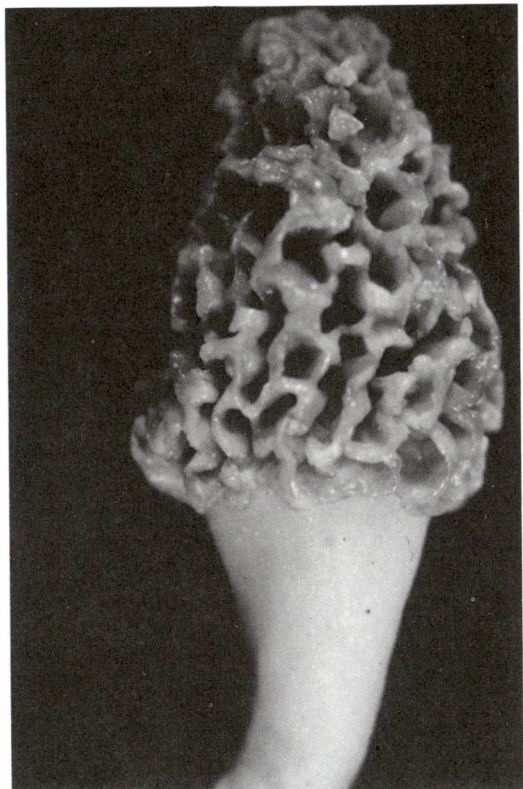

Fig. 3-10. A morel mushroom. The surface of the mushroom is lined with asci.

Brightly colored, cup-shaped saprophytic fungi on decaying wood and the morel mushrooms of gastronomic fame are the reproductive structures of other familiar Ascomycetes. Microscopic examination of their surfaces reveals masses of asci containing ascospores.

We should now consider why *M. ulei* destroyed the South American plantations but did not destroy the rubber trees in the jungles. The answer lies solely in the density of planting. All plants are parasitized by relatively harmless leaf spot fungi. If a rubber plant in the jungle becomes infected by *M. ulei* and spores are produced, the probability of those spores landing on another rubber plant is very small. The low density of the rubber trees, combined with the dense foliage of surrounding plants, protects rubber plants from frequent infections. In plantations, however, there is no blocking foliage of nonhost plants, and plants exist in high densities, creating the perfect environment for an epidemic, especially in the continuous warmth and moisture of the tropical climate. It is not surprising that this previously minor parasite became a deadly pathogen in plantations.

The plantations of Southeast Asia were, and are, just as vulnerable to

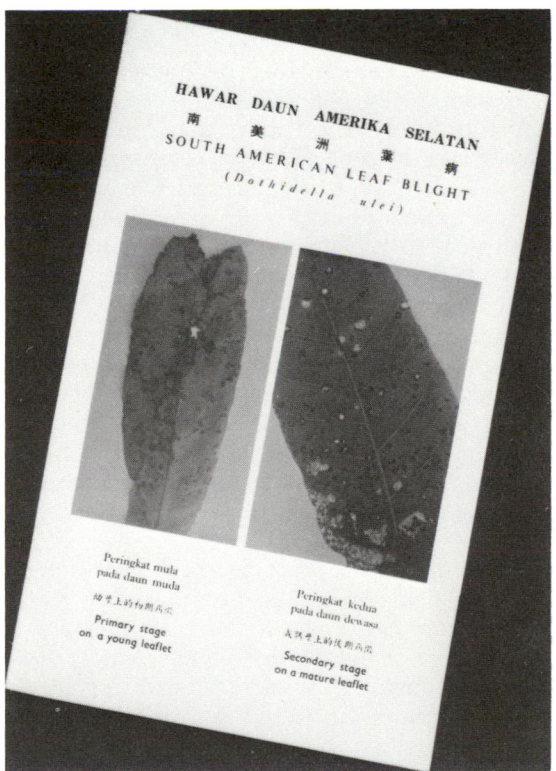

Fig. 3-11. Bulletin published by the Rubber Research Institute of Malaysia in 1966 to alert growers to the danger of South American leaf blight.

destruction, but they have been able to exist on a continent without the leaf blight fungus, the mirror image of coffee production. Rubber plantation workers are trained to recognize the lesions of *M. ulei*, and continual vigilance has kept the Asian plantations blight-free so far. Back in Brazil, only repeated fungicide applications ensure protection against the continuous supply of spores in a plantation environment. Genetic resistance has been investigated, but trees resistant to the fungus yield less rubber than the more susceptible ones. Disease-resistant tops can be grafted onto high-yielding root stocks, but the process is costly. Brazil, the site of origin of the rubber tree, is now an importer of natural rubber. Asian plantations, created from the fungus-free seeds smuggled out of South America, produce 90% of the world's natural rubber. Only a strict quarantine and many miles of ocean have prevented devastation of the Asian plantations.

The Need for Quarantines

Both coffee rust and South American leaf blight of rubber are examples of how monoculture agriculture increases the danger of major epidemics. In today's agriculture, most crops are placed in similarly precarious situations. A field or orchard planted with hundreds or thousands of plants of a single species is a generous invitation to any parasite. In perennial and tropical environments, the danger is even greater because susceptible plant tissue may be continually available, and winter stresses do not reduce the parasite population. For both rubber and coffee production, escape to another continent was the most effective solution. Rubber remains safe from leaf blight in Asia, but coffee producers must now learn to coexist with the rust fungus throughout the world.

Three important plant disease epidemics that occurred in the early part of the 20th century led to the first U.S. quarantine legislation in 1912. In two of the epidemics, native plants were killed by introduced parasites to which they had little resistance. The chestnut tree, *Castanea dentata*, a dominant species throughout the Appalachian mountain range, was nearly eliminated by a fungus from Asia, and the eastern white pine, *Pinus strobus*, suffered tremendous losses to a fungus introduced from Europe. Citrus is an introduced crop in America, but a serious bacterial disease, citrus canker, was also introduced to the citrus orchards at about the same time. All three diseases have important histories and will be discussed in greater detail in later chapters.

Following these epidemics, it became clear that open borders and unrestricted import of plants and plant products such as fruit, nuts, and lumber threaten important native and agricultural plants. It was also clear that the continued import of plants and plant products was necessary for agriculture and the economy. The world map showing the origin of many food crops (Fig. 3-12) makes it obvious that our agricultural production would be excessively limited if it were restricted to plants native to North America.

Microscopic parasites in plant tissue are very difficult to detect, unless the infection is well established so that lesions of dead cells or other symptoms of disease are evident. Some infected plants remain symptomless, and virus diseases in woody plants may take years to exhibit symptoms. Thus, the successful detection of parasites by government plant inspectors at borders falls short of that necessary to completely exclude threatening pathogens. In fact, most inspectors look for the larger insect pests and the general soundness of the plant material. Early infections or the presence of microscopic spores and other resting structures of parasites are impossible to detect visually.

How can parasites be stopped at the borders if they can not be quickly and economically detected? One important way is to simply forbid the import of any plant material by travelers. Anyone who reenters the United States is required to dispose of all plant material. They may even be checked by Department of Agriculture dogs specially trained to detect plant material. These dogs are selected to be docile around the milling crowds of travelers trying to pass through customs. One dog described in a recent issue of *The New Yorker* had been trained to sniff out citrus fruit to prevent the reintroduction of the citrus canker bacterium. He was very accurate except when confused by lemon-lime scented shaving creams. Like all laws, the success of border quarantines lies not in enforcement but in cooperation by educated travelers. Each year uninformed citizens threaten American agriculture with their botanical souvenirs.

People in the business of agriculture may import plants, seeds, and plant products by special permit according to the restrictions of quarantine

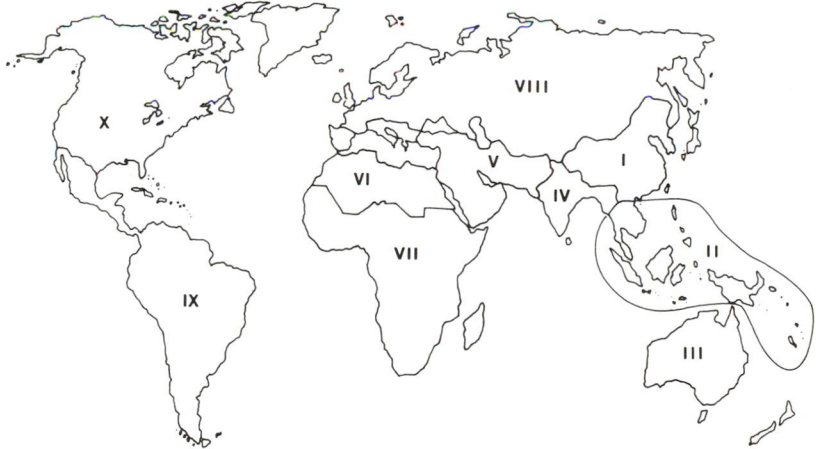

Fig. 3-12. Regions of genetic diversity and their associated crops. I, Chino-Japan: soybeans, oranges, rice, tea; **II,** Indochinese: banana, coconut, yam, rice, sugar cane; **III,** Australian: none; **IV,** Hindustanean: jute, rice; **V.** West Central Asiatic: wheat, barley, grapes, apples, linseed, sesame, flax; **VI,** Mediterranean: sugar beet, cabbage, rapeseed, olive; **VII,** African: oil palm (oil, kernel), sorghum, millet, coffee; **VIII,** Euro-Siberian: oats, rye; **IX,** Latin American: maize, potato, sweet potato, cocoa, cassava, tomato, cotton (lint), cottonseed (oil), seed cotton (meal), tobacco, rubber; **X,** North American: sunflower.

legislation. Three main categories exist for the importation of plants. The broadest category is called "restricted" and applies to most plants and seeds. They must be inspected and possibly subjected to chemical treatment before release into the United States. A second category includes many vegetatively propagated crops such as fruit trees, flowers, and woody ornamentals that must be maintained for a postentry period of about two years before their general release. During this period, the plants are carefully inspected and subjected to laboratory tests to determine whether they carry parasites that could threaten plants already present in the United States. Such plants are often maintained in agricultural experiment stations designated as entry stations. A third category exists for plant material that may carry parasites that pose such a hazard to U.S. plants that they are completely prohibited from import. This category is often the most controversial because growers may feel that they are unable to obtain plant material necessary for competition in world markets. Careful surveys to determine that the pest

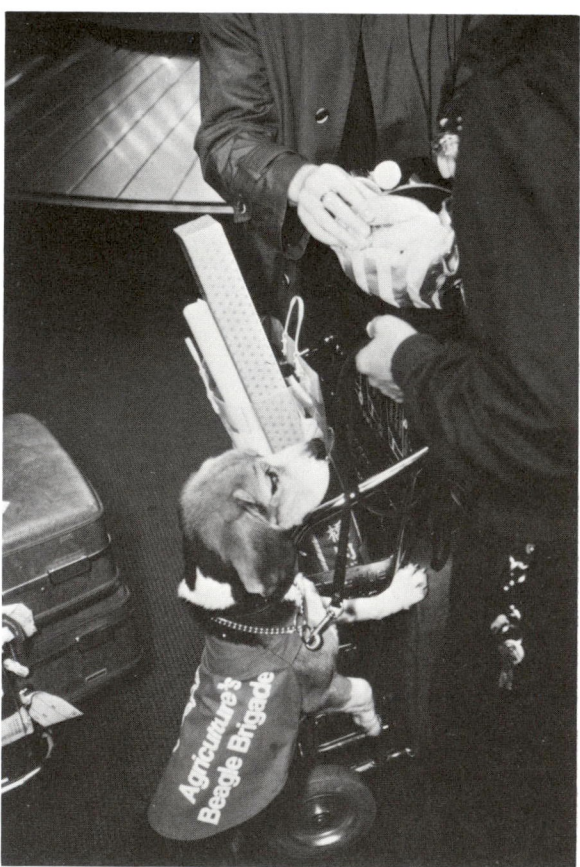

Fig. 3-13. The U.S. Department of Agriculture trains beagles to help detect hidden plant and animal products carried by travelers.

is not already present in the United States and periodic reevaluation of the threat are needed.

Many changes have occurred since the first U.S. quarantine legislation of 1912. The **Federal Plant Pest Act of 1957** includes restrictions on importation of a broad range of plant pests and parasites. It allows the Federal government to restrict interstate movement of plant material and permits destruction of diseased or infested material in emergency situations. The **Agricultural Plant Health Inspection Service** (APHIS) monitors imported plants and the distribution of pests and parasites within the United States. Individual states may have their own protective legislation. For instance, citrus from Florida is prohibited from California, Arizona, and other citrus-growing states.

The earliest quarantines were created out of fear of exposure to diseased people even before the nature of contagious or infectious disease was under-

Fig. 3-14. Part of the material confiscated at JFK International Airport in New York in a single day.

stood. The word **quarantine** is derived from the Latin word for 40 because ships coming into port with diseased passengers were required to wait 40 days before disembarking. Presumably all contagious people would have died or recovered by then. Plant quarantines began in Europe in the late 1800s to prevent the introduction of the Colorado potato beetle from North America. Today, it is necessary to justify quarantine regulations biologically because they are expensive to administer and restrictive to trade.

Some countries are geographically isolated. Plants in such countries have been biologically isolated from outside organisms and are thus particularly vulnerable. Such a country is Australia, where arriving travelers find themselves sprayed with an insecticide while still on the airplane and questioned in detail by border officials about plant and animal products and any recent visits to farms. Countries with ocean borders have the luxury of easy border control compared to land-locked countries. Those that share borders and climates often have relaxed restrictions with their neighbors because natural dispersal of pests and parasites is inevitable and will circumvent any human attempts to restrict their spread. Between neighboring countries, such as the United States and Canada, restrictions are usually confined to soil or machinery that might carry soilborne pests. In many countries, it is difficult to adequately protect the borders from invading pests. Language problems, prohibitive costs, and lack of adequate training of border guards contribute to the rapid spread of new pests. Because of these problems, over three fourths of all countries have no restrictions on plant material carried by travelers.

At best, quarantine restrictions represent a delay in the spread of a parasite. Restrictions are least effective where natural dispersal by air movement is possible. They are most effective for isolated areas where long expanses of ocean separate land masses and for parasites that are strictly soilborne. These delays can be of great significance in plant protection because they give plant pathologists time to determine the best practices to reduce epidemics and, more importantly, time to select disease-resistant cultivars before the parasite arrives.

Fusarium Wilt of Bananas

Earlier in this chapter, we saw that air-dispersed fungal parasites caused coffee production to move to the Western Hemisphere and rubber production to center in Asia. In both cases, it was necessary to cross oceans to avoid the easily dispersed spores. Such parasites are the most difficult to restrict by quarantines. Soilborne pathogens are dispersed not by air but by water and soil movement on machinery, people, animals, etc. Restriction of the movement of soilborne pathogens is theoretically easier although not necessarily more successful in the real world. Diseases caused by soilborne pathogens are no less devastating than those caused by airborne pathogens either. As seen with many *Phytophthora* diseases, protection of underground plant parts is particularly difficult. Thus, once a pathogen is introduced into soil, it continues to multiply on the host plants until susceptible plants can no longer be grown on that site.

Bananas (*Musa* species) are an example of a crop where the spread of soilborne pathogens has caused tremendous losses worldwide and necessitated the abandonment of many acres of contaminated land. Cultivation of the banana by human beings has an ancient history, probably originating in southeast Asia or Africa and spreading throughout tropical lands. Production of the sweet dessert bananas is greatest in Central America, but the starchy cooked banana, or plantain, is the more important food for tropical people. Plantain cultivation places Africa in first place for banana production, despite the extensive exports of Central American bananas to the United States, where bananas are the most popular of all fresh fruits.

The wild progenitors of the cultivated bananas are not known. Commercial cultivars are mostly sterile **triploid** plants, which means that their nuclei contain three sets of chromosomes. Because banana fruits contain no seeds, they are propagated vegetatively like the potato. New plants grow from the buds of upright, underground stems called **corms**. Like potatoes, these plants are genetically identical and thus equally susceptible to parasites. Banana plants are not trees but large herbaceous plants without woody tissue. Within one year the **inflorescence**, or flower stem, emerges from the plant, and on these the clusters of banana fruits will develop. Fruit

Fig. 3-15. Banana plant (*Musa* species).

production occurs only once on a plant, so new plants must be cultivated after each crop. As cultivation continues, soilborne parasites multiply as long as the crop is grown in the same land. If corms are transplanted to new soil, they are likely to carry the parasites with them in infected tissue or in soil clinging to their surfaces.

One of the most important diseases of banana, **Panama disease**, is named for the country where the first epidemics occurred. The disease is caused by the fungus *Fusarium oxysporum*, which is now well established throughout banana-growing areas of the world. The leaves of banana plants infected by *F. oxysporum* turn yellow, beginning with the lower leaves, until eventually all leaves appear wilted and brown as if unable to draw enough water from the soil. A cross-section of the stem reveals a brown discoloration of the water-conducting tissue, or **xylem**. Microscopic examination of the tissue reveals tubular xylem vessel cells filled with the mycelium and spores of the invading fungus.

Such a disease is called a **vascular wilt** and results in the plugging,

Fig. 3-16. Panama disease (*Fusarium oxysporum* f. sp. *cubense*) of banana. Note the numerous wilted lower leaves of the diseased plants.

degradation, and collapse of xylem vessels by a pathogen adapted for invasion of the xylem. The fungus produces enzymes capable of degrading the cell walls of the xylem vessels to release nutrients for its own growth. Enzyme degradation weakens the cell walls and plugs the vessels with gums and polysaccharides. The fungus also produces toxins that poison the living parenchyma cells adjacent to the water-transporting vessels. These

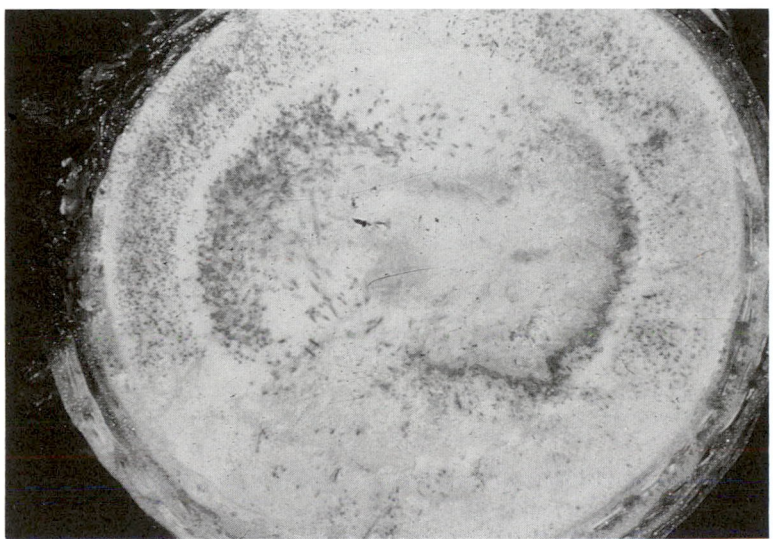

Fig. 3-17. Discoloration of infected xylem in the stem of a banana plant with Panama disease.

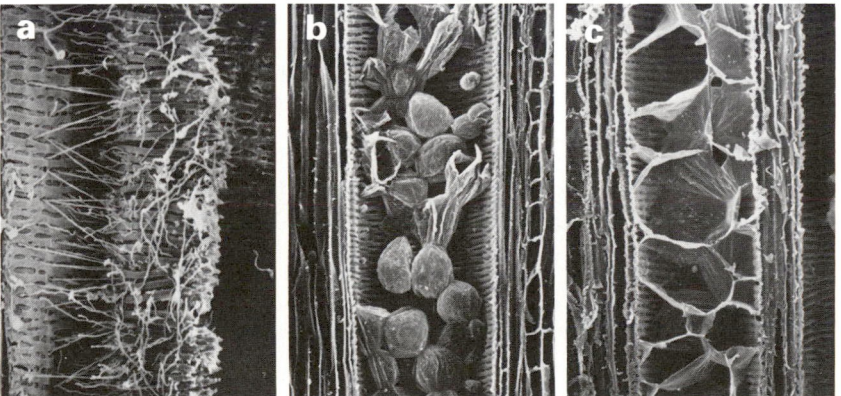

Fig. 3-18. Blockage of the xylem: **a,** Longitudinal section of a xylem vessel invaded by *Fusarium oxysporum*; **b,** tylose formation that contributes to xylem blockage; **c,** complete blockage of a xylem vessel by tyloses.

parenchyma cells balloon into the vessels through pits in the cell walls, forming **tyloses** (bulges) that further reduce water transport. The hyphae and conidia of the fungus itself also help block the xylem. Thus, the symptoms of the disease reflect the activities of the invading fungus.

The most obvious symptom of a vascular wilt disease is water stress or wilting. The distinct yellowing is partially due to nitrogen deficiency because the xylem fluid also carries important mineral nutrients from the soil to the plant. The lower leaves exhibit symptoms first because the fungus invades the plant below the ground by means of germinating spores. In many cases, the planted corms may already be infected, so the disease develops even more quickly.

Once the hyphae reach the xylem, they produce small oval **microconidia** that can be carried rapidly upward in the xylem fluid. While the plant is still alive, the fungus remains confined to the xylem. After the plant dies, *F. oxysporum* begins to colonize all stem tissue until it reaches the surface of the plant. On the surface, two other kinds of conidia are produced, sickle-shaped macroconidia capable of new infections and thick-walled **chlamydospores** for survival during adverse conditions. It should be noted that no sexual spores have been described, and none are known. Like certain other fungi, *F. oxysporum* is classified in the **Deuteromycetes**, or **Imperfect Fungi**, because it apparently lacks a sexual or "perfect" stage in its life cycle. Studies of closely related fungi have demonstrated that fungi in the genus *Fusarium* are probably Ascomycetes.

The naming and classification of Imperfect Fungi is particularly confusing to the novice because many important plant pathogens have two or more Latin names. This is because many of them rarely, if ever, produce a sexual state. Once a sexual state has been discovered, or manipulated to develop under laboratory conditions, a Latin name based on its characteristics is designated. This becomes the official name of the organism. But plant pathologists may see the sexual stage only late in an epidemic or under unusual laboratory conditions. For practical reasons, they need to be able to identify a fungus early in an epidemic based on its asexual, or "imperfect," conidial characteristics, so another set of Latin names has been created for this purpose. Of course, it is the indistinguishable mass of vegetative mycelium that is actually causing disease in a host plant, but we can usually only differentiate fungi using their sexual and/or asexual reproductive characteristics.

When the sexual stage is common and well known, the Latin name descriptive of that stage is used. When the sexual stage is uncommon or unknown, the Latin name(s) based on conidial characteristics are used. Fungi produce only one sexual stage, if at all, but may produce several conidial stages. *F. oxysporum*, for example, produces no sexual stage but three different types of conidia: microconidia for colonization of the xylem, macroconidia for infection of the roots, and thick-walled chlamydospores for survival.

A further complication in our ability to classify and identify fungi lies in the host specialization that many fungi exhibit. *F. oxysporum* causes vascular wilt disease in many kinds of plants besides bananas, but the strain of the fungus that attacks bananas is different from the one that attacks, for instance, tomatoes. These differences are not apparent in the appearance of the fungus, so the various strains are classified as one species. The group of plants that a strain can infect, its **host range**, is then designated by naming a "special form" (*forma specialis* in Latin) of the fungus. An isolate of *F. oxysporum* cannot be identified with respect to its ability to invade a plant based on its appearance. Only test inoculations on various host plants can determine the forma specialis (abbreviated "f. sp.") to which it belongs. Panama disease of bananas is caused by *F. oxysporum* f. sp. *cubense*. Other formae speciales of *F. oxysporum* cause vascular wilts of such economically important crops as tomatoes, cucumbers, and cotton. Although we are unable to see their differences, the special forms are obviously very different physiologically and genetically because of their distinct host ranges.

For many years, the most popular fresh fruit banana cultivar was Gros Michel. Corms of this cultivar were distributed to tropical regions around the world. As early as 1890, Fusarium wilt was recognized in Panama and has since been discovered in all banana-growing areas. Once the pathogen is established in the soil, trees begin to die. Since there is no cure or control for the disease, the only solution is to move on to new lands. Over 100,000 acres (40,000 hectares) have been abandoned in Central and South America. New land is usually cleared from the jungles at a cost of $1000–2000/acre. The fungus can be introduced to new areas on

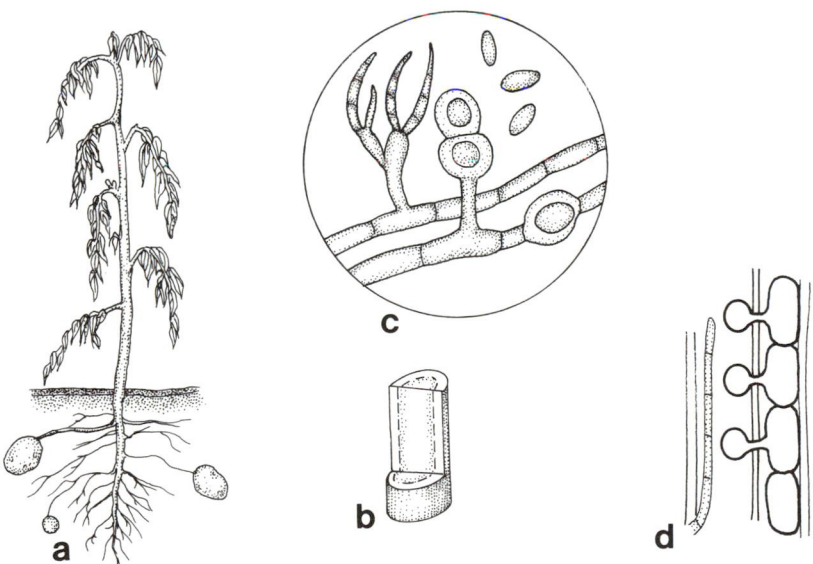

Fig. 3-19. Fusarium wilt: **a,** plant exhibiting wilt; **b,** vascular discoloration; **c,** spores of *Fusarium oxysporum* (macroconidia, microconidia, and chlamydospores); **d,** hyphae and tyloses in infected vessels, longitudinal section.

infected corms, on contaminated soil on feet and equipment, and by flowing water. These areas will also have to be abandoned when disease losses become too great. For some years, in places where new lands were not readily accessible, such as in Asia, the resistant banana cultivar Cavendish was grown. Fusarial wilt was discovered on this cultivar in Taiwan in 1967. A new race of *F. oxysporum* f. sp. *cubense* was determined. Inoculations of plants of various banana cultivars and close relatives of the banana have demonstrated that the fungus exists as several different races that vary in their host range.

At present, no commercially acceptable banana cultivar is resistant to all known races. This represents a great challenge to banana breeders who wish to select wilt-resistant bananas. Since cultivated bananas are sterile triploids, traditional breeding crosses between them or crosses to wild diploid relatives are impossible. Sources of genetic variation may be derived using tissue culture techniques and genetic engineering, but short-term disease control lies in the techniques appropriate for many soilborne pathogens. Pathogen-free corms should be planted in new land with special care not to contaminate the land with soil or water from infested areas. Where contamination already exists, crop rotation with paddy rice can reduce the fungus population because of the anaerobic (oxygen-free) conditions of the flooded soil.

Bananas, like potatoes, are a genetically uniform, vegetatively propagated crop. Eventual contamination of land by *F. oxysporum* f. sp. *cubense* leads to serious losses. In Costa Rica, banana production was 11 million bunches in 1923 but was reduced by Fusarium wilt to only 1.4 million in 1941. In the Western Hemisphere, jungle destruction and cultivation of new land remain the most effective means for escaping the fungus. Genetic uniformity and monocultural plantations have made it possible for even a soilborne fungus to hinder production of an immensely important tropical food crop in soils around the world as contamination spreads.

Selected Readings

Dowling, C. F., Jr., Graham, A. E., and Alfieri, S. A., Jr. 1982. Plant inspection and certification. Plant Disease 66:345-351.

Schieber, E., and Zentmyer, G. A. 1984. Coffee rust in the Western Hemisphere. Plant Disease 68:89-351.

Su, H., Hwang, S., and Ko, W. 1986. Fusarial wilt of Cavendish bananas in Taiwan. Plant Disease 70:814-818.

Thurston, H. D. 1984. Tropical Plant Diseases. American Phytopathological Society, St. Paul, MN. 208 pp.

Waterworth, H. E., and White, G. A. 1982. Plant introductions and quarantine: The need for both. Plant Disease 66:87-90.

Weber, N. S. 1988. A Morel Hunter's Companion. Two Peninsula Press, Lansing, MI.

Yarwood, C. E. 1970. Man-made plant diseases. Science 168:218-220.

Bacteria

CHAPTER 4

In Central and South America, the sight of brown and wilted banana plants is all too common. In many cases, the xylem vessels are infected with the vascular wilt fungus, *Fusarium oxysporum* f. sp. *cubense*, the Panama disease discussed in the previous chapter. Sometimes the source of the problem is different, and the cut tissue is filled with the sticky ooze of the vascular wilt bacterium, *Pseudomonas solanacearum*.

It is often difficult to diagnose the causal agent of a plant disease based only on the symptoms exhibited by the host plant. Both fungi and bacteria can cause leaf spots, blight, wilt, and other plant disease symptoms. For instance, the bacterial vascular wilt disease of banana just mentioned, **Moko disease**, may result in symptoms similar to those of Panama disease because both the fungus and the bacterium infect the xylem and therefore cause the same water stress symptoms.

Similar problems arise in the diagnosis of human and animal diseases. People often ask their physician for an antibiotic prescription for a severe sore throat. In most cases, however, the physician will wait for laboratory results from a throat swab to determine whether the sore throat symptom is the result of a viral or a bacterial infection. The symptom, a sore throat, is nearly identical in both cases, but antibiotics will be effective only if a *Streptococcus* bacterium is the causal agent ("strep throat"). If a virus is the causal agent, one can only wait for antibody production by one's own body to eliminate the infection because antibiotics are not effective against viruses.

Like other organisms, bacteria receive Latin binomials. *Pseudomonas* is a genus consisting of many bacterial species. The specific epithet of the causal agent of Moko disease, *solanacearum*, may seem surprising since it refers to the plant family, Solanaceae. It reflects the fact that one race of this bacterial species can cause vascular wilt disease in members of the Solanaceae, such as tobacco, tomato, and potato, when they are grown in warm climates. The species consists of several races that vary in their host range. A different race of *P. solanacearum* is responsible for the Moko disease in bananas.

The sticky ooze in infected plant tissues consists of millions of rapidly dividing bacteria. When these bacteria are present in plant tissue, even in low numbers, any cutting is likely to spread the bacteria to healthy plants. In the process of cutting banana corms or potato tubers for planting, even one infected piece can result in the spread of bacteria to cut surfaces of many healthy corms or tubers. Careful inspection and discard of infected propagative parts are required to prevent massive losses in new plantings.

Because visual inspection is insufficient for detection, vascular wilt diseases due to *P. solanacearum* are common in warm, humid climates. Bacterial wilt, caused by *P. solanacearum* Race 1, is a limiting factor in potato production in warm lowland tropics. Moko disease, caused by *P. solanacearum* Race 2, is an important banana disease in the Western Hemisphere, but it has not yet been found in Africa, the world's largest area of banana production.

Biology of Bacteria

Some very important plant diseases are caused by bacteria. As with fungi, the ability of bacteria to function as parasites is linked to their biology. Even though infection by fungi or bacteria can cause similar plant disease symptoms, bacteria are very different from fungi. One immediately obvious difference is that bacteria are much smaller than fungi. A typical bacterial cell is about 1 micron (micrometer) in diameter, whereas an average fungal spore might be 50–200 microns in diameter. Hyphae are commonly 5–10 microns wide and may be many millimeters long. It is even possible to see individual hyphae growing across a culture plate, whereas bacteria are visible only using the magnification of a compound microscope.

Bacterial cells are also very different from the cells of plants, animals, and fungi. These other life forms are all highly evolved eukaryotic organisms, whereas bacteria are **prokaryotes**. Fossil prokaryotes (*pro* = before, *karyo* = nucleus) appear in rocks about 3.5 billion years of age, and prokaryotes remained the only form of life on earth for over 2 billion years. Fossils

Fig. 4-1. Internal symptoms of the Moko bacterial disease (*Pseudomonas solanacearum*) at the base of a banana fruit, showing ooze and degraded tissue.

of eukaryotic organisms do not appear in rocks older than approximately 1.5 billion years. In fact, one theory suggests that the membrane-limited organelles of **eukaryotes** (such as **mitochondria** and **chloroplasts** that function in respiration and photosynthesis, respectively) originated from free-living prokaryotes that came to exist symbiotically within what was to become the eukaryotic cell. The mitochondria and chloroplasts divide independently of mitotic cell division in eukaryotic cells. They also contain

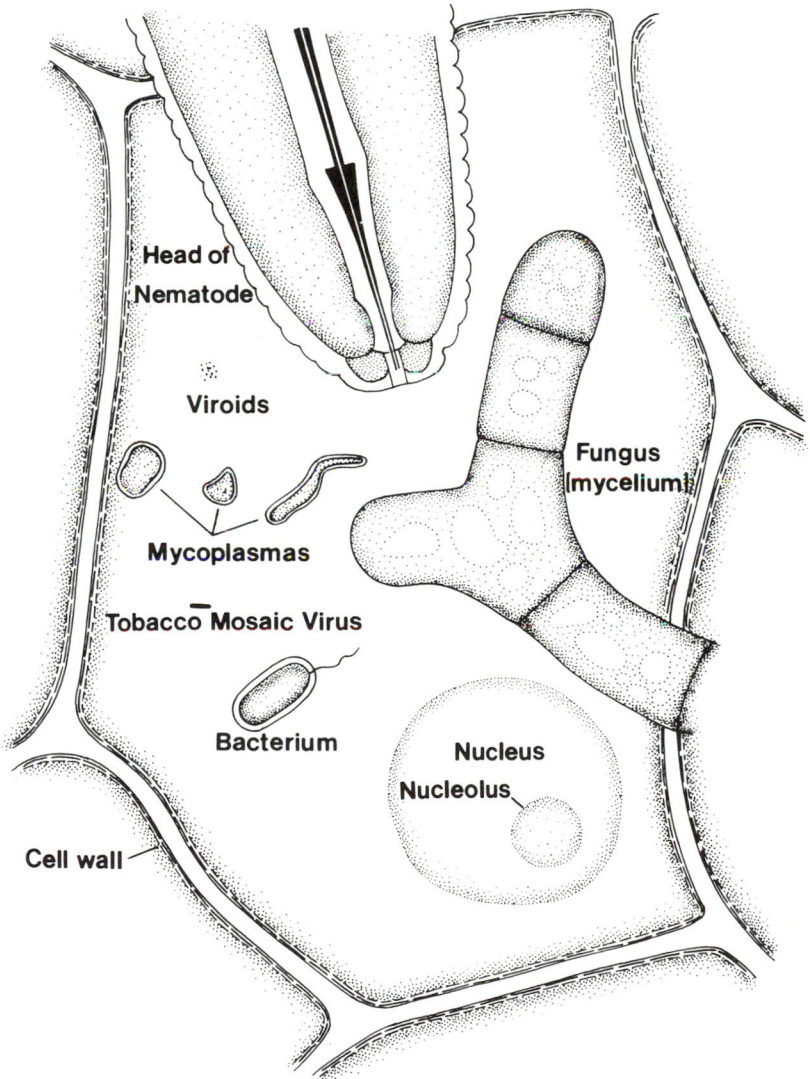

Fig. 4-2. Schematic diagram of the shapes and sizes of certain plant pathogens in relation to a plant cell.

their own genetic information in a form similar to that of bacteria.

Recent research has shown that the separate genetic information contained in organelles is a significant component of the total genetic makeup of a eukaryotic organism and can influence the interactions of plants and their parasites. During sexual reproduction in some eukaryotes, organelles are all contributed by the female because the male contributes only nuclear genes to the next generation, whereas the female contributes the cytoplasmic contents of its sex cell as well as nuclear genes. The genetic information associated with these organelles is variously described as **extrachromosomal** (outside of the nuclear chromosomes), **maternal** (contributed only by the female parent), or **cytoplasmic** (in the cytoplasm, i.e., outside the nucleus) **inheritance**. We will return to the significance of these genes in plant disease in Chapter 6.

Prokaryotic cells contain no nucleus, other organelles, or large, highly organized chromosomes. They are usually unicellular, although sometimes a few cells maintain a loose connection. Some prokaryotes are **autotrophic**, which means they require only simple nutrients and CO_2 and synthesize their own organic carbon. For example, cyanobacteria can photosynthesize and were previously called the blue-green algae, but the important cellular differences between prokaryotes and eukaryotes make them more closely related to other bacteria than to other algae, all of which are eukaryotes.

Most bacteria are **heterotrophs** and require organic carbon from outside sources. Most are also **saprophytes**, obtaining nutrients from nonliving organic matter, and so they play an extremely important role as decomposers. Of the approximately 25,000 bacterial species, only a relatively small number are actually pathogens causing disease in animals, people, or plants. Except for a few species that may be able to infect humans under extremely unusual conditions, the bacterial species that cause plant diseases threaten only plants.

Bacteria commonly exist in three general shapes: 1) **bacilli**, which are rod-shaped, 2) **cocci**, which are spherical, and 3) **spirilli**, which are long and coiled. Common bacteria have a cell wall, although some prokaryotes,

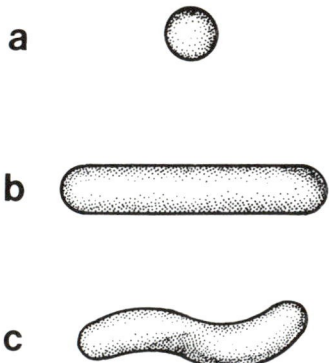

Fig. 4-3. The fundamental shapes of bacteria: spherical (a), rod-shaped (b), and spiral (c).

called mycoplasmas, do not. Some mycoplasmas are plant pathogens (see Chapter 12). The bacteria with cell walls that are plant pathogens are all rod-shaped bacilli.

Some bacteria move by means of **flagella**, although their small size limits movement to very short distances. The number and arrangement of flagella on a bacterial cell are important taxonomic characteristics. The flagella are so small and move so rapidly that special techniques must be employed if they are to be visible under the microscope.

Stains may be used to make the bacteria themselves more visible when viewed with a microscope. Dark-colored stains retained by the cells make them more distinct on a light background. An alternative method is to stain the background a darker color so that the light-colored bacteria appear more prominent. An important staining procedure was discovered by the Danish microbiologist **Hans Christian Gram**. He was looking for a stain that would be absorbed by all bacterial cells. Using his method, however, only some types of bacteria stained purple. This accidental discovery actually represented a means of separating bacteria into two biologically different groups. Those bacteria that stain purple, called **Gram-positive**, have a relatively thick, uniform cell wall. Those that do not retain the stain, called **Gram-negative**, have a thinner cell wall with an additional outer layer of polysaccharides and lipids. The layer containing the lipids protects the bacteria from absorbing certain substances.

In medicine, these differences can be important. For instance, the common **antibiotic** penicillin is ineffective against Gram-negative bacteria. A Gram stain of bacteria cultured from a patient can guide a doctor in choosing a correct prescription drug. Of the plant-pathogenic bacteria, only species in the genus *Clavibacter* (formerly in the genus *Corynebacterium*) are Gram-positive, which limits the usefulness of this technique in identification of

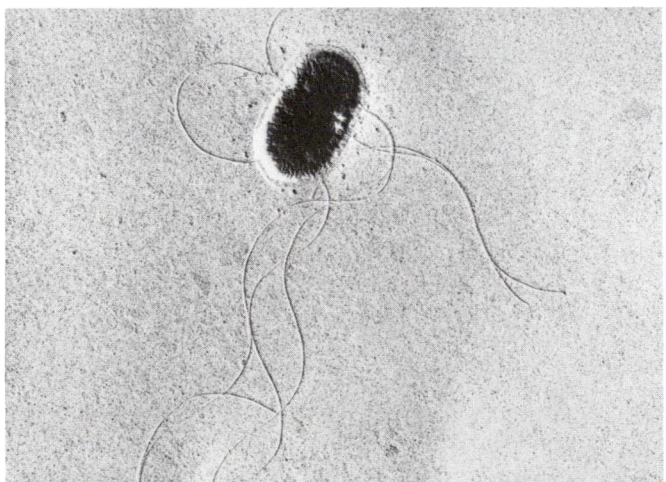

Fig. 4-4. A bacterium, *Erwinia amylovora*, with flagella (×30,000).

plant pathogens, because all the other genera are Gram-negative.

Bacterial cells divide by a process called **fission**, in which the cell membrane and cell wall grow inward to divide the cell in two. Most of the genetic information of a bacterium exists in a single circular chromosome that replicates, and each daughter cell receives a copy. Under optimal conditions, some bacterial cells divide in less than 20 minutes, so that a single cell can quickly become a population of billions of cells. This enormous reproductive capacity accounts for the ability of bacteria to quickly colonize new food sources and to rapidly cause infection and disease.

Bacteria lack the complex life cycles of fungi, but genetic recombination is possible. New genetic material can be obtained by bacteria in three ways.

1) Bacterial cells may temporarily fuse and exchange genetic material. This process is called **conjugation** and sometimes, inaccurately, "bacterial sex."

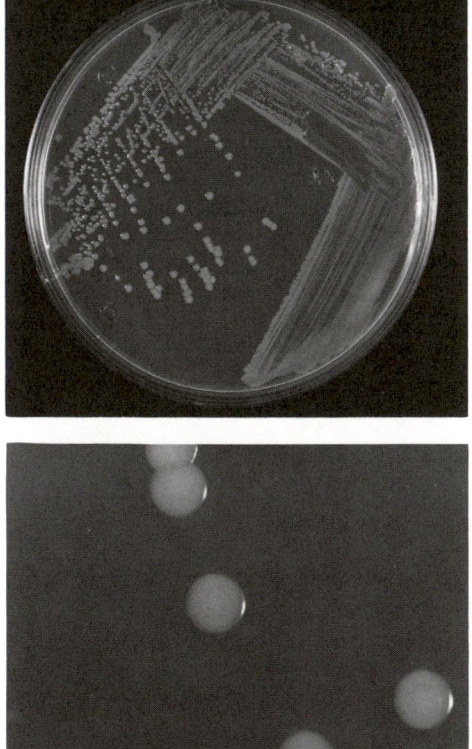

Fig. 4-5. Bacterial colonies on a nutrient agar plate. **Top,** dilution streaking to obtain individual colonies; **bottom,** close-up of colonies.

2) Fragments of DNA in the environment, released from destroyed bacterial cells, can be absorbed and become part of the genetic information of a living bacterium. This process is called **transformation** because the genetic information of the bacterium has been changed.

3) A third process, **transduction**, occurs when bacterial cells become infected by viruses called **bacteriophages**. The viruses may carry genetic information from previously infected bacterial cells to newly infected cells.

Although these three means of genetic recombination are important, they are also relatively rare in bacteria. The most important source of genetic variation in bacteria is mutations, or "mistakes," that occur during replication of the bacterial DNA during fission. Mutations are rare events, with perhaps one mutant cell per 10 million individuals for any given gene. In organisms such as plants and animals that reproduce only once or a few times a year, mutations occur only rarely. Even though the chance of a mutation occurring is no greater in bacteria, mutations are likely to be found in some members of the massive population of bacteria that can result from a single bacterium during a 24-hour period. Bacteria multiply so rapidly, and exist in such high populations, that they are able to adapt rapidly to changes in their environment through the genetic variation provided by mutation.

Bacteria as Plant Pathogens

Some bacteria produce resistant endospores that allow the cells to survive adverse environmental conditions such as heat, drought, and cold. It is fortunate that no plant-pathogenic bacteria form endospores, so they remain quite vulnerable and are particularly susceptible to desiccation. Because no endospores are produced, these bacteria are unable to disperse long distances aerially as many fungi do. Some plant-pathogenic bacteria can survive in the soil as saprophytes, but many die when plant tissue is removed, especially in the dry conditions of winter soils. In temperate climates, bacteria commonly survive in **cankers** on perennial plant parts and in or on seeds and other propagative parts such as tubers, corms, bulbs, and rootstocks.

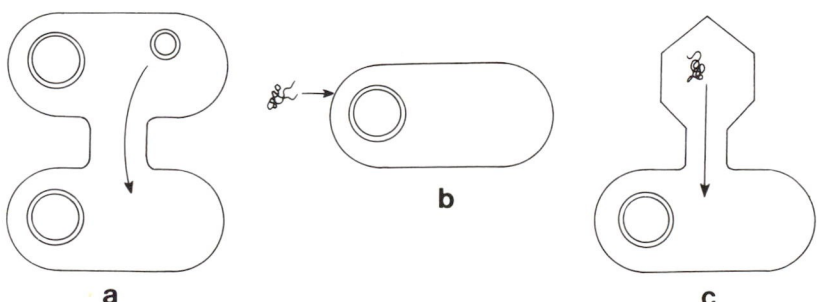

Fig. 4-6. Types of genetic recombination in bacteria: conjugation (a), transformation (b), and transduction (c).

The vulnerability of plant-pathogenic bacteria to destruction limits their ability to survive in the absence of a host plant and to disperse rapidly and to distant places. For bacteria to move from one plant to another, they must rely on passive dispersal by water, wind-blown water, insects, birds, and movement of soil on tools, machinery, or agricultural workers. Humans are responsible for most long-distance movement of plant-pathogenic bacteria. Plants, seeds, fruits, bulbs, and soil clinging to various tools and machinery have carried bacteria to all parts of the world during trade and commerce.

It should not be surprising that bacterial plant diseases are much more common in warm and humid climates. When soils do not freeze, many bacteria can survive as saprophytes in plant debris and in soil. When dew and rain are frequent, the bacteria can be splashed and spread by wind, tools, and machinery. In the United States, such growing conditions are primarily in the southeastern states and in greenhouses.

Some important plant diseases are caused by bacteria that are seedborne. For instance, snap beans are susceptible to three seedborne bacterial diseases, common blight, brown blight, and halo blight. Hot water treatment of seed is commonly used to kill the bacteria, but the process often has adverse effects on the viability of the seeds. Many seed production farms now exist in the drier western areas of the United States, where bacterial diseases are much less common. European countries that lack arid areas obtain vegetable seed grown in drier areas of Africa or the western United States. Under drier conditions, bacteria-free seed can be harvested, tested, and **certified** as meeting acceptable standards of freedom from pathogens for use in more humid areas. Many bacterial diseases of annual flowers and vegetables have been greatly reduced by the use of such certified seed.

Penetration of Plant Tissues

Besides being vulnerable to adverse environmental conditions, bacteria face another important obstacle as plant pathogens. Healthy plant surfaces

Fig. 4-7. Some means of pathogen dispersal.

are well protected from invasion by any pathogen, but especially from tiny single-celled bacteria. Woody tissues with multiple layers of bark cork cells offer an impenetrable barrier to a microorganism, but the surfaces of leaves and green stems may be more vulnerable.

A cross-section of a leaf of a typical broad-leaved plant demonstrates its anatomy (Fig. 4-8). A single layer of **epidermal cells** lines both the upper and lower leaf surfaces. In addition to the cell wall of the these cells, a **cuticle** of water-repellent compounds, primarily cutin, protects the exposed surfaces of cells. There is usually also a distinct **wax** layer on, and blending into, the cuticle that varies in thickness depending on the plant species. Usually the wax and cuticle layers are thicker on the upper surfaces of leaves. The wax and cuticle are also thicker on leaves of plants that are adapted for water conservation, such as evergreen and desert plants, and on leaves of many plants in warm and humid tropics.

The epidermal layer is periodically interrupted by specialized **guard cells** around openings called **stomata**. The guard cells in the epidermis are capable of opening and closing the stomata for control of gas exchange and transpiration (water loss). Stomata are usually more numerous on the lower leaf surface.

Between the epidermal layers is the **mesophyll**, which simply means the middle of the leaf. It is filled primarily with thin-walled living parenchyma cells of various shapes. Those just below the upper epidermis, in one or a few rows, are column-shaped parenchyma cells called the **palisade parenchyma**. Most of the photosynthesis of the leaf takes place in these chloroplast-filled cells. Below the palisade layer lies more loosely packed **spongy parenchyma**. There are air spaces between these irregularly spaced cells.

Within the mesophyll are also the **vascular bundles**, commonly called the leaf veins. The vascular bundle consists of **xylem** tissue for water and mineral transport and **phloem** tissue for food transport. In addition, support tissues, such as **collenchyma** and **sclerenchyma**, are often associated with the vascular bundles and leaf margins. Familiar examples of such tissues

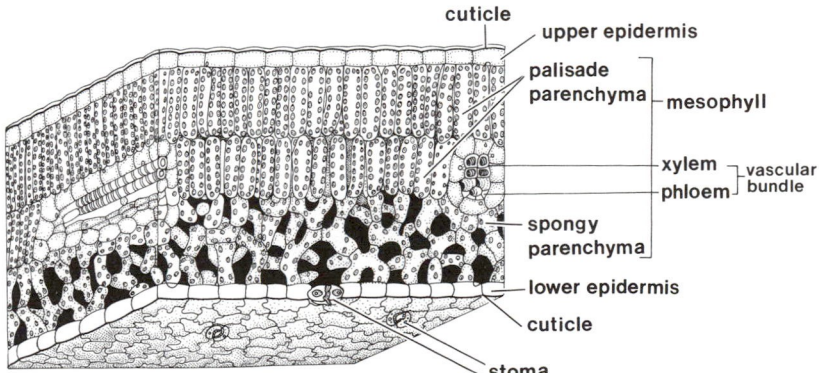

Fig. 4-8. Diagram of section view of a leaf, showing cells and tissues.

are the tough strings in celery and the gritty stone cells in pear fruits, respectively. The thickened cell walls of these tissues lend strength to leaves and nonwoody stems.

Among flowering plants, leaf anatomy exhibits tremendous variation. Epidermal cells may produce various types and densities of leaf **trichomes**, or hairs. The internal pattern of parenchyma cells is variable. The number, pattern, location, and morphology of stomata vary depending on the environment to which the plant is adapted and the botanical family to which it belongs. However, all leaves have a protective, and relatively **hydrophobic** (water-repellent) surface with natural openings at the stomata. The cuticle is least thick on the cells in the stomata, on newly expanding leaf tissue, on **nectaries** and flower petals, and in **hydathodes**, which are openings along the sides of leaves for exudation of excess water.

Even after penetrating the protective cuticle, parasites are faced with the multiple layers of the cell walls of the cells themselves. The cell wall consists primarily of **cellulose, hemicellulose**, and **pectin**, in layers. All of these components are **polymers**, or long chains, of simpler molecules. Cellulose is the predominant component of cell walls and is a polymer of **glucose**, a simple sugar. Hemicellulose and pectin polymers are chemically similar. The various pectin polymers are chains of galacturonic acid, a derivative of glucose. The intercellular layer that cements the cells together, called the **middle lamella**, consists primarily of pectin.

How can a fungal spore or a bacterium penetrate such well-protected plant tissue? Let us first consider the much larger fungus spore. Most spores begin to germinate if free water is present on the leaf surface for an hour or longer. After water absorption, **germ tube** growth begins. Water must typically be present for a total of 8–12 hours for infection to be successful because the germ tube takes some time for growth and then remains sensitive to desiccation until penetration takes place. The time necessary for spore germination varies with fungal species, temperature, and other environmental variables and is one of the most important factors in a disease epidemic.

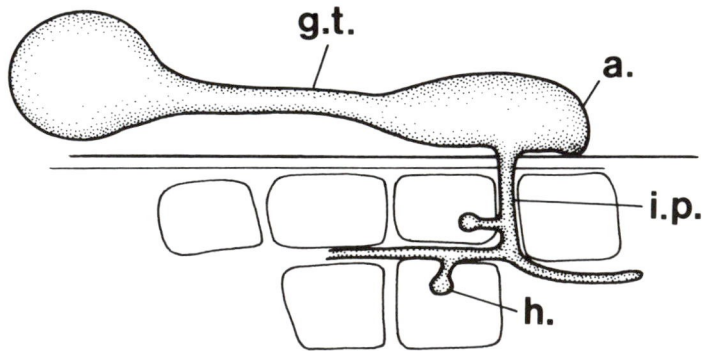

Fig. 4-9. Diagramatic view of spore germination. g.t. = germ tube, a = appressorium, i.p. = infection peg, h = haustorium.

Many plant-pathogenic fungi have evolved elaborate mechanisms to penetrate host plants. Many produce a flattened structure called an **appressorium** at the infection site. The appressorium often produces a sticky substance that helps attach the germ tube to the leaf. From the appressorium is produced a narrow **penetration peg**. This peg penetrates the cuticle with a combination of mechanical pressure and enzyme production that softens and degrades the cuticle and epidermal cell walls. Many fungi produce various enzymes such as **cutinase**, which degrades the cutin in the cuticle, plus **pectinases** and **cellulases**, which degrade pectin and cellulose in the cells walls. (Note that the suffix -*ase* is used to designate an enzyme that degrades the named substance.) Pectinases are particularly important because they degrade the middle lamella, which, for reasons not clearly

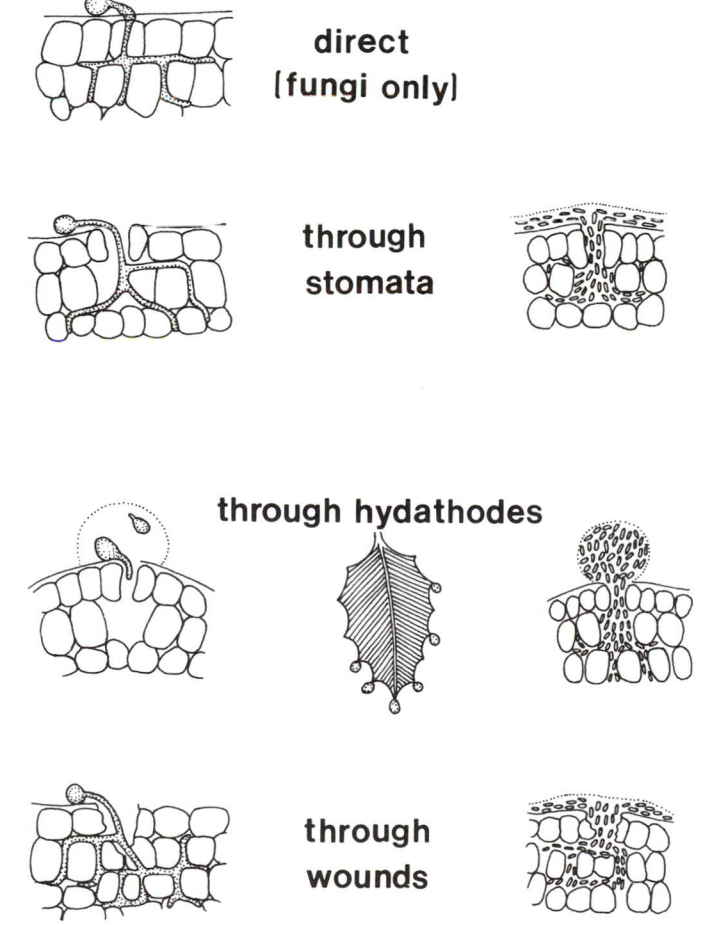

direct (fungi only)

through stomata

through hydathodes

through wounds

Fig. 4-10. Various modes of penetration and invasion of a leaf by fungi (left) and bacteria (right).

understood, contributes to the death of neighboring cells. Such enzyme production not only expedites penetration but provides important nutrients for fungal growth. Enzyme degradation of both pectin and cellulose releases glucose, an important energy source. As cells begin to die, the cell membranes become very permeable, and other important nutrients begin to leak out of the cells.

Some fungi penetrate plants only via stomata by entering the space between the guard cells to internal cells where the waxy layer is absent and the cuticle is thinner. Others penetrate between epidermal cells, while still others penetrate epidermal cells directly. Many fungi take advantage of wounds caused by mechanical damage, insect feeding, or hail to facilitate penetration of plant surfaces. Infection by some fungi results in visible lesions of dead cells only a few days later, whereas other fungi, such as the rusts, absorb nutrients without triggering immediate cell death. The hyphae of such fungi grow between the parenchyma cells of the leaf and absorb nutrients through **haustoria**, which are projections of the hyphae that push into the cytoplasm without breaking through the cell membrane.

The relatively large size and mechanical pressure that fungal spore germ tubes can bring to bear give fungi advantages over bacteria as terrestrial parasites. A single bacterium has essentially no chance of penetrating a healthy plant cell. It is unable to exert sufficient mechanical pressure for penetration. Many pathogenic bacteria exist in a saprophytic state on leaf surfaces until conditions are conducive for infection. In most cases, the bacteria must increase in number so that the enzymes and toxins that they produce are present in sufficient quantity to have an effect on the nearby plant cells. Bacteria can only enter healthy plant tissue through "natural openings" or points of vulnerability where the cuticle is very thin or non-existent, such as newly expanding leaf tissue or parenchyma cells exposed in **lenticels**, the lens-shaped air-exchange openings on bark and stems.

Bacterial penetration of plant tissues is much more common at wounds. Wounds not only expose vulnerable plant cells to the bacteria but contain torn cells that provide nutrients for the rapid multiplication of the bacteria and the subsequent increase in the enzymes and toxins they produce.

Many plant-pathogenic bacteria produce pectinases that degrade cell walls and the middle lamellae and contribute to the death of nearby cells. Bacteria may also produce toxins that poison plant cells. As the plant cells die, nutrients are released through the cell membrane for use by the multiplying bacteria. Bacteria lack the penetration ability of many fungi and usually progress through plant tissue intercellularly. The same processes that create wounds in plants can also serve to disperse the bacteria. Wind-driven rain drops can carry bacteria, and the sticky cells can also be carried easily by insects and agricultural implements such as pruning shears.

Isolation and Identification of Bacteria

Bacteria are ubiquitous and are found on most plant surfaces. Diseased plant tissues are rapidly colonized by saprophytic bacteria as soon as dead cells are present, so the accurate diagnosis of a plant disease is often confounded by the presence of such "secondary invaders." It is often difficult to distinguish pathogenic bacteria from the saprophytic species that are also present.

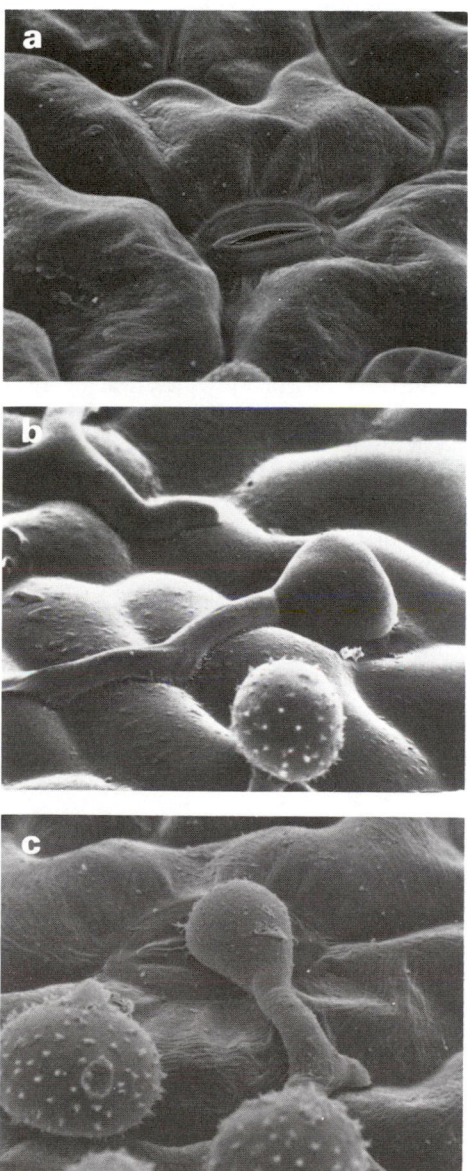

Fig. 4-11. Rust fungus (*Uromyces phaseoli*) spore germination and penetration of a bean leaf. **a,** Stoma in leaf surface; **b** and **c,** spore germ tube crosses epidermis and produces appressorium over stoma before penetration.

In 1866, shortly after fungi had been shown to cause disease in plants, the French scientist **Louis Pasteur** first proved, in experiments with wine, that bacteria are not produced by spontaneous generation. In the next few years, the role of bacteria in human and animal diseases became clear, and Koch's postulates allowed pathogenicity to be proven (see Chapter 1). Although it became generally accepted that bacteria were animal pathogens, fungi were considered the agents of plant disease. Studies by American plant pathologist **Thomas Burrill** in 1878 implicated a bacterium as the disease agent of the "fire blight" disease in North America that was causing the death of apple and, especially, pear trees. **J. C. Arthur**, using pure cultures, conclusively demonstrated the role of a bacterium in fire blight in 1885. The fire blight bacterium is now called *Erwinia amylovora*. The genus *Erwinia*, which includes many species, honors **Erwin F. Smith**, an important American plant bacteriologist who contributed greatly to our understanding of plant diseases caused by bacteria, beginning in 1895.

In the late 1800s, laboratory techniques were too primitive for scientists to accurately identify bacteria. One important advance involved sterilization, usually by heat, of nutrient media in which bacteria could grow. Early experiments were performed using various food broths, wine, urine, and other liquids. After the liquids were heated to kill bacteria already present, samples containing bacteria to be studied could be introduced and would multiply. Most of the samples were mixed populations of bacterial species. One important limitation to the use of liquid nutrient media is that, in a mixture, all the bacteria grow together.

The addition of a gelling substance to the nutrient medium provides a moist, semisolid surface on which individual colonies of bacteria can be grown. Nutrient media today often include **agar**, derived from seaweed, that gels after it is boiled and allowed to cool. When bacteria are streaked across the sterile, gelled surface of the medium, individual bacterial cells multiply to form small circular individual colonies, usually in a day or two. This technique was an important advance in the study of bacteria in pure cultures of a single species.

How can pathogenic bacteria in infected plant tissue be separated from the saprophytic secondary invaders? One simple technique (Fig. 4-12) relies on the assumption that pathogenic bacteria are probably present in higher concentrations than contaminating saprophytes. First, infected tissue is surface-disinfested with alcohol or a dilute bleach solution to kill contaminating microorganisms. It can then be macerated in sterile water to release bacteria from the infected tissue into the liquid. The sample of liquid containing bacteria is diluted with more sterile water in a series of clean sample bottles. The resulting set of sample bottles contains a series of dilutions from the original sample of 1:10, 1:100, 1:1000, and higher if necessary. A small amount of liquid from each bottle is then streaked across the surface of a **petri plate** of nutrient agar medium, using a sterile tool such as a bent glass rod. Petri plates are the common covered circular dishes found in laboratories everywhere that were invented by the 19th century microbiologist and student of Robert Koch, **R. J. Petri**, to prevent contamination from airborne microbes. In 24–48 hours, individual bacterial cells usually produce circular colonies that are visible on the medium. Plates containing samples from the lower dilutions often have many colonies of variable colors and morphologies. In plates containing more dilute samples, fewer colonies should be present. These colonies are most likely to be the pathogenic species because the contaminating bacteria are reduced, and perhaps eliminated, by dilution.

The tiny circular colonies grow from single cells (Fig. 4-5) and should therefore be pure cultures of the causal agent. Of course, it is only presumptive evidence that the bacterial colonies that appear after extensive dilution from the original sample are actually the pathogen. Following Koch's postulates, a pathogenicity test must be done on a healthy plant to prove that the isolated bacterium is the causal agent of the disease.

Even after a pathogenic bacterium has been isolated, it is often difficult to identify. The commonly used bacterial characters, shape and Gram stain,

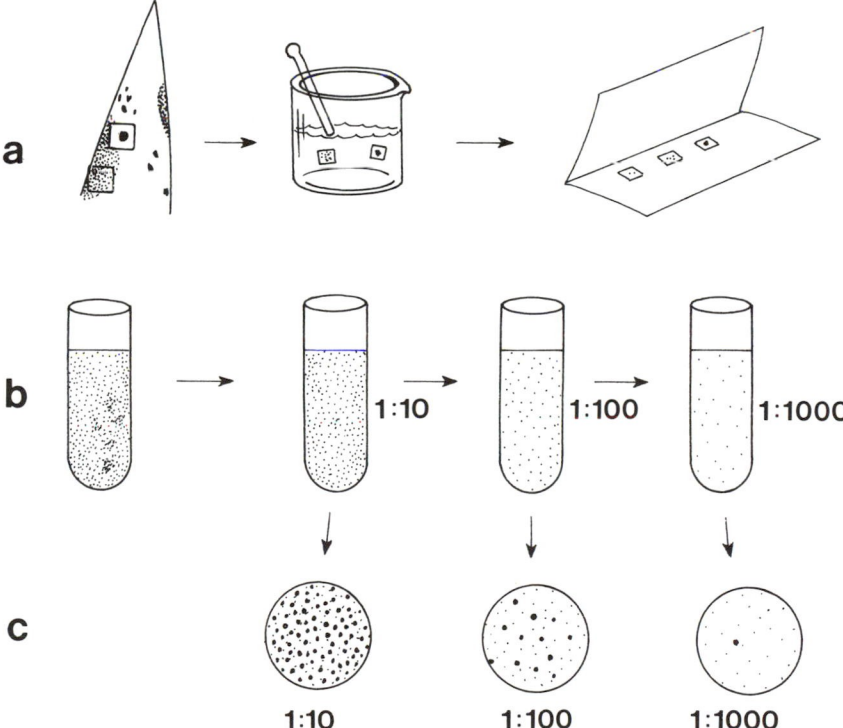

Fig. 4-12. Isolation of bacterial pathogens from infected plant tissue. **a,** A portion of infected tissue is surface-sterilized and blotted dry. **b,** The tissue is macerated in sterile water to release the bacteria. The sample is diluted several times to reduce the numbers of background contaminants. **c,** A sample of each dilution is spread on a nutrient agar plate. After incubation, pure colonies of bacteria may be present. A pathogenicity test must be performed to determine whether the isolated organism is the causal agent of the original disease.

are of limited utility for plant-pathogenic bacteria since all are rod-shaped and most are Gram-negative. Plant-pathogenic bacteria are commonly identified to genus based on their ability to grow on so-called **selective media**. These media contain various antibiotics and other nutrient mixtures known to allow the growth of bacteria of some genera but not others.

Bacteria of the genus *Erwinia*, for instance, can grow in the absence of oxygen. Bacteria that do not require oxygen are described as **anaerobic**. Because *Erwinia* species grow better in the presence of oxygen but can survive in its absence, they are considered **facultative anaerobes**. If mineral oil, or another agent that blocks the entry of oxygen, is layered over media inoculated with bacteria, only facultative anaerobes will be able to grow, whereas other plant-pathogenic genera will not. Other identifying characteristics include the ability of some species to fluoresce, or glow, under ultraviolet light and the appearance of others as gummy or yellow in culture. Most diagnostic laboratories complete a number of such tests routinely to arrive at the correct genus name for an isolate.

In the early years of microbiology, plant-pathogenic bacteria were listed in an entirely separate section of bacterial taxonomy. As knowledge increased, it became clear that their biological relationships to other groups of bacteria were more important than their role as plant pathogens. The plant-pathogenic species are now to be found scattered among many orders and families and in the genera *Erwinia, Clavibacter* (formerly included with *Corynebacterium*), *Pseudomonas, Xanthomonas,* and *Agrobacterium.* Taxonomic changes that divide some of these into new genera are still

Fig. 4-13. Early symptoms of fire blight (*Erwinia amylovora*).

being considered by bacteriologists as genetic and biochemical studies increase our understanding of their relationships.

Although a genus name for a bacterium is usually accepted by most scientists, identifying bacteria to species is quite another matter. For many years, a plant-pathogenic bacterum was identified to genus and then often given a species name that reflected the host plant from which it had been isolated. Many bacteriologists believed that the number of species was getting too large because discovery of a new disease often led to a new bacterial species. They did not believe that differences in host range necessarily reflected enough biological differences to justify a separate species designation.

In 1980, a new taxonomic category was created to try to solve some of the problems of naming new species, subspecies, strains, and races. When two strains of bacteria appear to be identical with respect to morphology and biochemical and physiological tests and vary only in host range, they are differentiated by the term **pathovar,** abbreviated "pv." For example, begonia leaf spot is caused by *Xanthomonas campestris* pv. *begoniae.* Many other pathovars of *X. campestris* exist that cause disease in other plant species, such as *X. c.* pv. *pruni,* which causes bacterial spot of stone fruits. This has greatly reduced the number of plant-pathogenic species and more accurately reflects their known biology as currently understood. Notice that when writing about pathovars, scientists often abbreviate the genus and species after they have written out the names once.

Fig. 4-14. Fire blight. **Left,** canker (dark streak) on trunk of apple tree; **right,** bacterial ooze from trunk of an infected apple tree.

Fire Blight

Many important plant diseases are caused by bacteria, including the previously mentioned **fire blight** of pome fruits. Pome fruits include apple, crabapple, and pear. *Erwinia amylovora* is a common bacterial parasite of many North American native plants in the Rosaceae, the rose family. Disease development is usually quite limited on native hawthorns, mountain ash, serviceberry or shadbush, and other closely related plants. It became an economically important pathogen as a result of the intercontinental movement of plants and agricultural practices. When European colonists arrived in North America, they brought with them apple and pear trees, which were very susceptible to *E. amylovora*.

Pear trees are especially susceptible, and whole trees can die in a single season, leaving a tree with blackened foliage and twigs that give rise to the name "fire" blight. The first recorded epidemic was in New York State in 1780, long before the role of bacteria in plant disease had been discovered. Epidemics were limited in the beginning because orchards were small and often planted from seeds, so there was more genetic diversity among the trees.

The bacteria survive winter in the margins of cankers in woody branches and trunks. In the spring, they begin to multiply and ooze out of the canker through cracks in the bark. The sticky bacteria become attached to insects that are attracted to the sweet ooze. When the insects visit the blossoms of fruit trees and their wild relatives, they deposit the bacteria. The bacteria in ooze are also distributed in splashing and wind-blown rain. Recent studies have demonstrated that the bacteria are able to survive on flower surfaces until environmental conditions are optimal for infection. Most infections occur in the flowers and on young leaves where insects, humans, or storms

Fig. 4-15. Effects of fire blight in a pear orchard in Colorado, 1914. Infected trees in the foreground were cut back due to infection. They had been the same size as the trees in the background.

cause wounds. After an infection begins in the flower, the bacteria multiply quickly and probably produce toxins and enzymes that cause the tissues to blacken and die. The infection spreads down the stem of the flower spur and, unless pruned off, may eventually reach the trunk of the tree. Extensive trunk cankers can cause tree death.

Fire blight epidemics of eastern pear orchards caused farmers to abandon most production, and pear orchards became established in the western states as agriculture developed there. But *E. amylovora* was also carried west in infected tissues of the pear and apple trees brought to establish the new western orchards. From 1900 to 1910, before control measures were known, fire blight epidemics destroyed 95% of the pear trees in the San Joaquin Valley of California and other areas throughout the western states. Pear orchards are nearly impossible to maintain in humid areas, so U.S. production continues to center in the drier western states, where careful pruning of cankers and the use of chemical sprays help to protect them from fire blight. Fire blight is also an important disease of apple trees, but they are less susceptible than pear trees and can be maintained in humid regions with appropriate management.

Copper sprays are toxic to bacteria but can also damage plant tissue and developing pear fruits. An alternative type of spray is an antibiotic such as streptomycin. It is relatively expensive to spray, and, if it is used frequently, the bacterial population may become resistant to the antibiotic and cause disease despite its application. When a single type of antibiotic is used repeatedly against a bacterial infection in humans, animals, or plants, resistant strains of bacteria may arise. The penicillin resistance of some strains of gonorrhea bacteria is a classic example. The ability of bacteria to develop resistance very quickly is one reason that medicinal antibiotics are available by prescription only.

Most antibiotics are produced by soil microorganisms against other soil microorganisms. They may provide an advantage in the highly competitive soil environment. Even though scientists continually search for new antibiotics, it is important to use the safe and well-known types with care to prolong their effectiveness. Life-threatening situations can develop and the cure is delayed when an antibiotic is ineffective because the bacteria are resistant and other antibiotics must be tried.

Many people fear that unrestricted agricultural use of antibiotics may result in antibiotic-resistant strains of many pathogens, including human ones. How can this occur? As described previously, it is possible for bacteria of different species to exchange genetic material. Antibiotic sprays applied to an orchard, for instance, may increase the proportion of antibiotic-resistant bacteria on or in the plant, in the soil, and in the human applicators. The genetic resistance factor may eventually be passed on to a bacterial pathogen of humans. In Europe, agricultural antibiotic sprays are restricted to use by special permit only or are not allowed at all. In the United States, antibiotic sprays are usually restricted to commercial growers by their cost and their relatively short effectiveness when used repeatedly. Some scientists favor banning antibiotic sprays altogether and restricting the use of antibiotics for plant disease to certain uses only, such as the injection of trees infected with mycoplasmas (see Chapter 12).

Careful pruning of the overwintering cankers remains the most effective management practice for fire blight. Even with careful pruning, pear production is difficult unless the climate is quite dry. In recent years, *E. amylovora* was transported to the Eastern Hemisphere. In 1919, fire blight was first reported outside North America in New Zealand. It was discovered in England in 1957, perhaps carried on infected fruits or shipping crates from New Zealand. It then spread to continental Europe, appearing first in wild hawthorns on islands off the Dutch coast in 1966, which suggests that the bacteria were carried by birds that found the hawthorn fruits attractive. These first areas of infection were burned in an attempt to eradicate the disease, but it now seems firmly established in The Netherlands. Although pear orchards can be carefully pruned, the common hedgerows of hawthorn can remain a significant source of fire blight bacteria carried by insects to the orchards. In The Netherlands, hawthorns near orchards are required by law to be pruned to prevent blooming to reduce the spread of the bacteria by insects. The hawthorn laws have caused controversy between orchardists and environmentalists, who believe that hawthorns are ecologically important in natural areas.

Fire blight continues to spread and was detected in France in 1978. It is now present in many other European and Mediterranean countries, where it threatens not only pome fruits but many rosaceous ornamentals. Certainly, fire blight caused by *E. amylovora* will continue to have important economic effects for years to come.

Fig. 4-16. First observations of fire blight in Europe.

Citrus Canker

In the fall of 1984, Americans heard reports of a serious bacterial disease problem in citrus nurseries in Florida. In that state, the citrus industry

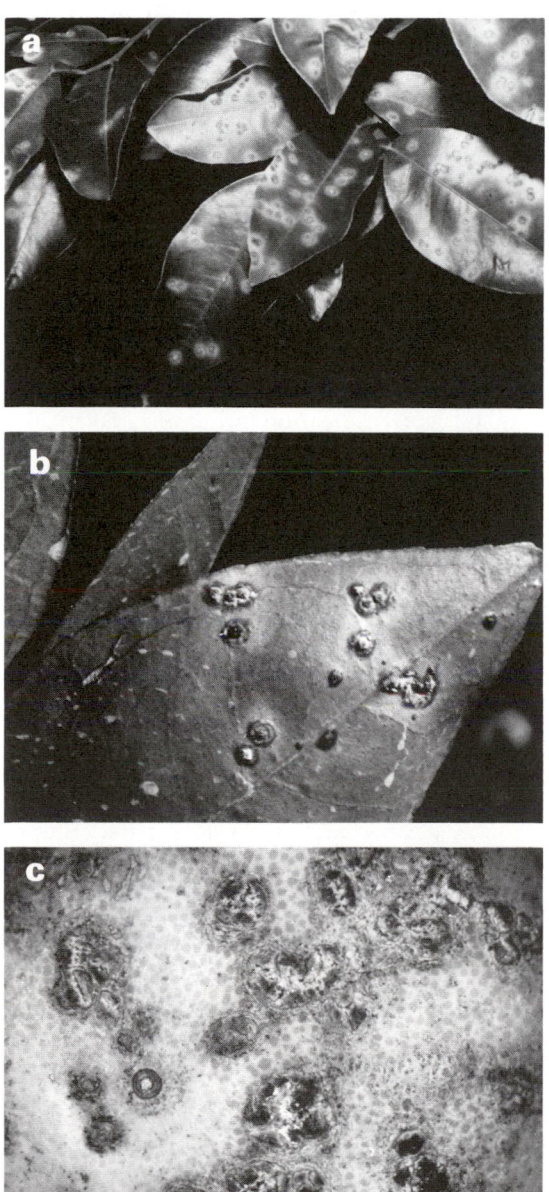

Fig. 4-17. Citrus canker symptoms. **a,** Upper leaf surfaces; **b,** lower leaf surfaces; **c,** close-up of fruit lesions.

has traditionally worried about two possible dangers, frost and bacterial canker disease. **Citrus canker** was not really a new disease to the United States. It had been previously introduced in the early 1900s and was reported to have been eradicated from Florida in 1933 following the burning of millions of citrus trees at a cost of about $6 million in that state alone. The eradication of citrus canker from the citrus-growing states of the United States is the only successful eradication of a well-established pathogen ever reported. No citrus canker had been again reported in any citrus-growing state until 1984. Stringent border inspections of plants and fruits had been maintained to prevent the reintroduction of citrus canker. Even so, many scientists feared that the quarantine would only be temporarily successful. Between 1973 and 1978, citrus canker on infected fruits and plants carried by travelers was intercepted at borders 2,603 times.

Coming seemingly from nowhere, bacterial citrus canker was reported not in a commercial orchard but in a nursery, where workers take particular care to protect plants from pathogens. The symptoms of the disease were atypical, but the bacterium was identified as *Xanthomonas campestris* pv. *citri* using standard methods. State and federal regulatory agents were required by law to initiate eradication programs. Over 20 million citrus trees have been destroyed at a cost exceeding $25 million.

Compared to many parts of the United States, the warm, humid climate of Florida is very conducive to the survival and spread of bacteria. In addition, citrus is a vegetatively propagated crop, produced primarily by bud grafting. The propagation techniques require very careful precautions to prevent the spread of bacteria to new plants. There is also considerable concern that the genetic uniformity of citrus plants makes them particularly vulnerable to major epidemics. Most of U.S. citrus cultivars are susceptible to infection by *X. campestris* pv. *citri*. Grapefruit, in particular, is highly susceptible.

The quarantine restrictions initially required that infected trees and all plant material within 125 feet (41 meters) be burned. Citrus workers wear clothing that can be removed at the site, and all tools, machinery, and vehicles are thoroughly cleaned after work in a grove. Shipments of plant material and fruit are restricted to noncitrus-growing states.

As studies of the outbreak continued, more precise characterization of the bacterium was possible. Before the 1984 Florida outbreak, four different citrus diseases had been reported to be caused by various strains of *X. campestris* pv. *citri*. Besides citrus canker disease (A group strains), there are more limited diseases including cankrosis or false citrus canker (B group strains), Mexican lime cankrosis (C strain), and Mexican bacteriosis (D group strains). The disease first reported in Florida in 1984, initially called citrus bacterial canker and now called citrus bacterial spot, represents a possible fifth strain. However, bacteriologists have reevaluated the relatedness of these strains, using many standard methods as well as modern techniques that compare the similarity of the bacterial DNA. In 1989, their conclusions were published in the *International Journal of Systematic Bacteriology*. They reinstated the A strains that cause citrus canker disease as a separate species, *X. citri*, separating it from the various pathovars of *X. campestris*. The remaining strains that cause the other citrus diseases have been divided into several pathovars of *X. campestris*, including *X. c.* pv. *aurantifolii* (B, C, and D strains) and *X. c.* pv. *citrumelo* (E strains). *X. c.* pv. *citrumelo* causes the nursery disease and was named for the susceptible cultivar Citromelo rootstock on which the disease is commonly found.

Does this mean that 20 million young trees were destroyed needlessly? A number of lawsuits have been filed by growers who sustained significant economic losses. The regulatory officials, with the advisement of a 16-member council made up of growers, nurserymen, scientists, and regulators, were compelled to act, however, based on the diagnostic and biological evidence available at that time. Although the details of bacterial systematics may not fascinate the average person, our ignorance of the true relation between the various bacteria that cause disease on citrus or any other host can be of critical economic importance.

Fig. 4-18. Top, burning of small trees in citrus nursery originally believed to be infested with Asian canker; **bottom**, destruction of grove infested with Asian canker.

The citrus canker that growers had most feared, the A group strain, did appear in 1985 and 1986 in a number of locations, mostly on mature trees at residences and in one citrus grove. Regulatory agents enforced quarantine and eradication procedures at these sites also. Efforts are currently underway to once again eradicate citrus canker, but the cost will be high. Some scientists question whether citrus canker was truly eradicated previously and have also questioned the destructibility of the bacterium in the Florida climate. Although certain scientists argue that eradication should be a priority, others suggest that the fear of canker was partly a political ploy to restrict the import of competitive citrus from countries where canker is endemic. In those countries, copper sprays are used to reduce infections, and windbreaks help reduce the spread of the bacteria. Although the cost of an eradication program is high, management costs may also rise, especially in highly susceptible grapefruit groves. In the meantime, the destruction of nursery stock has been a particular burden to growers trying to replant orchards killed by the frost and freezes of recent years. The reintroduction of citrus canker to the vegetatively propagated, genetically uniform citrus crop grown in a warm and humid climate will remain a significant problem for Florida's $2.5 billion citrus industry in the coming years.

Crown Gall

As a final example of an important bacterial plant disease, let us consider **crown gall** disease caused by *Agrobacterium tumefaciens*. Unlike the previously described diseases, crown gall symptoms do not include cankers

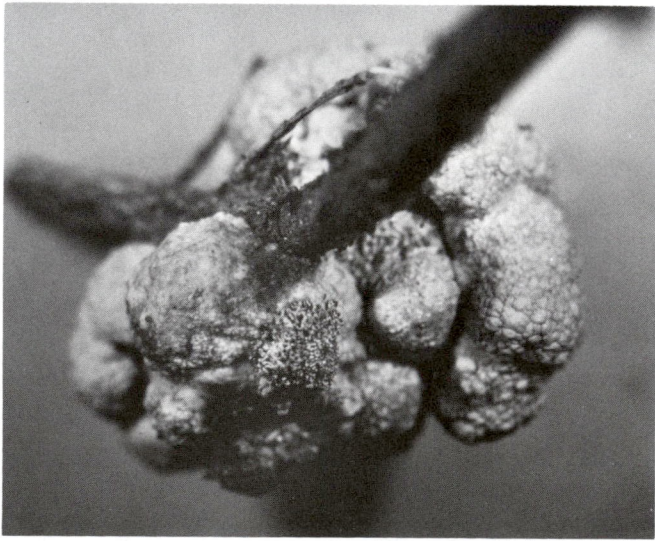

Fig. 4-19. Crown gall (*Agrobacterium tumefaciens*).

and dying tissues but tumors, or **galls**, that were referred to in the past as "plant cancer." Since these masses of undifferentiated plant cells are often seen at or near the soil line on the "crown" of the plant, the disease is called crown gall.

A. tumefaciens lives in the intercellular spaces of the gall tissue. It survives well in soil as a saprophyte for many years and enters susceptible plants through wounds when they are planted in infested soil. The host range of this pathogen is very wide; susceptible plants include grapes, roses, apples, raspberries, and many species of woody and nonwoody ornamentals. After infection, galls appear and continue to grow, eventually crushing the vascular tissue and distorting the growth of the plant. Galls can be removed for aesthetic improvement of the plant, but that will not eradicate the infection. Special care must be taken to prevent the spread of crown gall bacteria between plants during pruning or grafting.

A commercially successful **biological control** exists for this disease. A closely related strain of the bacterium, *A. radiobacter* Strain 84, can be applied to root systems of many susceptible plant species before they are planted. This bacterium produces an antibiotic that is specifically antagonistic to the crown gall bacterium. In addition, since this nonpathogenic bacterium becomes established on the root system or graft union first, it effectively competes with the crown gall bacterium in the soil to help prevent infection.

Some years ago, *A. tumefaciens* attracted considerable research attention from scientists who were trying to understand animal cancer. Although that research did not elucidate any new understanding of animal cancer, much was learned about crown gall disease itself. For instance, studies showed that infected plant cells were transformed into uncontrollably dividing gall or tumor cells that continued to divide even when the bacteria was removed from the plant tissue. It is now known that *A. tumefaciens* is a naturally occurring "genetic engineer." The bacterium inserts some of its genetic information into the nucleus of the plant cell, causing it to divide repeatedly and form a gall. At first, the phenomenon seemed to be a biological curiosity. Today, the crown gall bacterium is the workhorse of scientists who wish to insert foreign genes into plants. The potential results of such work have caused both excitement and fear and are the subject of the next chapter.

Selected Readings

Aldwinckle, H. S., and Beer, S. V. 1978. Fire blight and its control. Horticultural Reviews 1:423-474.

Billing, E. 1987. Bacteria as Plant Pathogens. Aspects of Microbiology 14. American Society for Microbiology, Washington, DC.

Brock, T. D. 1988. Robert Koch: A Life in Medicine and Bacteriology. Science Tech Publishers, Madison WI.

Campbell, C. L. 1983. Erwin Frink Smith: Pioneer plant pathologist. Annual Review of Phytopathology 21:21-27.

Campbell, R. N. 1979. Fire blight. Natural History 88(5):62-69.

Moore, L. W. 1988. Use of *Agrobacterium radiobacter* in agricultural ecosystems. Microbiological Sciences 5:92-95.

Schoulties, C. L., Civerolo, E. L., Miller, J. W., Stall, R. E., Krass, C. J., Poe, S. R., and DuCharme, E. P. 1987. Citrus canker in Florida. Plant Disease 71:388-395.

Starr, M. P. 1984. Landmarks in the development of phytobacteriology. Annual Review of Phytopathology 22:169-188.

Genes and Genetic Engineering

People have long recognized that certain traits are passed on from parents to children, from animals to their young, and from plants to seedlings. Selective breeding has resulted in larger farm animals, faster race horses, and higher-yielding crops. The means by which traits are inherited by successive generations, although recognized and exploited for centuries, has only been understood in recent years. New discoveries in this field are now occurring so rapidly that it is difficult to predict the possible innovations that may occur even in the next few years. Because the genetic information in the deoxyribonucleic acid (DNA) of an organism contains the blueprints for all cellular control, some basic understanding of genes and their structure is necessary for an appreciation of modern biology.

Before we examine the molecular structure of DNA, let us consider a **gene** simply as a locus, or site, on a chromosome. The description of meiosis in Chapter 2 indicated that offspring receive one set of chromosomes from each parent. Thus, for any particular genetic trait, the offspring receives a gene on one chromosome from one parent and another gene for that trait on the homologous, or similar, chromosome from the other parent. When genes exist in at least two alternative forms, the forms are called **alleles**. Offspring may receive different alleles of the same gene on the homologous chromosomes contributed by each parent. An individual that receives two identical alleles of a gene is called **homozygous**, and an individual that receives two different alleles of a gene is called **heterozygous**.

A familiar example of gene alleles is eye color in humans. The brown eye allele is **dominant** over the blue eye allele, which means that an individual will have blue eyes only if both parents contribute blue eye alleles. Brown-eyed individuals may be homozygous for brown eye alleles or they may be heterozygous, that is, possessing one brown eye allele and one blue eye allele. The actual genetic constitution of an individual is called the **genotype**, but the resulting appearance of an individual is called the **phenotype**. Brown-eyed individuals may have the same phenotype, brown eyes, but different genotypes: homozygous for brown eye alleles or heterozygous, with one brown eye allele and one blue eye allele. Thus, blue-eyed parents will always produce blue-eyed children, but heterozygous brown-eyed parents have a 25% probability of producing a blue-eyed child.

Mendelian Genetics

The beginning of modern genetics can be traced to 1865, when the famous studies of **Gregor Mendel** were first published. As a monk in an Austrian monastery, he made specific crosses between garden peas with different genetic traits. He recognized that certain traits are inherited in a predictable manner. Traits such as flower color, seed color, and smooth or wrinkled seed coat exist as alleles of genes on different chromosomes as described above for eye color in humans. Mendel chose pea plants and carefully controlled the pollination of flowers, so he would know the phenotype of each parent. He then observed the phenotypes of the resulting progeny and determined that certain traits exist as distinct units that are inherited independently. His experiments explained how factors, which we now recognize as genes, are passed on from parents to offspring to maintain distinct characteristics.

People sometimes comment that certain traits "skip a generation." This is a somewhat inaccurate phrase for a common genetic occurrence. Mendel observed the phenomenon of dominant and **recessive** traits, which explained for the first time how characteristics might disappear in one generation and reappear in the next. If two parents are homozygous for different alleles, they will produce only offspring with the phenotype of the parent homozygous for the dominant allele. For example, since red flower color is dominant over white flower color in peas, a cross between a homozygous red-flowered parent and a homozygous white-flowered parent would produce only heterozygous red-flowered offspring. The white-flowered trait would seem to have disappeared in that generation.

If those heterozygous offspring were crossed, however, one fourth of the next generation would be homozygous for the recessive allele and thus have the phenotype of the recessive allele. That is, one fourth of the offspring would have white flowers. The other three fourths of the offspring would

	B	**b**
B	**BB** brown	**Bb** brown
b	**Bb** brown	**bb** blue

Fig. 5-1. The brown eye color gene is dominant over the blue eye color gene in humans. Individuals with the brown-eye phenotype may have homozygous (*BB*) or heterozygous (*Bb*) genotypes. Two brown-eyed individuals may have a blue-eyed child if each is heterozygous for eye color. Blue eye color is recessive, so the blue-eye phenotype results only from a homozygous genotype (*bb*).

have a red-flowered phenotype consisting of two different genotypes, homozygous and heterozygous (Fig. 5-2). Identical patterns were observed for all the other traits he studied.

Mendel recognized that genetic traits can be discrete factors that are governed by pairs of genes contributed randomly by the parents. This is called the **principle of segregation**. Because the genes for the traits he studied happened to be on different chromosomes, he was also able to observe a second important phenomenon, the **principle of independent assortment**. This describes the fact that the inheritance of one genetic trait may be independent of the inheritance of other traits. Thus, the alleles for seed color are inherited independently from alleles for flower color and the other traits he studied. In these simplest cases, he was able to identify the important principles of inheritance long before the roles of chromosomes and DNA had even been discovered. However, his work remained virtually ignored until the early 20th century, when cell biology had progressed to the point where his studies could begin to be explained.

Of course, all inheritance phenomena are not as simple as the ones described by Mendel. For instance, many alleles are not strictly dominant or recessive. In some plant species, a cross between red-flowered plants and white-flowered plants produces pink flowers in the heterozygous offspring, a phenomenon referred to as **incomplete dominance**. Also, genes

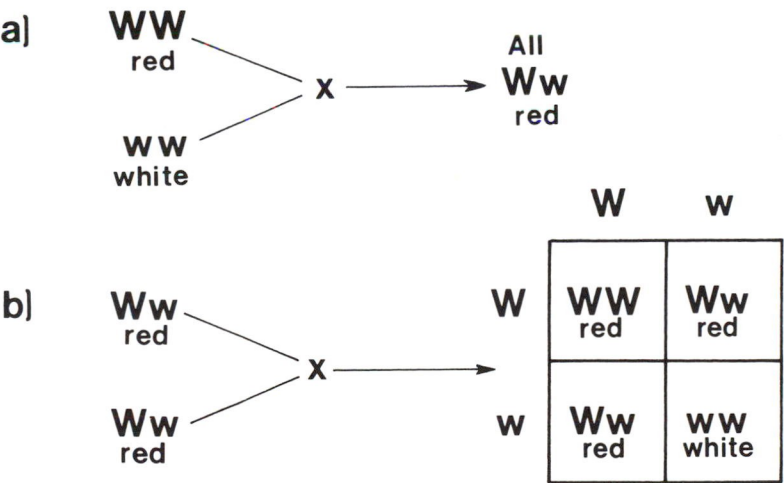

Fig. 5-2. If homozygous red-flowered peas (*WW*) are crossed with homozygous white-flowered peas (*ww*), the next generation is all heterozygous and red-flowered (*Ww*) (**a**). If these heterozygous progeny are crossed, the next generation will appear in a 3:1 ratio of red-flowered plants to white-flowered plants (**b**). One fourth of the plants will be homozygous for red flowers and half of the plants will be heterozygous for red flowers. The gene for red flowers is dominant over the gene for white flowers. White flowers is a recessive trait that appears only in the homozygous condition.

may be very close to each other on a chromosome, and so they do not exhibit independent assortment. Such genes are termed **linked** because they are inherited identically due to their close proximity on the chromosome. In addition, many phenotypic traits are not governed by a single gene but by several or many genes. Such **polygenic** traits govern many important and complex characteristics of plants and animals. The discrete nature of genes and the process by which genetic traits are passed to the next generation, as first recognized by Mendel, are fundamental to our understanding of **genetics**, the science of inheritance and evolutionary change. His selection and/or luck in choosing a plant with such distinct characters and his genius in recognizing the patterns of inheritance gave us the elementary principles of genetics that govern even the more complex phenomena.

DNA and the Genetic Code

While important genetic studies continued throughout the first half of the 20th century, the concept of the gene itself remained something of a "black box." Chromosomes, composed of protein and nucleic acid, had already been recognized as the source of genetic information. The processes of mitosis and meiosis had explained how each cell receives its genetic information and how chromosome numbers remain stable between generations. Many detailed genetic studies revealed inheritance patterns of various traits, both independent and linked, and the occurrence of mutations, or mistakes, had been observed. But the way in which so much complex genetic information could be stored, retrieved, and replicated was totally unknown.

Many clues suggested that the molecule DNA was probably the key element in the mystery. For instance, haploid gametes (sex cells) contain half as much DNA as the diploid cells of the parent. Chemical analyses had also revealed that different species contain varied amounts of **purines**

	W	w
W	**WW** red	**Ww** pink
w	**Ww** pink	**ww** white

Fig. 5-3. In snapdragons, the red flower color gene is incompletely dominant over the white flower color gene. Heterozygous individuals have pink flower color.

and **pyrimidines**, the nitrogenous bases that are components of DNA. In 1953, **Francis Crick** and **James Watson** published their proposed model of DNA, based on the experimental evidence compiled by many other scientists. The discovery of the structure of DNA must be considered one of the most important in all of biology because its structure suggested how genetic information is coded, how it is replicated, and how mutations, or mistakes, can occur.

DNA is a very long, double-stranded, helical molecule composed of many **nucleotide units**. Each unit consists of a **sugar** (deoxyribose), a **phosphate group**, and a **nitrogenous base**. Four nitrogenous bases are found in DNA: the purines, **adenine** (A) and **guanine** (G), and the pyrimidines, **cytosine**

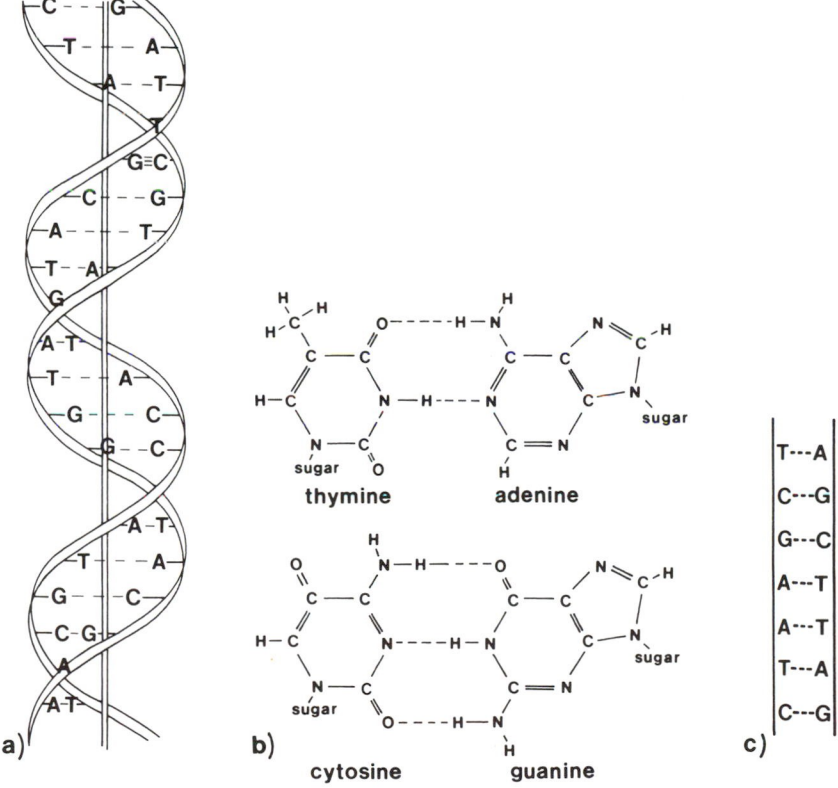

Fig. 5-4. The double helix. **a,** Deoxyribonucleic acid (DNA) is a double-stranded, helical molecule composed of nucleotide units. Replication of DNA occurs through the pairing of nucleotide bases to form new strands. **b,** Diagram of the molecular structure of the base pairs of DNA. "Sugar" indicates the bond to deoxyribose. Phosphates connect the sugars to make the "backbone" of DNA. A nitrogen base plus deoxyribose and a phosphate group constitutes a nucleotide. Adenine (A) always pairs with thymine (T), and cytosine (C) always pairs with guanine (G). **c,** Diagram showing pairing of the nucleotide bases in a short DNA segment.

(C) and **thymine** (T). In the double strand of DNA, the nitrogenous bases pair in a distinct pattern; A always pairs with T and G always pairs with C to form the rungs of the DNA helical ladder (Fig. 5-4).

When the DNA of an organism replicates, the double-stranded molecule opens, and each strand can then serve as a template (or mold) to form a new, complementary strand. The process is very complex and requires a number of different enzymes to open the double strand and complete the replication, but the means of replication itself was deduced almost immediately from the Watson-Crick model of the structure of DNA. Since chromosomes contain enough DNA to account for millions of base pairs, the discovery of the structure of this molecule also revealed a means for the storage of tremendous amounts of genetic information in a code based on the pattern of base pairs.

Further studies demonstrated how the genetic information stored in base pair patterns of the DNA molecule can be transformed into cellular products. Scientists already knew that proteins play very important roles in cellular control, particularly the specialized proteins called enzymes. **Enzymes** are complex molecules that catalyze chemical reactions in living organisms. They are responsible for controlling the synthesis and degradation of the numerous compounds necessary for life. Scientists also knew that complex proteins are made from chains of building blocks called **amino acids**. It seemed reasonable that the chains of nucleotides in the DNA somehow coded for the chains of amino acids of proteins, many of which were the enzymes that governed the chemical reactions in living organisms.

Production of Proteins

How do the chains of purines and pyrimidines in DNA code for the patterns of amino acids found in proteins? Since there are 20 different amino acids and only four different DNA bases, groupings of at least three bases are necessary to provide enough different combinations to code for the various amino acids. If combinations of two bases were used for the code, only 16 combinations would be available, which is not enough to code for the 20 different kinds of amino acids. Sixty-four combinations can be made using various groups of three bases. As seen in Table 5-1, each amino acid is coded for by more than one three-base group, and three of the combinations are stop signals. The three bases that code for an amino acid are called a **triplet codon**. This pattern of nucleic acid codons that match particular amino acids is the "genetic code" and is almost identical in all living organisms.

How can the information coded in the base pairs of the DNA in the nucleus become translated into the amino acid chains of the cellular proteins? An important intermediary molecule has this function. It is called **ribonucleic acid**, or RNA, and is closely related to DNA. RNA is different from DNA in two important ways. Although RNA has essentially the same nucleotide unit structure as DNA, the sugar component is **ribose** rather than deoxyribose, and **uracil** replaces thymine as one of the nitrogenous bases. Adenine (A) of RNA and DNA pairs with uracil (U) rather than thymine

(T). Several types of RNA exist in the cell, and their function is to mediate the translation of the DNA nucleotide code into the amino acid chains that become proteins.

Messenger RNA (mRNA) uses a section of DNA as a template to transcribe the DNA code into a mirror-image chain of RNA nucleotides (Fig. 5-5). Just as the double strands of DNA match each other with base pairings, mRNA matches the pattern of a segment of DNA. After transcription of the DNA segment into mRNA, the mRNA leaves the nucleus and becomes associated with ribosomes in the cytoplasm.

In the **ribosomes**, two other kinds of RNA participate in the production of proteins. **Ribosomal RNA** functions to bring together mRNA and the third type of RNA, **transfer RNA** (tRNA). tRNAs are small and composed of two important regions. One region binds to a specific amino acid. The other region is composed of a triplet codon. The tRNAs carry their amino acids to the mRNA, matching their triplet codon region to the correct codon on the mRNA. The amino acids line up into a chain exactly in the order originally specified in the nuclear DNA. The resulting chain of amino acids is used to build a protein. All these processes are carefully controlled by a complex series of enzymes that determine the type and amount of amino acid chains created. Not all of the bases of the DNA are transcribed into mRNA and then translated into proteins. Some of the DNA encodes the other various types of RNA, and other parts serve as starting and ending segments for transcription.

In prokaryotic bacteria, a sequence of DNA is simply transcribed into mRNA and then translated into a chain of amino acids. Thus, in bacteria,

Table 5.1. The Genetic Code[a]—Triplet Codons in Ribonucleic Acid (RNA) and Their Corresponding Amino Acids

Codon	Amino Acid	Codon	Amino Acid	Codon	Amino Acid	Codon	Amino Acid
UUU	Phenylalanine	CUU	Leucine	GUU	Valine	AUU	Isoleucine
UUC	Phenylalanine	CUC	Leucine	GUC	Valine	AUC	Isoleucine
UUG	Leucine	CUG	Leucine	GUG	Valine	AUG	Methionine
UUA	Leucine	CUA	Leucine	GUA	Valine	AUA	Isoleucine
UCU	Serine	CCU	Proline	GCU	Alanine	ACU	Threonine
UCC	Serine	CCC	Proline	GCC	Alanine	ACC	Threonine
UCG	Serine	CCG	Proline	GCG	Alanine	ACG	Threonine
UCA	Serine	CCA	Proline	GCA	Alanine	ACA	Threonine
UGU	Cysteine	CGU	Arginine	GGU	Glycine	AGU	Serine
UGC	Cysteine	CGC	Arginine	GGC	Glycine	AGC	Serine
UGG	Tryptophan	CGG	Arginine	GGG	Glycine	AGG	Arginine
UGA	None (stop signal)	CGA	Arginine	GGA	Glycine	AGA	Arginine
UAU	Tyrosine	CAU	Histidine	GAU	Aspartic	AAU	Asparagine
UAC	Tyrosine	CAC	Histidine	GAC	Aspartic	AAC	Asparagine
UAG	None (stop signal)	CAG	Glutamine	GAG	Glutamic	AAG	Lysine
UAA	None (stop signal)	CAA	Glutamine	GAA	Glutamic	AAA	Lysine

[a]U = uracil, C = cytosine, G = guanine, A = adenine. The codons in deoxyribonucleic acid (DNA) are complementary to the RNA codons. C and G are complementary in both RNA and DNA. U in RNA is complementary to A in DNA.

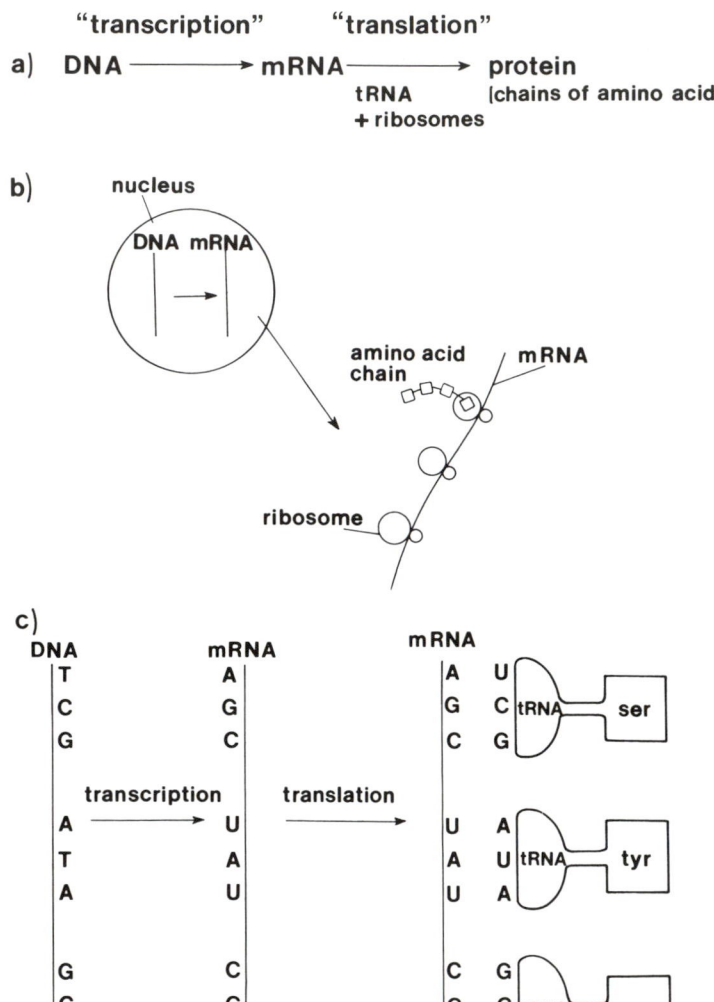

Fig. 5-5. The translation of a deoxyribonucleic acid (DNA) genetic message into a protein. Ribonucleic acid (RNA) differs from DNA in having ribose sugar units rather than deoxyribose and uracil (U) rather than thymine (T) as a nucleotide base. **a,** Overview of transcription of DNA to messenger RNA (mRNA) and translation to a chain of amino acids (protein) with the help of transfer RNAs (tRNAs). **b,** After transcription from DNA, mRNA leaves the nucleus and moves to the cytoplasm. tRNAs carry amino acids and "match" with a triplet codon on the mRNA to create a chain of amino acids in the order coded for in the DNA. **c,** Diagrammatic view of transcription and translation. Nucleotide bases: C = cytosine, G = guanine, A = adenine. Amino acids: ser = serine, tyr = tyrosine, arg = arginine.

a single gene can be identified as the section of DNA bases that codes a particular protein. In eukaryotes, the sequence of DNA that codes for a particular protein does not usually exist as a single linear sequence. It consists of a number of shorter segments along the DNA that are interrupted by sections of bases that do not function in the coding of the amino acid sequence. Long segments of RNA are transcribed from the DNA and then processed by enzymatic action into the actual mRNA that will move into the cytoplasm for translation into an amino acid chain. Thus, the concept of a gene as a section of DNA bases that codes for a particular protein is more complex and more difficult to identify in eukaryotes. Eukaryotic DNA contains large sections of nontranslated bases and many identical copies of certain base sequences, the function of which are not yet completely understood.

Despite these complexities in eukaryotes, the basic genetic code and its translation into proteins via RNA is essentially the same for all living organisms. The new genetic engineering techniques, to be described shortly, have shown that DNA sequences from one kind of organism can function in a totally unrelated organism. This remarkable phenomenon illustrates that the genetic code of DNA is the common thread that connects all forms of life.

Mutation and Evolutionary Change

The discovery of the structure of DNA led to a clearer understanding of how genetic information is stored, replicated, and translated into proteins for cellular control. It also helped explain how mistakes, or **mutations**, might occur. The idea of mutations was first suggested in 1901 by the Dutch scientist, **Hugo DeVries**, to describe new genetic characteristics that arose suddenly and had not been previously observed in the parental lines. Studies since 1951 have shown that mutations may occur in many different ways. Some of the simplest can be explained from the structure of DNA itself. Certain chemicals and physical damage may cause insertions, deletions, or mispairings during replication that affect one or a few nucleotides. Such small changes may cause shifts in the reading of the genetic code to create very different translations.

Larger changes in the structure of the DNA are also possible. **Barbara McClintock** received the Nobel Prize in 1984 for her identification of small mobile portions of chromosomes, called **transposons**, that can randomly move to different sites on the chromosome, causing changes in the genetic information. In addition, chromosomes are known to break and reattach in ways that also change the genetic information. For instance, a piece of a chromosome may break off and reenter the same chromosome with its gene sequence "backwards" from its previous orientation. This phenomenon is called an **inversion**. Chromosome segments may also attach to different chromosomes in a process called **translocation**.

All these various "mistake" mechanisms have been discovered in the course of investigations of sudden phenotype changes in offspring that could not be explained by the previous genetic histories of the parents. Mutations, although generally rare, are an important source of genetic variation that allows organisms to adapt to their changing environment. Mutations may also have very deleterious effects, such as contributing to increased susceptibility to diseases such as cancer. Many human cancers have been linked to specific gene mutations or to a series of mutations that accumulate over a period of time. Such mutations may reduce the production of an important enzyme or cause a different enzyme to be produced, which may affect control of the growth and development of cells and tissues.

People are warned to avoid exposure to X-rays and ultraviolet light because such radiation is capable of causing mutations. The ozone layer in the upper atmosphere helps shield the earth from ultraviolet light, and its possible destruction by industrial activities may result in increased mutations among living organisms. Chemicals to which people are exposed in their work and diet may also act as **mutagens**. Many compounds have been banned or restricted to protect the genes of human beings.

The recently developed **Ames test** is used to detect the ability of a chemical to increase the rate of mutation in bacteria. The test can be performed more quickly and inexpensively than animal tests because bacteria multiply so rapidly. Ames test results are used as preliminary evidence that a particular chemical may cause increased mutation, and perhaps increased cancer, in humans.

The Gene-for-Gene Theory

Let us now consider how this improved understanding of genetics and gene structure is useful in agriculture and, specifically, in plant pathology. For years, farmers have been selecting crop plants for their various desirable traits, including yield, quality, and pest resistance. The rediscovery of Mendel's work at the turn of the century led plant breeders and pathologists to the study of specific genes for desirable traits. In the area of disease resistance, many examples of resistance to various parasites were discovered in which a **single dominant gene** was the source of the resistance. As mentioned in Chapter 1, *Solanum demissum* served as a source of resistance genes that were crossed into cultivars of *S. tuberosum*, conferring resistance to *Phytophthora infestans*. The effect of such resistance genes was specific to *P. infestans* and to no other parasite of the potato. The success of the resulting resistance was, at first, so complete that many scientists believed that late blight had been eliminated. Unfortunately, the continued use of single-gene resistance often selects for new races of the parasite population, resulting in the failure of the resistance.

We have already referred to the genetic interactions between hosts and parasites. Although the actual interaction between organisms that allows parasitism to occur is very complex, some specific factors have been discovered. For instance, some parasites produce enzymes or toxins that contribute to their ability to attack the host plant. A resistant plant may produce a chemical that blocks the function of a parasite enzyme. The toxin-binding site may be different in the cells of a resistant plant, so that a parasite toxin is no longer effective. With time, genetic change in the parasite, through sexual reproduction or mutation, may result in an enzyme of a slightly different shape that cannot be blocked or a toxin with improved binding ability, so that host resistance is overcome. This greatly over-simplified description suggests the kind of stepwise evolutionary interactions that occur between hosts and parasites in natural communities.

H. H. Flor was the first plant pathologist to simultaneously study the genetics of a plant, *Linum usitatissimum* (flax), and its parasite, *Melampsora lini* (a rust fungus). Based on his studies, he proposed the **gene-for-gene theory**, in which a successful infection is governed by genetic factors in both the host and the parasite. In the example of potatoes and *P. infestans*, genetic studies have shown that potato plants with a resistance gene can be infected only by certain races of *P. infestans*. When a whole field of potatoes with a particular resistance gene is planted, the initial effect is a tremendous reduction in the amount of disease because most genetic strains of the fungus are eliminated. However, the surviving races capable of infecting the resistant plants infect and multiply on the plants. "Resistant" potato cultivars eventually appear just as susceptible as the original "susceptible" ones. Farmers often say that resistance has "broken down," when, in fact, the parasite has "overcome" the resistance.

Genetically uniform monoculture agriculture is very vulnerable because it puts selective pressure on the parasite to adapt to the genetic makeup of the host. The parasite must "adapt or starve." Flor's work explained why the use of single-gene resistance in an agricultural crop is often only effective temporarily. Gene-for-gene systems have been identified in plant interactions with many kinds of plant pests, including bacteria, viruses, nematodes, and even insects.

One of the most difficult aspects of these genetic interactions is that parasites generally have the advantages of shorter life cycles and much greater reproductive capacity than the plants we wish to protect. The parasite may be able to quickly overcome genetic resistance in the host plant. This is particularly true in parasites that reproduce many times during a growing

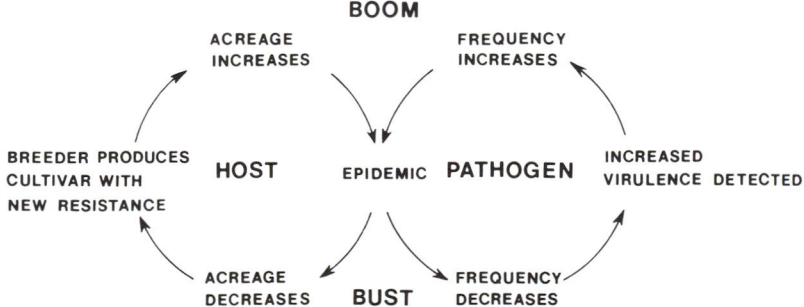

Fig. 5-6. The boom-and-bust cycle. As a resistant cultivar becomes popular, the selective pressure for the pathogen to overcome the resistance becomes greater. As the new pathogen race increases, the cultivar no longer successfully resists infection.

season and are air-dispersed, such as leaf spot fungi, rust fungi, and downy mildews.

Certain kinds of crops are more vulnerable to resistance failure. Vegetatively propagated crops, such as potatoes, are genetically uniform, which places very strong selective pressure on the parasite to overcome the resistance. In addition, perennial crops, such as grapes or asparagus, remain in the same soil for many years so that parasites have a much longer time to overcome resistance. In contrast, with annual crops such as corn or soybeans, farmers can alternate crop species or cultivars of the same crop from year to year. The longer the normal life of a plant, the more time the parasite has in which to overcome the resistance. For example, it is likely that the rapidly reproducing, airborne rust fungus, *Hemileia vastatrix*, might overcome single-gene resistance in a coffee tree before the tree could even mature enough to produce a harvestable crop.

Genetic Resistance to Disease

It has been said that the popularity of a new cultivar is both the reward and the doom of a plant breeder. The more widely a single genetic line of a crop is planted, the more likely it is that a devastating epidemic might arise from a new parasitic race. This has been dubbed the **boom and bust** cycle of plant breeding (Fig. 5-6). Although some spectacular resistance failures have plagued plant breeders, some single-gene resistance factors have been very effective for many years. For instance, resistance genes to various formae speciales of the vascular wilt fungus, *Fusarium oxysporum*, have been very durable because the pathogen produces only one generation of spores at the end of the growing season, and the spores remain soilborne. The reproductive capacity and dispersal ability of this fungus are obviously much less than those of a parasite such as *P. infestans*. The same genetic interactions exist, but the probability of selecting a new fungal race capable of overcoming the resistance is much lower due to the biology of the parasite. Tomato plants with single-gene resistance to *F. oxysporum* f. sp. *lycopersici* have remained resistant for over 30 years. Resistance first failed in places where growers planted tomatoes in the same land repeatedly for many years. Eventually, a new race appeared in the fungus population capable of infecting the tomatoes despite the resistance gene. Luckily, the dispersal of the new race was slow because the pathogen is soilborne. In most areas, where tomatoes are rotated with other crops and planted in the same land only every three to four years, resistance has remained much more durable. Even if some spores of the new race were produced, most would starve or be destroyed in the intervening years without a host plant.

Single-gene resistance can be useful for rapidly reproducing, airborne pathogens when it is used appropriately. It remains the most common means of combating rust diseases in our grain crops, but its use requires that plant breeders continually assess the races that exist in the rust population. They can then change the resistance genes in the grain cultivars from year to year according to the genetic status of the parasite population. This important subject is discussed in more detail in Chapter 10.

Even though single-gene resistance can be useful and effective, many plant breeders have sought a form of genetic resistance effective against all possible races of a parasite. Their goal is a resistance that cannot be overcome by genetic changes in the parasite. This type of resistance has been called by various terms such as *durable, general, nonspecific, field, horizontal*, or *polygenic* resistance. (Note: Despite the term *general*, this resistance is effective against all races of the parasite for which it has been selected but does not necessarily confer resistance to any other pests or pathogens.) The theoretical basis for such resistance is that it involves many genes that each contribute to the resistance in some small way. Because many genes are involved, it becomes highly unlikely that a parasite would be able to simultaneously make all the genetic changes necessary to overcome the resistance.

Such a resistance is identified in the host plant by a reduced amount of disease even when the plant is exposed to a variety of races of the pathogen. Disease might be less because lesions develop more slowly, spores take longer to appear on lesions, and fewer spores are produced on a lesion compared to these factors in a plant lacking the "general resistance." Plant breeders hope that with time they will be able to increase the levels of general resistance in various crops. Some potato cultivars are now available with levels of general resistance to *P. infestans* that reduce the amount of fungicide needed to protect the foliage from infection. General resistance in certain grain crops is referred to as *slow-rusting*; it can be combined with single-gene resistance for added protection against rust diseases.

Genetic resistance has always been a popular means of plant disease management because it requires no special inputs by growers during the growing season and is compatible with other agricultural activities. Research in the area of genetic resistance to plant parasites has received renewed interest in recent years for several reasons. In many parts of the world,

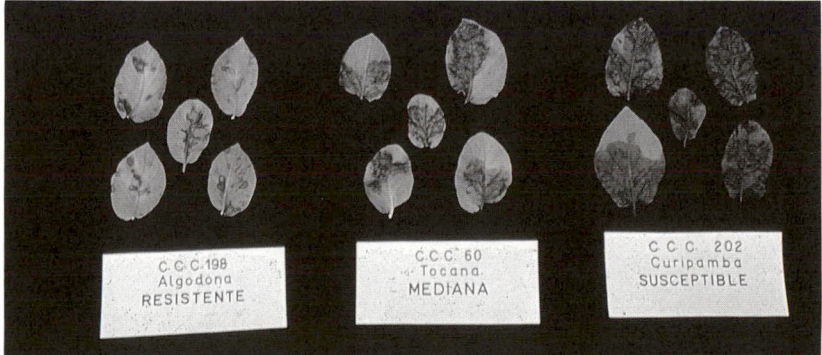

Fig. 5-7. Potato leaves from plants with varying degrees of resistance to *Phytophthora infestans*. General resistance reduces the rate of an epidemic. Lesions are smaller, develop more slowly, and produce fewer sporangia. It functions against all races of the pathogen.

subsistence farmers do not have the equipment or money for frequent pesticide applications. In addition, it has become clear that intensive pesticide use can have detrimental environmental effects. Also, advances in modern genetics and cell biology have greatly increased our understanding of the genetic interactions between fungi, bacteria, viruses, and their host plants. This improved knowledge allows plant breeders and plant pathologists to develop breeding programs based on scientific genetic principles rather than on the random selection of naturally occurring genetic variation.

Tissue Culture and Biotechnology

One of the major limitations for plant breeders is the reproductive isolation between plant species. Because of pollen incompatibilities, it is not usually possible to successfully cross plants unless they are very closely related. If pollen from an oak tree is transferred to the flower of a potato plant, no fertilization will occur and no seeds will be produced. However, once the structure of DNA and gene sequences had been discovered, scientists began to speculate about the possibilities of transferring specific genes between species. Such a technique is called **genetic engineering** and is particularly applicable to plants because of the way plants develop.

Each somatic (nonreproductive) cell of a living organism contains the full complement of DNA in its nucleus, and yet, as an organism grows and develops, cells become specialized. Only some of the total genetic information is used for the function of any particular cell. For example, leaf epidermal cells contain the same DNA information as root cortex cells. Yet, the appearance and function of each is very different. Various growth

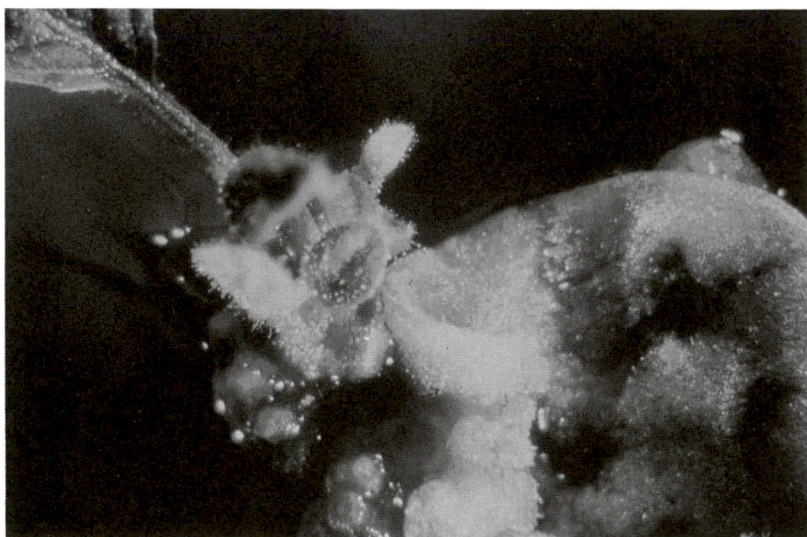

Fig. 5-8. The regeneration of a tomato plant from callus growing on a nutrient medium.

hormones and other complex controlling mechanisms determine the genetic information that is used in a specific cell. Unlike animal cells that become predestined for a particular function early in their development, many plant cells retain the ability to change their function and take on new roles at different times.

Cuttings from many plants develop roots when placed in water. As early as 1902, the German botanist **Gottlieb Haberlandt** studied these possibilities and described plant cells as having **totipotency**. That is, any plant cell should be capable of developing into a whole mature plant. His beliefs remained theoretical until 50 years later, when the American plant physiologist, **F. C. Steward**, grew cells from mature carrot root phloem in coconut milk, which provided nutrients and growth hormones. He succeeded in growing whole new carrot plants from cells that had previously been in differentiated tissues of a mature plant. Steward's work opened an important door for genetic engineering. If a specific gene could be transferred to the nuclear DNA of a single plant cell, that cell could then be regenerated into an entire plant in which each cell would contain the new gene in its nucleus.

Besides specific gene transfer, the culture of individual plant cells in nutrient media has offered other nontraditional breeding methods to scientists. For instance, entire haploid plants have been grown from haploid cells cultured from pollen and ovules. One advantage of haploid plant culture is that the full genotype appears because recessive factors are not masked by dominant genes on homologous chromosomes. Since many agriculturally important plants are **polyploids**, having more than two sets of chromosomes, they are difficult to work with; haploid plant culture from pollen and ovules reduces the chromosome numbers by half and simplifies genetic studies. Because agriculturally important potato cultivars are tetraploids, having four sets of chromosomes in each nucleus, haploid potato culture provides a means of reducing *S. tuberosum* to a diploid state, so that it can more easily be crossed with wild diploid *Solanum* species.

Through careful degradation with enzymes, cell walls can be removed from plant cells, leaving only the **protoplast** surrounded by the cell membrane. Such protoplasts can be grown like single-celled microorganisms

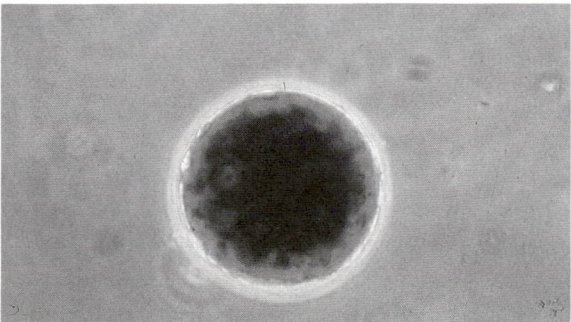

Fig. 5-9. Cucumber protoplast stained with neutral red. The cell wall has been removed enzymatically, leaving only the cell membrane.

in liquid nutrient media. Some protoplasts have been induced to fuse together, and successful nuclear fusions have even occurred. New plants have been regenerated from the fused protoplasts. Protoplast fusion presents a possible means of combining plants that are unable to be crossed by traditional breeding methods. For example, cultivated bananas are sterile triploids. Protoplast fusion between banana cultivars can be used to increase genetic diversity in plants that no longer undergo sexual reproduction. Large numbers of protoplasts can be exposed to chemicals such as parasite toxins, and the survivors can be regenerated into whole plants that may then be resistant to the parasite that produces the toxin. The ability to manipulate plant cells and tissues in culture and subsequently regenerate whole plants is now an extremely important area of plant biotechnology that may result in important plant improvements in the coming years. Commercial success in plant biotechnology has already occurred in the rapid propagation of pathogen-free planting material from tiny pieces of plant tissue, which is discussed in Chapter 12.

Recombinant DNA and Genetic Engineering

The ability to regenerate whole plants from single cells has stimulated great interest in the genetic engineering of plants. Plant breeders using traditional breeding methods must often perform crosses between plants for many generations to create new plant cultivars with a desirable genetic trait. At this time, classical breeding methods remain the most efficient way to introduce new genetic traits into plants that are sufficiently closely related for hybridization. However, genetic engineering can potentially be used to transfer any genetic sequence to the nucleus of any plant cell. It is becoming an important means of genetic improvement.

There are three important steps in such a procedure: 1) identification of a specific gene sequence that controls the desired trait, 2) removal of a functional gene sequence from the donor organism, and 3) transfer of the gene sequence to the nucleus of the receptor organism, followed by expression of the desired trait.

Identification of a gene sequence is simplest in prokaryotes, in which genes are coded in a linear sequence. For eukaryotic genes, it is often necessary to remove and further process large DNA sequences to ensure that an entire message is transferred. To begin genetic engineering, the DNA from the donor organism is isolated and purified of other cellular components. Removal of a gene sequence and its transfer is accomplished with special enzymes discovered in the 1970s. **Restriction enzymes** cut DNA at particular nucleotide sequences so that a section of DNA containing the desired sequence can be removed. **Ligases** are enzymes that seal the cut ends of DNA so that the transferred sequence can be inserted into another piece of DNA (Fig. 5-10).

The earliest genetic engineering experiments were accomplished with bacteria. The first step was to determine the actual base sequence of DNA molecules in bacteriophages (viruses that infect bacteria) and in bacterial plasmids. **Plasmids** are relatively short, circular pieces of DNA in bacterial cells that are independent of the main chromosome. After repeated degradation and analysis of the DNA, scientists were able to make genetic maps of the base sequences of these relatively short DNA pieces. Once they knew the base sequences, they could use enzymes to insert new gene sequences into the virus or plasmid. The new DNA containing the introduced sequence is known as **recombinant DNA** (rDNA). When "normal" bacteria are then exposed to the genetically engineered viruses or plasmids, the gene can be transferred to a few cells by the natural means of genetic recombination described in the previous chapter. When the recipient cell absorbs the recombinant DNA and begins to express the gene by producing the protein it coded for, the cell is said to be "transformed"; the genetic engineering is complete.

Many such experiments have been performed since 1973, and the results have been very interesting. It is quite remarkable that a gene from a eukaryotic organism can function in a bacterium. For instance, the genetic information necessary for the production of insulin in humans has been

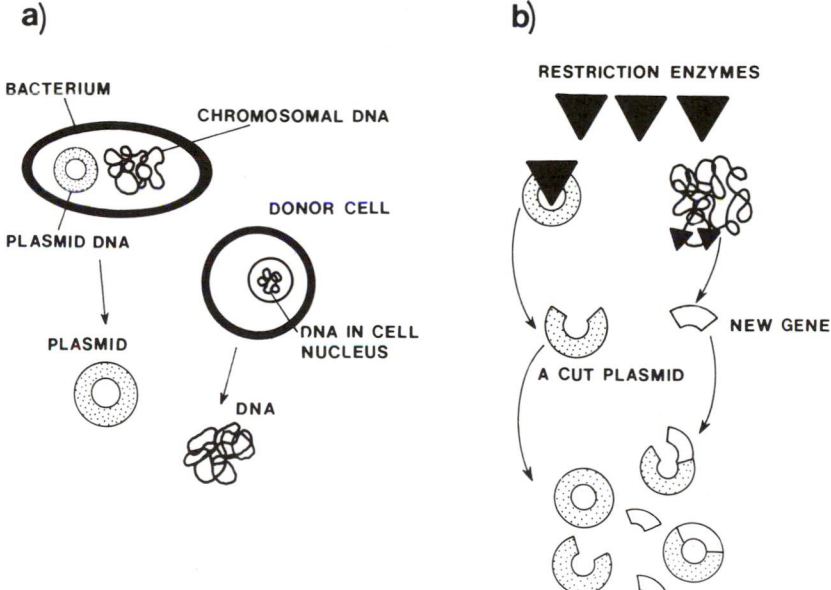

Fig. 5-10. Recombinant deoxyribonucleic acid (DNA) technology, or "gene splicing." **a,** The plasmid of a bacterium and the DNA of a donor cell (plant, animal, or human) are removed from their respective cells. **b,** Restriction enzymes are used to cut open the plasmid and to cut out a gene from the DNA of the other organism. The cut ends of the plasmids and the cut ends of the new genes are chemically "sticky," so they will attach to each other to form a new loop containing the inserted gene. This method has been used to insert the human gene responsible for insulin production into the plasmid of a bacterium. The bacterium is then able to produce human insulin.

transferred to a bacterium so that inexpensive insulin can now be produced by bacteria, reducing the need for its isolation from butchered pigs and cows. Because the insulin produced by bacteria is identical to that produced by human cells, allergy problems are eliminated for diabetic individuals who are sensitive to insulin derived from animals.

In a relatively short time, the techniques for the identification and transfer of gene sequences have become well known. Out of the enormous number of important research projects involving genetic engineering of bacteria and viruses, let us consider one of agricultural significance. Studies have revealed that plant surfaces support many kinds of bacteria. Dr. Steven Lindow, a plant pathologist working at the University of Wisconsin (and now at the University of California at Berkeley), noted that some leaf surface

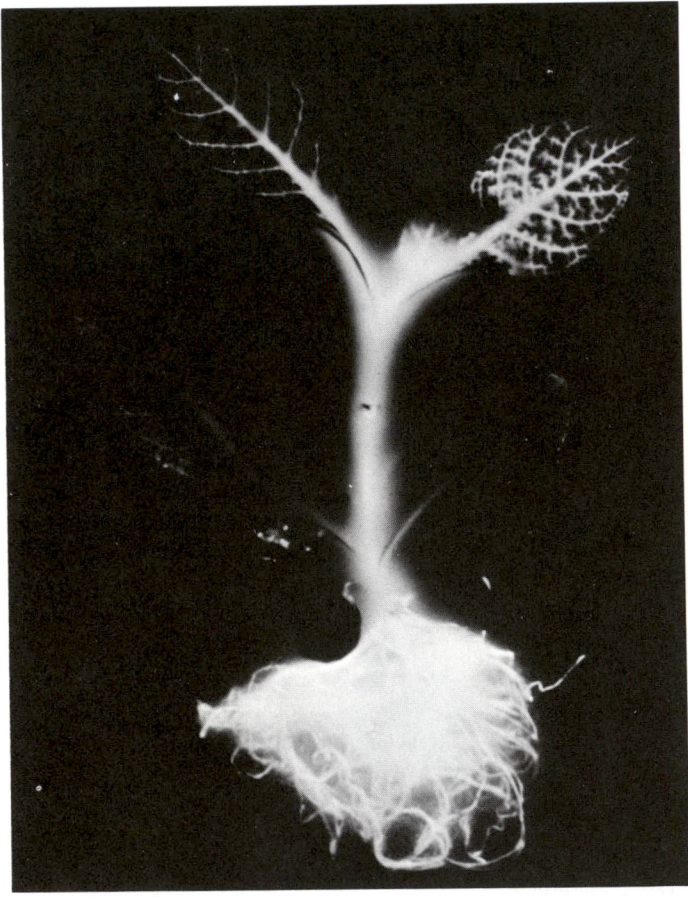

Fig. 5-11. A genetically "transformed" tobacco plant that received the luciferase gene responsible for the luminescence of fireflies. The photograph was taken over a 24-hour period.

bacteria produce a particular cell wall protein that serves as a nucleation site for (acts as a nucleus for) ice crystal formation, while other bacteria do not. In fact, ice-nucleating bacteria, marketed under the trade name Snowmax, are mixed with water to increase artificial snow production for ski slopes. When many ice-nucleating bacteria are present on a leaf surface, frost damage is greater due to ice crystal formation. When ice-nucleating bacteria are removed from the leaf surfaces, plants can survive to −5°C (23°F) without suffering frost damage.

Tremendous crop losses are attributable to frost damage each year. Damage occurs most often in spring, when leaf and flower buds are expanding, and at harvesttime, so this discovery was of considerable significance. How could the ice-nucleating bacteria be removed or reduced so that plants could be protected from frost damage? Antibiotics and other chemical sprays were effective in experiments but were not suitable for commercial use. Non-ice-nucleating bacteria were sprayed on plant leaves in an attempt to displace the ice-nucleating bacteria and reduce their number—but without great success. Lindow reasoned that the ice-nucleating bacteria already present on leaves were well-adapted residents of the leaf surfaces and would be difficult to displace with bacteria less well adapted to that environment. He then used genetic engineering techniques to delete the genetic information that controlled the formation of the ice-nucleating protein. The resulting bacteria (designated "ice-minus") were well adapted for survival on leaf surfaces but could not serve as ice-nucleating sites. Controlled laboratory and greenhouse experiments proved that plants sprayed with these genetically engineered bacteria were able to survive lower temperatures without frost damage. In 1987, the first controlled outdoor

Fig. 5-12. Protection of potato plants from frost damage using genetically engineered "ice-minus" bacteria. A naturally occurring bacterium, isolated from the plant leaf surface, was modified so that it lacks the single gene that triggers frost formation. When the bacteria are applied to potato seedings, the seedlings are protected from frost damage (right). The plant on the left is noninoculated.

experiments with these bacteria were permitted under considerable public protest. We will return to the controversial aspects of such a release at the end of this chapter.

Crown Gall Bacteria—The Natural Genetic Engineers

The techniques for genetic engineering of bacteria are well established, which leads us to consider the problem of inserting a gene into a plant cell. The previous chapter described the crown gall bacterium, *Agrobacterium tumefaciens*, as a natural genetic engineer. During the same years that the bacterial genetic engineering techniques were being developed, it was discovered that *A. tumefaciens* possesses a plasmid that transfers a segment of its DNA into the nuclear DNA of the infected plant host cell. This plasmid was subsequently named the **tumor-inducing**, or **Ti**, plasmid because the presence of a segment of the Ti plasmid in the plant nucleus caused the uncontrolled cell division and tumor growth characteristic of the disease. After the plasmid gene transfer, gall formation could continue even if the bacteria themselves were removed. Scientists began to realize that crown gall bacteria had great potential for gene transfer to plants. *A. tumefaciens* is now the workhorse of plant genetic engineering.

It was discovered that only a small section (the T-DNA) of the Ti plasmid is transferred into the plant nucleus. In addition, only the two border "ends" of the gene sequence are needed to control gene transfer. The nucleotide sequence between the ends can be removed to prevent the bacteria from causing crown gall disease in the plant but without loss of the genetic engineering ability of the bacteria. So the tumor-inducing sequences were replaced by a smaller, well-known plasmid from the common bacterium, *Escherichia coli*. To transfer a gene to a plant cell, the DNA sequence of the gene is first inserted into the *E. coli* plasmid. Recombination between the plasmids results in the transfer of the desired gene into the Ti plasmid. When plant cells are exposed to the Ti plasmid, the gene is inserted into a chromosome in the nucleus of a susceptible plant cell. Using tissue culture techniques, the plant cell can then be regenerated into a whole plant in which each cell contains the new gene.

The Ti plasmid is a very efficient and genetically predictable means of gene transfer in plants. It is limited, however, to plants that are susceptible to *A. tumefaciens*, which eliminates many important crops, including the cereals. For nonhosts of the crown gall bacterium, other means of gene transfer have been used, although they are not as efficient nor as genetically predictable. One procedure, called **electroporation**, involves using a brief electrical current to open temporary pores in cell membranes of plant protoplasts to allow entry of foreign DNA. Electroporation of pollen tubes is also being investigated as a means to introduce foreign DNA into a plant during fertilization. **Microinjection** of foreign DNA has also been attempted, although this technique is more commonly used for introduction of foreign DNA into animal cells. Although these techniques circumvent the host range problem in the use of the Ti plasmid for gene transfer, it is still difficult to regenerate plants of some species, including most cereal

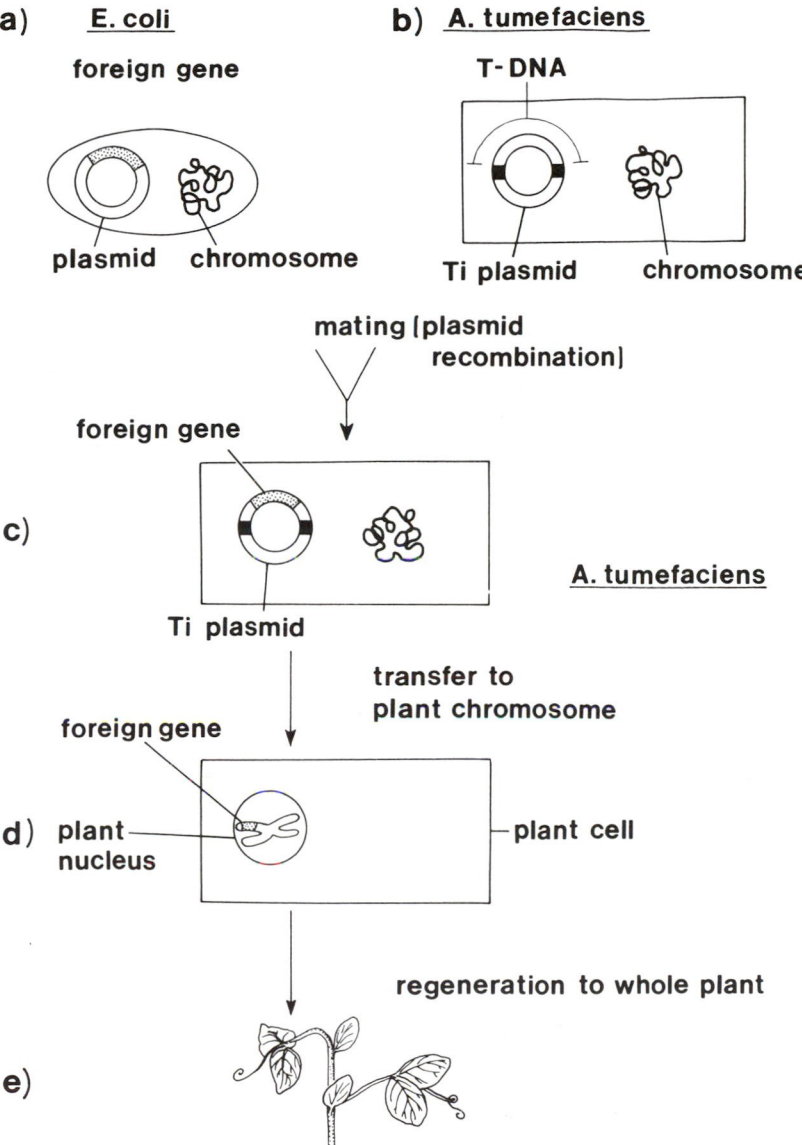

Fig. 5-13. Plant genetic engineering using the tumor-inducing (Ti) plasmid of *Agrobacterium tumefaciens*, the crown gall bacterium (**b**). Using the recombinant-DNA techniques described in Fig. 5-10, a foreign gene is inserted into a plasmid of *Escherichia coli* (**a**). Through plasmid recombination, the foreign gene is inserted into the T-DNA section of the Ti plasmid (**c**). *A. tumefaciens* transfers the foreign gene to a chromosome in the nucleus of a plant cell (**d**). Each cell of a plant regenerated from the transformed cell contains the foreign gene (**e**).

crops. However, rice has already been regenerated from protoplasts, and such technical problems will likely be overcome in other crops in the near future.

The potential for plant improvement using genetic engineering is vast. It may become possible to improve the amount and quality of protein in various food crops and perhaps even insert genes that control the fixation of atmospheric nitrogen into a form usable by plants, reducing fertilizer needs. As our understanding of resistance mechanisms in plants improves, the transfer of genes to make plants more resistant to disease and pests will become possible. Of course, this will not eliminate the need for traditional breeding methods to improve and maintain more complex characteristics, but genetic engineering is an important new tool for rapid transfer of specific gene sequences. For example, tomato plants have been genetically engineered to resist infection by the tobacco mosaic virus by the transfer of the gene for the virus coat protein to the nucleus of a tomato cell, using the Ti plasmid. Virus-resistant plants can be regenerated from the genetically engineered cells. Herbicide-resistant plants have also been created by transfer of a resistance gene from bacteria via the Ti plasmid. Plants have been created that produce a toxin against insect pests. The identification of useful gene sequences and their functional transfer to plants is the goal of many current research projects.

Regulation of Genetic Engineering

The various types of genetic engineering and cell manipulation described above are grouped under the broad term **biotechnology**. As always happens

Fig. 5-14. Transgenic tomato plants are resistant to tobacco mosaic virus. The plants on the left have been transformed using the Ti plasmid. The coat protein gene of the virus was transferred to the DNA in the nucleus of the plant cell. Virus-resistant plants were regenerated from the transformed cells. Both sets of plants were inoculated with the virus, but only the plants on the right became diseased.

when rapid change of any kind occurs, many people feel that such activities are unnatural and dangerous, whereas others find the potential improvements too great to ignore. The products of genetic engineering, dubbed **transgenic** organisms, present potential dangers that must be evaluated. Everyone agrees that the release of products of genetic engineering must

Fig. 5-15. Field testing of genetically engineered organisms. Dr. Steven Lindow and his assistant wore protective clothing in the 1987 small-scale field test of "ice-minus" bacteria, as required by regulations of the Environmental Protection Agency.

Fig. 5-16. An example of a field plan used to "track" genetically engineered organisms. Note the bare soil beyond the edge of the test plots, the additional "trap" plants at the edge of the test area, and the secure fencing of the experimental area.

be done cautiously and only after careful study. As described in the previous chapters, the introduction of a "foreign" plant or microorganism to an ecosystem can have devastating results. Many important weed pests in the United States are actually native to Europe and other continents but rapidly invaded the new lands when inadvertently transferred by emigrants. So, it is only reasonable that we closely supervise the new genetic strains of plants and microbes created through genetic engineering.

Experiments using what is called "recombinant DNA" are now permitted only in special laboratories. Genetically engineered plants and microbes are grown in special containment facilities where all waste is carefully sterilized before disposal. Of greater controversy is the eventual outdoor release of the products of such experiments. Some people believe that the organisms produced in recombinant DNA experiments represent a special danger greater than that posed by genetic recombination through the so-

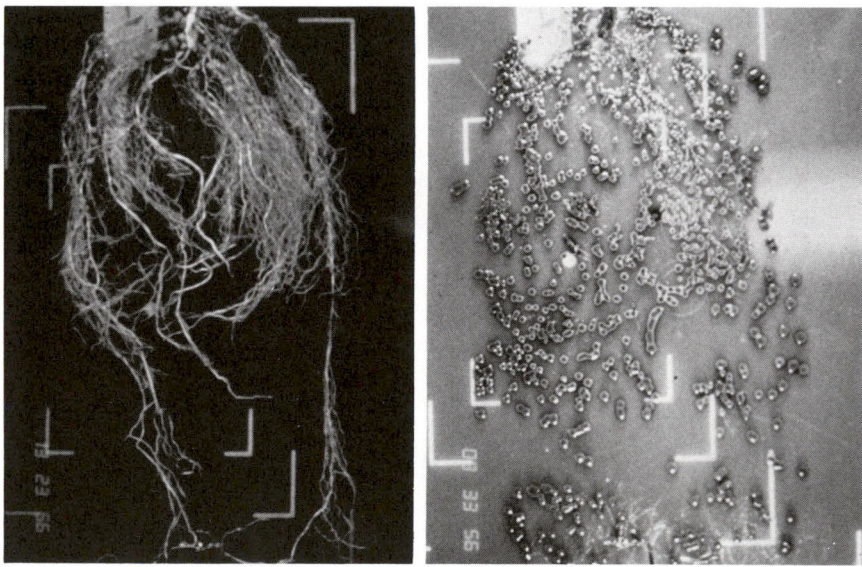

Fig. 5-17. The use of live microbial agents requires knowledge about their long-term performance in the field and potential effects on the environment. This is particularly important for microbes that have been genetically modified by recombinant DNA techniques. This photograph depicts the use of a method developed at Monsanto Company to monitor the presence and spread of bacterial biocontrol strains. Scientists at Monsanto have introduced two specific marker genes, called the lacZY genes, which allow the detection of a single microbe in a gram of soil. The left panel is a soybean root that was inoculated at the time of planting with a marked biocontrol microbe. After 4 weeks of growth in field soil, the seedling was harvested, shaken free of soil, and placed on a square dish containing a medium selective for the engineered microbe. The plant was then removed and the dish incubated. After 2 days, the marked microbe is clearly visible as colonies in the right panel. The colonies have a distinctive blue color.

called "natural means" discussed previously. Recent studies have shown that natural genetic exchange between various species via bacterial plasmids and viruses is probably more common than was thought even 20 years ago. Yet, genetic engineering certainly poses potential dangers that must be respected.

A number of scientific panels composed of molecular biologists, microbiologists, plant pathologists, ecologists, public health scientists, and toxicologists carefully considered the experimental work done over the past 20 years from laboratories throughout the world. The U.S. Department of Agriculture ruling, published in the *Federal Register* on June 16, 1987, concluded that organisms produced through recombinant DNA techniques are of no greater danger than organisms that might be produced through naturally occurring processes. In 1989, a distinguished group of ecologists published ecological considerations and recommendations for the planned release of genetically engineered organisms. They emphasized that risk is associated with the phenotype of an organism, or its actual activity and survival in the environment, rather than with its genotype, or the source and combination of its genes. This does not imply that the products of genetic engineering should not be contained and controlled, because new genetic strains, no matter what their origin, can pose a danger. In particular, careful study is required before the release of organisms to the environment. Several important questions must be answered to assess the potential risk of the release of a particular organism, such as:

Fig. 5-18. Biocontrol of *Gaeumannomyces graminis* f. sp. *tritici*, the causal agent of wheat take-all disease. This photograph demonstrates the potential of live microbial agents to effectively control take-all disease in the field. Vertical rows of wheat were planted to areas infested by the pathogen (white head rows) or noninfested (green rows). In horizontal rows, a native anti-take-all microbe was placed on the wheat seed at the time of planting. This microbe has provided visible disease protection.

l) Will the organism be able to survive, multiply, and disperse in the environment?

2) Can the new genetic trait be transferred to other organisms in which it might pose a danger?

3) What are the ecological effects of the new organism on other organisms it might encounter?

4) Does the new organism pose any specific health threats to human beings, animals, or crop plants?

Before release, a genetically engineered organism may be given a tracing mechanism so its multiplication and dispersal can be followed. Extensive, contained laboratory and greenhouse experiments are also performed before outdoor release is contemplated.

The release of the genetically engineered, non-ice-nucleating bacteria described previously is considered by most scientists to be a very carefully tested and safe experiment. It was delayed by long court procedures brought about by individuals who do not consider any release of genetically engineered organisms to be safe. These people also fear that the potential profit from such experiments will encourage scientists to take short cuts that may increase risks. Although nearly all scientists favor appropriate regulations, some feel that excessively restrictive rules are biologically unjustified, which may encourage experimenters to ignore or evade them.

At this point, all regulations are in anticipation of possible problems that have, so far, not occurred. Because most of the general public has little or no understanding of the rapidly changing science involved in genetic engineering, it is difficult for people to appropriately assess the various opinions offered in the mass media. When release of recombinant DNA organisms is decided in the courts, the decisions may be affected more by the legal skills of the lawyers than by the scientific merits of the cases. Judges may find themselves poorly prepared in the scientific background needed to make a rational decision. The potential benefits of biotechnology are enormous, and genetic engineering will continue. It is now necessary to settle the legal entanglements so that the experiments can be performed under restrictions that are appropriately protective. Because these decisions affect us all, everyone should be encouraged to try to follow the progress of this rapidly developing area of science.

Selected Readings

Gasser, C. S., and Fraley, R. T. 1989. Genetically engineering plants for crop improvement. Science 244:1293-1299.

Keller, N. P., and Bergstrom, G.C. 1988. Genetic engineering and its applications to plant protection. Plant Pathology Extension Notes 88-1. Cornell University, Ithaca, NY.

Lindow, S. E., Panopoulos, N. J., and McFarland, B. L. 1989. Genetic engineering of bacteria from managed and natural habitats. Science 244:1300-1307.

Moses, P. B. 1987. Strange bedfellows. BioScience 37:6-10.

Root, M. 1988. Glow-in-the-dark biotechnology. BioScience 38:745-747.

Strange, C. 1990. Cereal progress via biotechnology. BioScience 40:5-9, 14.

Tiedje, J. M, Colwell, R. K., Grossman, Y. L., Hodson, R. E., Lenski, R. E., Mack, R. N., and Regal, P. J. 1989. The planned introduction of genetically engineered organisms: Ecological considerations and recommendations. Ecology 70:298-315.

U.S. Department of Agriculture, Animal and Plant Health Inspection Service. 7 CFR Parts 330 and 340. Introduction of organisms and products altered or produced through genetic engineering which are plant pests or which there is reason to believe are plant pests. June 16, 1987. Federal Register 52:22892-22915.

Plant Disease Epidemics and Their Management

"The history of mankind is the story of a hungry creature in search of food," wrote historian Hendrick Van Loon. The need for food has always governed the movement and activities of human beings. Early humans were foragers, sometimes following seasonal food supply patterns but always moving on as supplies became depleted. Low population densities could be supported this way without too much interference from neighboring groups. With the advent of agriculture, people were able to live in one place. As food production became more efficient, some people could produce more food than they needed for themselves, so that others had time to build cities, write books, create art, and pursue scientific studies. However, along with the threat of illnesses and the fear of invading armies, the potential for famine was always there.

Most readers of this book are living in a place and in a time when food surpluses seem more of a problem than food shortages. We have access to diverse food crops from many different regions so that localized epidemics and crop losses become only a momentary inconvenience, even though individual farmers may be ruined. Only a low percentage of the population (2–3% in the United States) is employed in agriculture, so most people are removed from direct experience with food production. However, highly mechanized agriculture and the rapid transit of crops in refrigerated trucks, railroad cars, and airplanes across continents is a very recent phenomenon. When the store shelves are well stocked with inexpensive foods, it is easy to lose sight of the precarious future of tiny seeds planted in a field, exposed to the vagaries of weather, insects, diseases, and soil fertility that will determine the harvest several months later. Even western Europe, with its highly technological agriculture, became self-sufficient in wheat production only about 1975.

Early farmers were nearly totally dependent on their own food production, so significant losses most likely meant hungry days or even starvation. With all agricultural tasks performed by human or animal power, production was limited. Since cultivation was by hand and herbicides were not used, many weeds competed with the crops for water and nutrients. We have available today a large variety of food crops, but until about 500 years ago, as seen in Chapter 3, each continent had access to only some of these important plant species. Food storage was certainly less successful before refrigeration became commonplace, so supplies could dwindle quickly, to be replaced only after the next growing season. In many Third World countries today, lack of refrigeration and storage facilities that can protect food from insects and decay limit human diets severely.

All agricultural crops are, by their nature, vulnerable to losses to plant pathogens. The planting of many individuals of the same plant cultivar, variety, or species over many thousands of acres (or hectares) invites epidemics. Plant disease epidemics have plagued humans since agriculture began, just as human diseases have caused tragedy in crowded cities. How frightening a crop epidemic must have been to those early farmers, and modern ones as well, who had to watch a crop wither or wilt!

Plant Disease in Traditional Agriculture

Were plant diseases more or less of a problem in those early times? A good argument could probably be made for either answer. Ancient writings and Biblical references describe plant disease epidemics, especially those that were particularly recognizable, such as mildews and rusts. Losses due to soilborne pathogens, such as damping-off and root-rotting organisms, were probably considered fertility problems ("tired land") rather than disease problems. Farmers do not always separate the causes of poor harvests, which can be due to flooding, drought, temperature extremes, insects pests, soil fertility problems, poor seed, or a variety of disease problems. All of these factors can play a role in the final harvest, so ancient crop protection probably involved a general approach to crop health—an early example of what modern scientists recommend today in their more complex programs of **integrated pest management**.

Because the understanding of agricultural biology was quite primitive, religious and superstitious activities were important in early attempts at crop protection. Ancient agricultural and historical records from throughout the world contain references to various sacrifices, prayers, and ceremonies created to ensure a good harvest. Many of these rites are described in *The Golden Bough* by Sir James Frazer. When certain animals were sacrificed, or blood was applied to the fields, or human sexual activities took place where crops were to be planted, the overall objective was a good harvest rather than control of a specific problem. Religious ceremonies for specific plant parasites did exist, however. Examples include the Roman rust gods, who received sacrifices to protect the grain crops; requests to St. Anthony and other saints for protection from the "holy fire" of the Middle Ages caused by the rye ergot fungus; and many complex ceremonies among tree worshipers involving the tree parasite, mistletoe. These ceremonies are discussed further in Chapters 9 and 10.

Agricultural methods from past centuries, some still practiced successfully by subsistence farmers in many parts of the world, may superficially appear to be primitive but in reality can reflect careful observation and selective choices by intelligent human beings over many generations. For instance, the role of certain environmental factors in plant disease epidemics have been known for many centuries. The Greeks and Romans of past civilizations recognized that wheat rust was worse in wet years and in low-lying areas of fields where moisture accumulates. Of course, they did not understand the role of water in spore germination and its subsequent effects on the development of an epidemic caused by a fungal pathogen, and they could not control the weather any more than we do today, but these early observations were important clues to future scientists who would have access to better tools for the study of pathogenic microbes.

In some ways, early agriculture was probably better adapted to the general prevention of plant disease than is the agriculture of modern times. If one imagines back many centuries, several important factors should come to mind quite easily. For example, ancient farmers worked in relative isolation from both their immediate neighbors and from farmers on other continents. Farming was done in small patches surrounded by wild vegetation that could help block the movement of spores and insects. Even though new kinds of plants, and inadvertently the pathogens they might carry, were a popular cargo of ancient explorers, the amount of plant material that might actually survive long ocean voyages was relatively limited. After new plants arrived in harbors, their dispersal on horse-drawn carts or in packs of walking travelers would be quite slow. This can be compared to the

Fig. 6-1. Mixed cropping in the highlands of Ecuador.

tremendous quantities of plant material confiscated at any international airport in a single day in modern times. Certainly much plant material carried by today's travelers goes undetected and finds its way to new habitats.

Consider, also, the genetic makeup of the crops planted in ancient times. They were not the seed of careful genetic crosses and many generations of inbreeding and back-crossing to obtain maximum yields but the result of many years of survivor selection by farmers. These were seeds that would grow into plants that could produce some yield despite competition from weeds, insects, pathogens, and poor soil fertility. The levels of general resistance to a multitude of pests in plants selected under such circumstances ensured at least some harvest, even if not the maximum possible. In addition, early crops were first grown in the centers of origin of the plant species so that natural cross-breeding with closely related wild species probably continued. Genetic diversity was also much greater in the past because most crops were planted from seeds rather than vegetatively propagated.

Genetic diversity was commonly quite substantial, within a crop species and among the many species planted. Cautious farmers planted many kinds of plants, when possible, in hopes that some food would be harvested even if certain crops did poorly. Genetic diversity between farms was probably quite substantial, since farmers would select seed from their own plants each year rather than purchasing it from a major seed company as modern farmers do.

Even centuries ago, problems of soil fertility and the buildup of soilborne

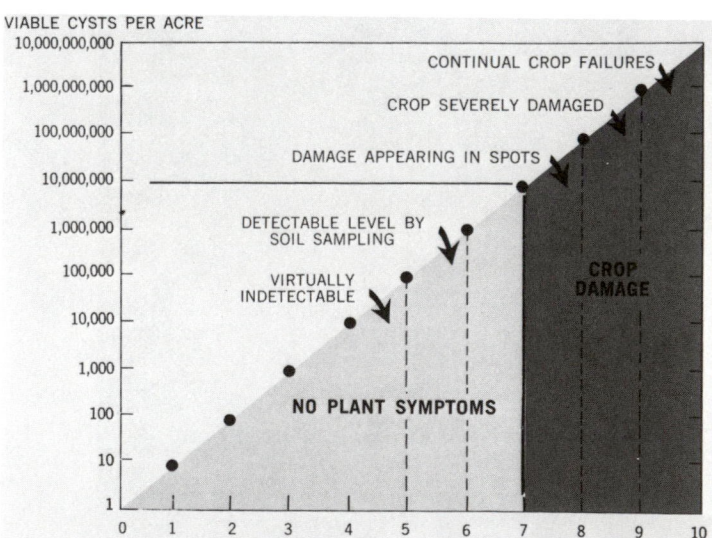

Fig. 6-2. Increase in the population of golden nematode (*Globodera rostochiensis*) over a 10-year period with continual cultivation of potato. The Incas proscribed the planting of potatoes in the same soil more than once in 7 years, which prevented the rapid increase in the nematode population.

pathogens were observed. When human populations were low and land was plentiful, early farmers simply moved on to new land when crop yields declined. **Slash and burn agriculture** represents one means of nomadic agriculture that is still practiced in many Third World areas. The heat from burning plant material reduces many insect and parasite populations and releases nutrients that can be used by future crop plants. When pests became too great a problem, or soil fertility declined, farmers moved on to new areas. Eventually, as population pressures increased in many areas, all suitable land was cultivated, and communities became more sedentary. The reuse of agricultural lands became more important, and **crop rotation** came into use.

The Incas of Peru developed detailed rotation laws for the production of potatoes and other crops that were enforced with the threat of death for those who did not follow the prescribed plantings. For example, potatoes were to be planted in the same soil only every seventh year, a requirement the biological significance of which was determined only recently, when the golden nematode, a parasitic worm native to South America, was discovered in Europe and the United States. One can obtain an acceptable yield if potatoes are not grown repeatedly in a field; this prevents the nematode population from reaching a damaging level. In the intervening years, the population of this obligate parasite declines, so it again becomes safe to plant potatoes. This rotation plan is biologically sound, even if not commercially acceptable to most modern farmers.

Although traditional farmers did not understand the intricacies of soil microbiology, their careful observations led them to effective practices to reduce soilborne pathogens. In the United States in the early 20th century, before prices of farm land became prohibitive and population pressures increased, farmers grew a greater diversity of crops in rotation patterns that traditionally included leaving land fallow, or uncultivated, for a year or two.

Over the years, trial and error practices by observant farmers have resulted in many cultural practices that help reduce disease. In some cases, these methods have turned out to be accurately connected with the life cycles of pathogens in ways that help protect crops even though the farmers knew little, if anything, about the microorganisms that cause plant diseases. Even now, subsistence farmers in many parts of the world cling to their traditional methods in the face of "modern agriculture," and it often turns out that their methods are appropriate for the climate and environment in which they live.

Frijol tapado, or "covered beans," is a name for the practice of planting bean seeds in carefully selected weed patches and then cutting and chopping the weeds until the seeds are covered by the resulting mulch. This practice, described as an established means of bean culture in Central and South America by the early Spanish explorers, protects the growing plants from splashing soil that carries the spores of the fungus, *Thanatephorus cucumeris* (asexual stage: *Rhizoctonia solani*), which causes a severe bean disease called web blight. In warm, moist tropical climates, beans planted the higher-yielding "modern way" in weed-free rows succumb quickly to this disease.

Traditional methods are not always more appropriate or suitable than newer methods. The Romans sacrificed red animals, such as cows and dogs, to their rust gods to protect their wheat fields. Unfortunately, there is no evidence that rust disease was subsequently reduced. On the other hand, folk wisdom led to laws in Europe and the United States banning the planting of common barberry bushes near wheat fields over 200 years before Anton deBary, the famous 19th century German mycologist, discovered the important role that barberries play in the life cycle of the wheat stem rust fungus. But we must admit that the multitude of potential pests and parasites that threaten our crops present a very complex problem. When we concentrate on solving one or a few specific problems, we often neglect other important factors and find ourselves no better prepared to protect the crops.

Fig. 6-3. Bean plants. **Top,** grown through *frijol tapado* mulch. Mulch prevents weed growth and conserves moisture. **Bottom,** web blight (*Thanatephorus cucumeris*) destroys beans grown without mulch.

Ecological Agriculture

The recent intense study of microbial ecology and the genetic interactions of parasites and plants has greatly increased our appreciation for the complexities of the agricultural ecosystem. An **ecosystem** includes all the living organisms in a natural community and their various interactions. **Agricultural ecosystems** are different from natural ecosystems because humans are trying to tip the ecological balance in favor of the crop plant to be harvested. Such an unbalanced system is difficult to maintain and often requires continuous inputs such as fertilizer, cultivation, and irrigation.

The concept of a modern agricultural ecosystem forces us to accept the fact that all agricultural activities have a multitude of effects on the interactions between the various members of the community. Plant pathologists often spend years studying the details necessary to understand even one aspect of a single host-parasite relationship. At the same time, entomologists are studying similar problems with insect pests, and soil scientists and agronomists are trying to understand the complexities of soil structure and fertility. Eventually, it is the farmers who must try to apply all this detailed information in a way that results in optimal crop production.

In this chapter, we will look at some of the various ways to protect plants from disease. These methods are discussed in a general way, even though some very important types of plant pathogens have not yet been introduced. Diseases mentioned in previous chapters are used as examples of the relative success or failure of various control methods in relation to the type of epidemic that occurs. The rationale behind these choices can then be applied to the diseases still to be presented. As the diseases and pathogens are introduced in the following chapters, it will be a useful exercise to imagine the types of controls that might be effective in controlling them. This will help us to better understand how plant pathologists determine their recommendations for growers.

Several terms are used to describe activities that reduce plant disease. The simple word **control** is often used for such activities. In recent years, use of the word **management** has become more popular. This small difference in word choice represents a great philosophical change on the part of plant pathologists. In the early years of plant pathology, disease-free plants were the ultimate goal of control programs, and scientists even hoped to be able to totally eliminate certain pathogens. As the science of ecology became more sophisticated, it became clear that the elimination of most pathogens from an agricultural ecosystem is neither realistic nor necessary. In addition, the economic and environmental costs of trying to achieve complete control of a particular disease usually cannot be justified.

The idea of disease management was created from a better understanding of the agricultural ecosystem. When a disease is properly managed rather than controlled, low levels of disease are acceptable because their elimination would cost more than the value of any additional yield that might result. Disease management schemes require detailed economic as well as biological studies and are thus much more difficult to develop than control programs that attempt to eliminate all disease, but they represent a much more ecologically sound means of crop protection.

In the following discussion, the word *control* is used to describe a specific means of disease reduction generally incorporated into a disease management recommendation comprised of a number of control measures. An entire page could be filled with a list of control measures, but this would not be very interesting or instructive. There must be a rational means of choosing controls that are appropriate for the crop and for the disease. When certain important aspects of the life cycles of both the host plant and the parasite have been determined, one can then create a reasonable strategy for disease management. These choices are closely tied to an understanding of how a plant disease epidemic develops in a particular crop and of the economics of crop production.

Epidemiology

Most people use the word **epidemic** to indicate rapid and widespread disease development. The word more accurately refers to the increase of disease with time, which can occur slowly or rapidly. *Epidemic* technically refers specifically to disease increase in a human population. Disease increase in animals is an *epizootic*, and disease increase in a plant population is an *epiphytotic*. Although some scientists prefer the more specific terminology, many use the more general and familiar term *epidemic* and the related word *epidemiology* to describe the study of disease increase in any kind of population.

To gain some appreciation for the importance of **epidemiology** in modern plant pathology, imagine a field of wheat in which some plants have suddenly appeared to be diseased. What can one expect to see happen over the next few weeks? Will all the plants die, leaving no grain to harvest? Will only currently infected plants produce less grain at harvesttime? Are the plants that appear healthy already infected but not yet symptomatic? Is the parasite air-dispersed and already producing propagules that will cause disease in neighbors' farms nearby? Or have spores from a neighbor's fields caused the disease? Is the pathogen air-dispersed, or is it carried in water droplets or by sticking to a passing insect? Perhaps the pathogen is soilborne and will spread slowly, causing patches of diseased plants surrounded by healthy plants. Can this same crop be planted next year, or might the epidemic begin even earlier and with greater intensity?

Epidemiology attempts to answer such questions by describing disease development patterns during a single season and from year to year. In some cases, epidemiological studies involve the creation of complex mathematical equations that model disease increase or changes in the parasite population. Such studies can lead to more general models that have broad application to a variety of disease situations.

Several general approaches to plant disease management can be taken. It may seem quite hopeless to even think that plants can be protected from the various rapidly reproducing microorganisms that are potentially or actually present. Remember from Chapter 1, however, that although parasites may always be present, disease development occurs only when the three major factors are interacting in a suitable fashion. As illustrated in the **disease triangle**, the three components necessary for disease are: a susceptible plant, a virulent pathogen, and a favorable environment. When we consider a plant disease epidemic, we can now expand our two-dimensional disease triangle to a three-dimensional **disease pyramid** that includes time as a fourth important factor. All three components must interact over a period of time for an epidemic to occur. The reduction of plant disease can be approached from two perspectives: 1) controlling the pathogen or 2) protecting the plant.

Pathogen Exclusion

If the pathogen has not yet reached the plant, there are ways to prevent its arrival. These control methods are grouped under the control category of **exclusion**. Without the pathogen there can be no disease, so exclusion is an excellent control strategy when successful. The general goal of exclusion is to control the **distribution** of the pathogen. If economics were not a factor in plant production, plants could be grown as sterile organisms in pathogen-free soil in sealed containers receiving only filtered air. In some very special cases, as discussed in Chapter 12, this extreme form of pathogen exclusion is employed, although obviously it is not practical for most crops.

Although total pathogen exclusion is not usually possible, the planting of pathogen-free seeds and propagative parts is an extremely useful means of disease control. To prevent the establishment of the pathogen in or on the seed, many seeds are tested and "certified" as meeting predetermined standards for freedom from pathogens. Since it is even more likely that large propagative parts might carry important pathogens that will cause severe losses, particularly systemic bacteria, fungi, and viruses, the production of pathogen-free propagative parts is an important agricultural

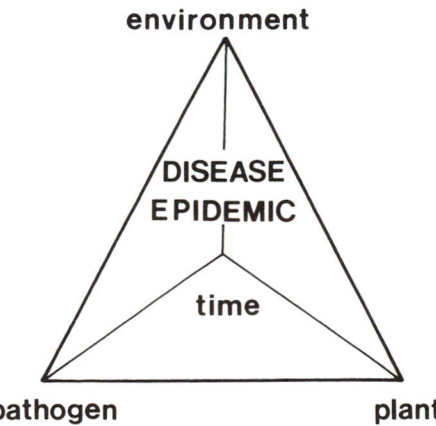

Fig. 6-4. The epidemic pyramid, in which time is an important factor in the development of an epidemic in addition to the susceptible plant, virulent pathogen, and favorable environment of the disease triangle.

industry. Examples have been discussed previously: potato tubers without mycelium of *Phytophthora infestans*, citrus nursery stock free of citrus canker bacteria, raspberries and many woody ornamentals free of crown gall bacteria, and banana corms free of *Fusarium oxysporum* f. sp. *cubense* and *Pseudomonas solanacearum*. Quarantines and agricultural embargoes between countries, regions, or states are also an important means of pathogen exclusion.

Pathogen Eradication

For most common pathogens, exclusion from planting material is helpful but generally temporary, because the potential always exists for infection from pathogens existing in the growing environment as dormant propagules or in a saprophytic state in the soil or as parasites of weeds and nearby crop plants. Thus, the second general approach to control of the pathogen is **eradication**, which attempts to control the **survival** of the pathogen. This does not usually mean total eradication, since, unfortunately, this is economically and biologically impossible to accomplish in most cases.

Total eradication may be the goal when quarantines fail, and extensive eradicative measures may delay the spread of a pathogen that has bypassed quarantine restrictions, but total eradication is very difficult. Previously mentioned eradication failures include fire blight in Europe and coffee rust in South America. The eradication of citrus canker from the United States early in the 20th century is perhaps the only example of successful eradication of an established pathogen, but its recent reintroduction leaves the future status of this pathogen uncertain.

The purpose of some seed treatments is to eradicate the pathogen before planting. Historically, seeds were soaked in hot water or chemicals and other harsh substances such as horse urine or salt water in an attempt to destroy pathogens. The present use of systemic fungicides, available since the 1960s, that can eradicate certain important seedborne pathogens without harming the plant embryo has been far more successful.

Eradication can also include the destruction of infected plants having an economic value less than that of the crop plants to which the pathogen

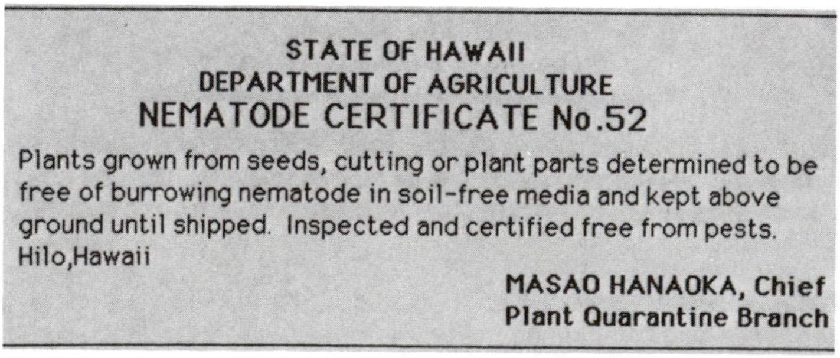

STATE OF HAWAII
DEPARTMENT OF AGRICULTURE
NEMATODE CERTIFICATE No.52

Plants grown from seeds, cutting or plant parts determined to be free of burrowing nematode in soil-free media and kept above ground until shipped. Inspected and certified free from pests. Hilo, Hawaii

MASAO HANAOKA, Chief
Plant Quarantine Branch

Fig. 6-5. Inspection certificate used in a program to reduce the movement of nematode-infested plant material from Hawaii to other areas.

may be transmitted. A good example is the infection of many common weed species by plant viruses that also infect crop species. Virus transmission from weed hosts to crop hosts can occur when the plants rub together or, more commonly, when sucking insects such as aphids and leafhoppers transmit viruses during feeding. This is discussed in more detail in Chapter 12.

Similar situations can occur with other types of pathogens. Hawthorns, which are closely related to many fruit tree species and grow wild in hedge rows, are eradicated in Europe because they serve as a source of fire blight bacteria that may be carried to nearby pear and apple orchards by wind-blown rain or insects. Many rust fungi require two unrelated plant species, known as the alternate hosts, to complete their life cycles. This amazing biological phenomenon, explained in Chapter 10, can be exploited for practical disease control. When one host is economically important, and the other is not, the removal of the host species of lesser value has been practiced to eradicate a rust fungus by preventing the completion of its life cycle.

Eradication of diseased crop plants themselves or infected parts of those plants is commonly practiced. Many growers inspect newly emerged plants and remove diseased individuals. The removal of virus-infected plants early in the season prevents them from serving as sources for infection of healthy neighboring plants. Cankers in woody plant parts serve as important reservoirs of bacteria and fungi that produce inoculum for new infections of healthy plant tissue. Pruning out cankers not only removes sources of inoculum for new infections but also eliminates the continued destruction of infected woody tissues by the growing canker pathogen. Dormant-season pruning is often most practical because deciduous plants are better exposed during winter and early spring for the inspection of perennial tissues, and most pathogens are in an inactive state and less likely to spread to new infection sites during the pruning activities.

A multitude of other agricultural practices, the purpose of which is

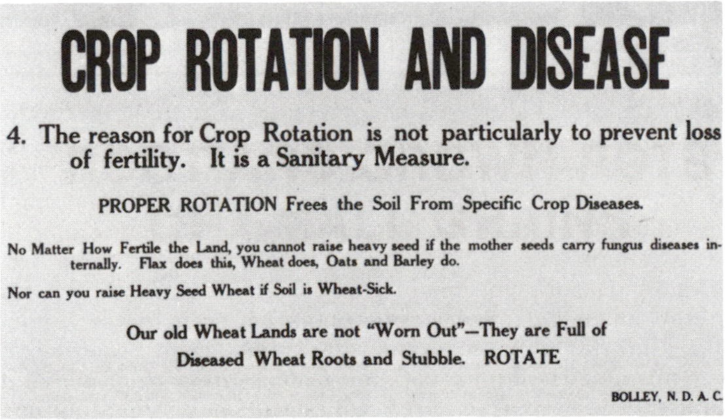

CROP ROTATION AND DISEASE

4. The reason for Crop Rotation is not particularly to prevent loss of fertility. It is a Sanitary Measure.

PROPER ROTATION Frees the Soil From Specific Crop Diseases.

No Matter How Fertile the Land, you cannot raise heavy seed if the mother seeds carry fungus diseases internally. Flax does this, Wheat does, Oats and Barley do.

Nor can you raise Heavy Seed Wheat if Soil is Wheat-Sick.

Our old Wheat Lands are not "Worn Out"—They are Full of Diseased Wheat Roots and Stubble. ROTATE

BOLLEY, N. D. A. C.

Fig. 6-6. A poster by H. L. Bolley from North Dakota in 1909, explaining that the benefits of crop rotation for wheat result from a reduction of soilborne pathogens.

pathogen eradication, can be grouped under the general term **sanitation**. Careful growers remove soil from tools and machinery before transporting them between farms or sometimes even between fields to reduce the movement of soilborne pathogens. Tools used for pruning, grafting, or other activities that cut plant tissues may be sterilized or disinfested between cuts to prevent pathogen spread. (Note: We use the term **disinfested** to describe the removal of pathogens from the surfaces of objects such as pruning shears or potting table tops or from the surface of seeds and other plant parts. We reserve the term **disinfected** to describe the removal of a pathogen from within infected living plant tissue.) Greenhouse benches and storage areas for harvested products or propagative parts, such as potatoes and flower bulbs, are carefully cleaned and disinfested to eradicate any pathogens left from the previous planting or storage. This can be accomplished by the use of steam heat or chemicals.

Careful removal and destruction of plant material at the end of the growing season is often useful to eradicate existing pathogens that might threaten plants in the next planting. In large fields this is obviously not a practical means of eradication, but crop debris and stubble can be plowed under to encourage decay of the remaining plant material. In moist soil, with its teaming microbial population, pathogens find themselves quickly faced with an intensely competitive soil environment after the plant debris has decayed. Without the safe haven of the host plant, the pathogen population can quickly decline because most pathogens are far less competitive than the saprophytic microflora of the soil. Farmers can thus exploit the natural competitions between soil microorganisms to help eradicate pathogens before planting the next crop.

Although the competition between soil microorganisms and plant pathogens can be exploited for crop protection, two aspects of modern agriculture limit our ability to make use of it. The first is **reduced tillage**. Soil tillage, or plowing, has been greatly reduced in some areas to help prevent soil erosion and to preserve soil moisture. Although the effects of reduced tillage vary depending on the crop and the pathogens, one can generally expect an increase in diseases caused by pathogens that survive best in crop debris. Increases in fungal diseases in grain crops, corn, and soybeans grown under reduced tillage have already been observed.

A second important aspect is the pressure to limit **crop rotation**. As mentioned earlier, many soilborne pathogens are host-specific to some degree. Thus, in choosing a crop to plant in a particular field, it is important to plant the crop that is least susceptible to pathogens that might still be present in the debris from the previous crop. Rotation recommendations often involve alternations between crop species from different plant families. More distantly related plants tend to have different parasites. For example, a grain crop might be followed by sunflowers or a forage legume such as alfalfa. Unfortunately, economic constraints have encouraged many farmers to abandon traditional rotation patterns in favor of more intensive monoculture. Expensive, specialized planting and harvesting equipment and the differential value of various crops pressure farmers to grow high-value crops more often. Due to reduced tillage and monoculture, we are less able to exploit the ability of competitive soil microbes to eradicate pathogens from the soil.

For some high-value crops or special disease situations, rotation and tillage do not eradicate pathogens sufficiently for economic crop production. In such situations, more rapid and complete pathogen eradication from the soil is necessary. In greenhouses, relatively small amounts of soil are used in plant production, so **chemical or heat pasteurization** methods are used to kill pathogens in the soil. Soil can be heated to temperatures that kill most pathogens without killing all the beneficial saprophytic competitors, so some protective competitors will still be present in the soil in case a pathogen is accidentally reintroduced. **Pasteurization** is the term used to describe the heating of soil to kill pathogens without destroying all micro-organisms. When soil is heated to very high temperatures that kill all organisms (**sterilization**), ammonia and salt toxicities are common problems. In many modern greenhouses, plants are also grown in soilless mixes such as composted bark and perlite and even liquid nutrient solutions (hydroponics) on rock wool to avoid soil pathogens.

Outdoors, soil pasteurization is expensive and temporary, because only small areas of soil can be treated, and pathogens can then recolonize treated soil from the edges of the treatment area. Such soil treatment is only done for high-value crops or in soils where a pathogen is present that can prevent an acceptable harvest if not eradicated. Heat treatment with steam is time-consuming and difficult in field situations, although it is done in nurseries

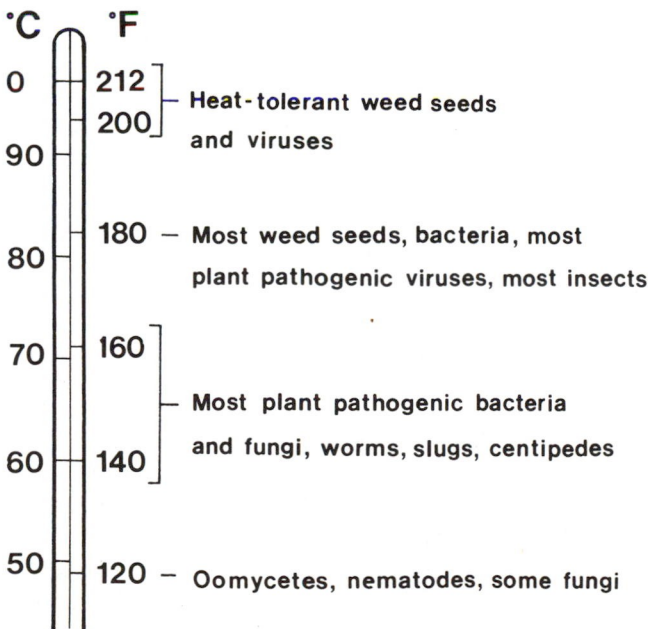

Fig. 6-7. Temperatures necessary to kill pathogens and other plant pests. Most of the temperatures are for a 30-minute exposure under moist conditions.

or seedbeds. Outdoor soils are more commonly sterilized with chemical fumigants. In climates where intense sunlight is available, pathogen populations can be reduced in soil by using clear plastic tarps to trap sunlight to heat the soil in a process called **solarization**. In hot and dry climates, repeated tillage may be practiced on fallow land to eradicate certain soilborne pathogens.

Protecting the Host Plant

We have seen that control of the pathogen can be approached in two general ways: exclusion, which controls the distribution of the pathogen, and eradication, which controls the survival of the pathogen. We now change our perspective and, assuming that the pathogen has not been completely excluded or eradicated, consider what might be done to defend the host plant. Certainly the use of **pesticides** to kill the pathogen comes quickly to mind. For some diseases, such as late blight of potatoes, frequent protective sprays with fungicides are necessary to produce a crop in humid climates. However, the use of pesticides for disease control is decidedly ineffective for many diseases, such as those caused by bacteria, viruses, and many soilborne pathogens, such as *Fusarium oxysporum*. Efficacy and economic constraints limit the use of pesticides for plant disease control to relatively few crops. The complex and controversial subject of using pesticides to control plant diseases is discussed in Chapter 7.

Plants may also be protected against infection by pathogens through **genetic resistance**. Genetic resistance is pathogen-specific, so it is unlikely that a particular cultivar will have genetic resistance to all the possible pathogens that can attack it. However, most modern crop cultivars have genetic resistance to at least one important pathogen, and many are resistant to several. At least 75% of all agricultural crops possess genetic resistance to at least one pathogen, and 98% of all grain and forage crops possess

Fig. 6-8. Steam being used to reduce soilborne pathogens in Massachusetts before chemical fumigants became available.

genetic resistance to one or more diseases. Genetic resistance may be race-specific or may be generally effective against all races of a pathogen. The choice of these different forms of genetic resistance depends on the pathogen, the crop, and the type of epidemic.

A final group of disease control methods can be grouped under the heading of **cultural practices**. These consist of agricultural activities that do not directly exclude or eradicate pathogens but serve to reduce plant disease by protecting the plant. In a very general sense, vigorous plants are less susceptible to disease than are nutrient-deficient plants grown with suboptimal water, poor soil drainage, or intense weed competition. Therefore, agricultural activities that improve plant growth and vigor will, at least indirectly, protect plants from disease. Many studies have shown the direct and indirect effects of specific soil elements on plant susceptibility. One important example concerns nitrogen, which results in increased infection by many pathogens when it is applied in high amounts that stimulate the growth of soft, succulent plant tissues. Insufficient nitrogen,

Fig. 6-9. Many cultivars of garden vegetables have resistance to several diseases. "Good 'n Early" tomatoes have resistance to Fusarium and Verticillium wilts, tobacco mosaic virus, and root knot nematodes.

on the other hand, may increase symptoms caused by pathogens that reduce nitrogen uptake, such as vascular wilt fungi.

The application of irrigation water can also affect plant diseases. Too little water puts plants under stress and contributes to susceptibility to some pathogens. Frequent, shallow watering encourages shallow, poorly developed roots that cannot withstand attack by root pathogens as well as a deep, well-developed root mass can. Overwatering, the most common cause of house-plant death, fills all soil air spaces with water and prevents root growth by depriving these living tissues of oxygen. Poorly growing roots are more susceptible to attack by root-rotting pathogens.

Aboveground effects of watering can be very direct because foliar (leaf) fungi require water on the leaves for spore germination and infection. Irrigation water should be applied early in the day when the leaves will dry quickly or during the night hours when leaves are already wet. In either case, the duration of leaf wetness is not extended. Foliar diseases increase greatly when water is applied to leaves in the late afternoon or early evening so that they remain wet throughout the night. Despite common misconceptions, application of water to plants in sunlight does not burn the foliage but actually helps cool the plants. Rapid drying of foliage can be accomplished by orienting the row with the wind direction, proper spacing of rows, thinning of plant density, and weed removal. In modern greenhouse production, irrigation systems apply water only to soil surfaces, keeping the leaves completely dry and preventing many foliar diseases. Many pathogens such as bacteria and fungi can be spread during normal activities such as pruning or cultivation when the foliage is wet. It is best to wait until the leaves dry before handling or working among plants.

Monocyclic and Polycyclic Pathogens

The preceding discussion has shown that plant diseases can be reduced by two approaches: control of the pathogen through exclusion and eradication, and protection of the plant through protective chemicals, genetic resistance, plant vigor, and numerous cultural practices. How does one create an appropriate disease management strategy from the many possible control measures? Epidemiology has given us some general principles that suggest a rational approach to this problem. The South African plant pathologist, **J. E. Vanderplank**, first introduced the concept of monocyclic and polycyclic pathogens and provided two simple models of plant disease epidemics that result from these two general types of pathogens.

Monocyclic pathogens infect host plants during the growing season, and no new inoculum for further infections is produced until the end of the season, usually when the host plant dies. One familiar example of such a pathogen is *Fusarium oxysporum*. If a graph is made of the "disease progress" of a monocyclic pathogen, one can observe that the number of infected plants increases until a plateau is reached (Fig. 6-10a). In the case of Fusarium wilt, conidia are present in the soil at a particular density that results in a certain number of plant infections. At the end of the season, new conidia are added to the soil to serve as inoculum for the next season.

Polycyclic pathogens, on the other hand, produce new generations of inoculum during the growing season so that infections can continue. A familiar example of a polycyclic pathogen is *Phytophthora infestans*, which produces new sporangia several days after each infection. If a graph is made of the disease progress of a polycyclic pathogen, one observes a slow increase in disease at first because only a few infections have occurred, but acceleration may be rapid because each set of infections results in the production of new propagules (Fig. 6-10b). The disease increase may level off later in the season if no new plant tissue is available for infection. This so-called S-shaped curve is typical for rapidly reproducing polycyclic pathogens such as many rusts, downy mildews, foliar fungi, and some bacteria and viruses.

Now we will use these two models to predict the amount of disease that might occur in a single growing season. For monocyclic pathogens, we should be able to predict how much disease will occur if we know how much **initial inoculum** (inoculum in existence at the beginning of the growing season) is present. The amount of disease at the end of the season should be proportional to the amount of initial inoculum. When inoculum levels are high, disease will be greater; low inoculum levels should result in less disease.

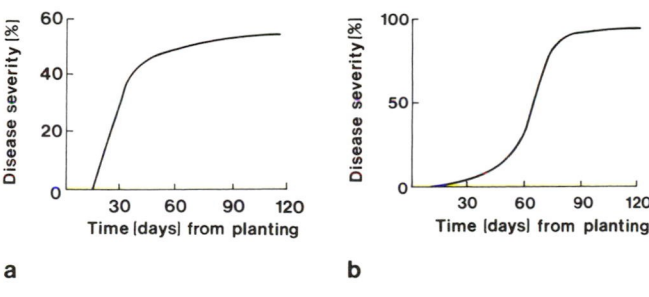

a   b

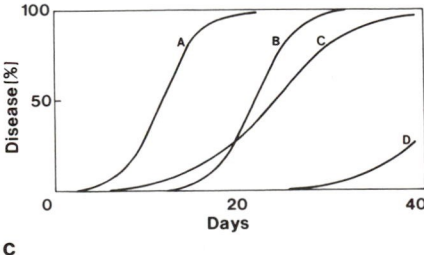

c

Fig. 6-10. Disease progress curves of plant disease epidemics. **a,** Monocyclic pathogen; **b,** polycyclic pathogen; **c,** polycyclic pathogen under different management stategies: A = no control, B = reduction of initial inoculum, which delays the epidemic but does not reduce the epidemic rate, C = reduction of epidemic rate, D = combination of B and C, i.e., the epidemic is delayed and the rate is reduced.

Can the same relationship be applied to the initial inoculum of a polycyclic pathogen? In most cases, the relationship is less direct between the initial inoculum of a polycyclic pathogen and the final amount of disease. The increase of the polycyclic pathogen during the growing season is governed by factors such as temperature, moisture, and crop plant resistance, which play a much more important role in the final amount of disease than the amount of initial inoculum. The length of time between generations of the pathogen, called the **latent period**, is governed directly by these factors. Inoculum produced during the growing season is called **secondary inoculum** and is responsible for the explosive development of epidemics of some polycyclic pathogens. Pathogens such as *P. infestans* increase so rapidly that different levels of initial inoculum can still result in total destruction of a potato field.

How can an understanding of the differences between monocyclic and polycyclic pathogens be useful in choosing appropriate control measures? Let us first consider the effects of reducing initial inoculum. A reduction in initial inoculum has a direct effect on the amount of disease in an epidemic caused by a monocyclic pathogen. Control measures that reduce initial inoculum include many of the pathogen eradication methods discussed previously. In addition, the use of specific resistance can be useful, because some races of pathogens will be incapable of causing infection, which reduces initial inoculum.

What are the effects of reducing initial inoculum on an epidemic caused by a polycyclic pathogen? The epidemic may be delayed in the beginning because less inoculum is available at first, but the geometric increase of the pathogen may still cause high levels of disease. If some inoculum is present, the rate at which the epidemic develops is governed by factors unrelated to initial inoculum and will be relatively unaffected by its reduction. This means that methods that reduce initial inoculum of a polycyclic pathogen may be insufficient to protect the crop. Reducing the **rate of the epidemic** is usually necessary as well. Thus, for polycyclic pathogens, reduction of initial inoculum is important and helpful but not usually sufficient.

In the case of potato late blight, initial inoculum is reduced by destruction of infected tubers and planting of pathogen-free tubers. The rate of the epidemic is reduced by the timing of irrigation so as not to extend leaf wetness, by frequent application of fungicides, and by the use of general resistance that is effective against all races of *P. infestans*. Other examples of diseases caused by polycyclic pathogens for which rate-reducing methods of disease control are necessary include downy mildews, coffee rust, and South American leaf blight of rubber. Of course, if all initial inoculum can be excluded or destroyed, polycyclic pathogens will not cause epidemics. South American coffee growers must now approach rust control with rate-reducing methods because the quarantine failed to totally exclude *Hemileia vastatrix*, but, so far, total exclusion of initial inoculum by quarantine has protected Asian rubber plantations from South American leaf blight.

In conclusion, disease management is most effective when composed of methods appropriate for the type of epidemic that results from a particular pathogen. Most practices that reduce initial inoculum are those described previously in the sections on controlling the pathogen, i.e., exclusion and eradication. Complete pathogen exclusion is obviously suitable for all plant pathogens but is not usually possible. In most cases, reduction of initial inoculum is most effective for monocyclic pathogens. Control methods that protect the host plant, such as cultural practices, protective chemicals, and genetic resistance, have their greatest effect in reducing the rate of an epidemic. Thus, these practices are most effective against polycyclic pathogens. Reduction of initial inoculum may delay an epidemic of a polycyclic pathogen, but it has no effect on epidemic rate. If both initial inoculum and epidemic rate can be reduced, the epidemic is both delayed and slowed, as illustrated in Figure 6-10c, curve D. This is the most desirable situation and explains why a complete disease management strategy for rapidly reproducing polycyclic pathogens includes methods that reduce both the initial inoculum and the epidemic rate.

These epidemiological principles are extremely useful, but one must not oversimplify the complexities of an agricultural ecosystem. Despite the comfort that such categories give the human brain, few things in nature can really be divided into two categories. Many pathogens are not easily labeled as monocyclic or polycyclic. Many produce two or three generations of new inoculum during a growing season. One must then determine whether reduction of initial inoculum is sufficiently effective or possible, or if rate-reducing activities are more appropriate. Sometimes a disease epidemic should be observed over several years rather than just in a single growing season. In a single year, the Dutch elm disease fungus could be called monocyclic, but its behavior is polycyclic when observed over several years (see Chapter 11).

One should also not conclude that polycyclic pathogens are necessarily more threatening or cause more damage than monocyclic pathogens. Many foliar fungi are polycyclic pathogens that cause relatively little damage to a host plant, whereas some soilborne monocyclic pathogens such as *F. oxysporum* and *P. cinnamomi* cause great losses. Studies have shown that *Verticillium albo-atrum* and *V. dahliae*, monocyclic vascular wilt fungi, can be present in the soil at levels that cause 5–10% of the plants to become infected and may reach levels capable of causing 80–90% infection in just 3 years.

Economic factors always play an important role in control choices, as well, and involve the cost of the control and the value of the crop. Economically sound choices must be based on accurate disease assessment that can be correlated with predictable losses. Figure 6-11 illustrates how difficult accurate **disease assessment** can be. Cover the numbers below the diagram that indicate the percentage of diseased tissue and try to guess how much tissue is diseased.

The 1970 Southern Corn Leaf Blight Epidemic

We conclude this chapter with a description of a relatively recent epidemic of southern corn leaf blight that occurred in the U.S. in 1970, causing over $1 billion losses in the corn crop. The pathogen responsible is a polycyclic leaf spot fungus, *Cochliobolus heterostrophus*, an Ascomycete. It is commonly known by its asexual or conidial stage, *Bipolaris maydis* (previously called *Helminthosporium maydis*). It produces dark, multi-cellular conidia on the surface of infected leaves several days after initial infection. These conidia are blown or splashed to new plant tissue. Infections are most common on leaves and husks of the developing ears of corn. This fungus can be isolated from spots on corn leaves in almost any season in all but the most northern states.

What is the common management strategy for such a pathogen? We can first say that theoretically this polycyclic pathogen is best managed by a combination of methods that reduce initial inoculum and the rate of the epidemic. The fungus overwinters primarily in stubble left from the

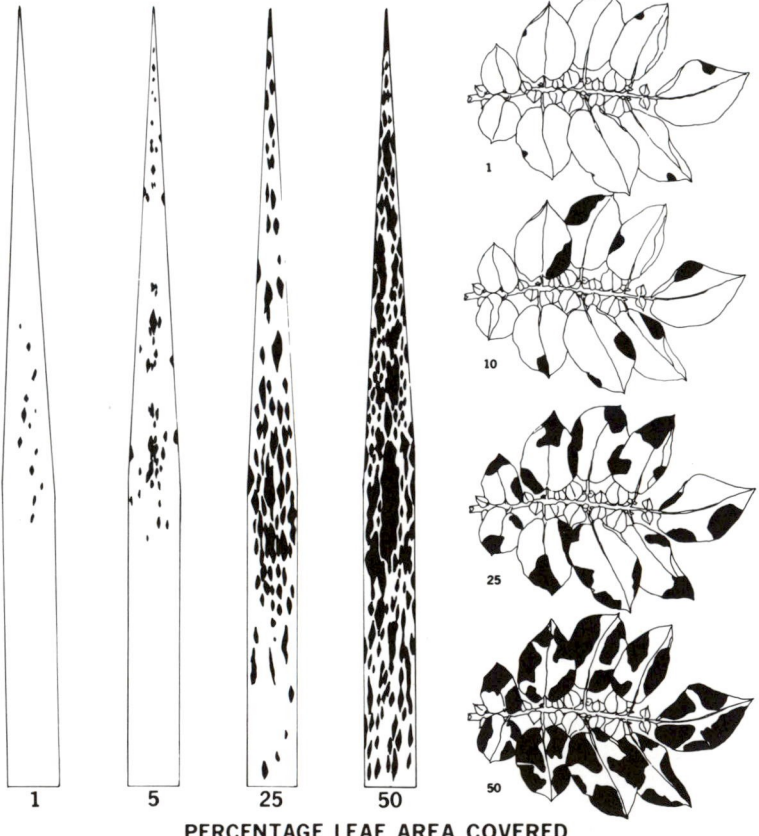

PERCENTAGE LEAF AREA COVERED

Fig. 6-11. Disease assessment keys allow visual comparison with diseased plants to improve the accuracy of evaluation of resistance, fungicide control, and cultural practices that reduce disease. **Left,** Septoria leaf blotch of cereals. **Right,** late blight of potatoes.

previous crop, so stubble should be plowed under to enhance decay, and corn should not be grown in the same field in consecutive years. What means are available to reduce the rate of the epidemic? Cultural practices that reduce leaf wetness consist of row orientation and spacing. Protection of the foliage with fungicides cannot be justified because of economic factors except possibly in seed-producing fields. The value of the corn crop is usually insufficient to offset the expense of chemical applications. Genetic resistance seems to be the most economically appropriate means of control. The available cultivars of corn appear to have high levels of general resistance to this polycyclic pathogen because infections occur but rarely cause significant loss. In 1969, southern corn leaf blight was one of several corn foliar diseases that together accounted for losses of about 2.3% each year. One must ask: What changes occurred that led to the tremendous loss in 1970?

This corn epidemic is an excellent example of the problems that can arise from genetic uniformity and a narrow view of an agricultural ecosystem. To understand what occurred, it is first necessary to review the production of seed corn. Early in the 20th century, breeders discovered that inbreeding of corn did not increase yield but rather reduced it. When two high-yielding parental lines were crossed, the initial progeny produced the highest yields, but the continued propagation of seeds from these progeny produced disappointing harvests. From these studies came the concept of **hybrid vigor**, also called **heterosis**, which says that the highest yields are produced in progeny from the crossing of two inbred parental lines that are genetically different for many loci on many genes. Similar observations have been made for many other types of plants, so seed catalogues offer "hybrid" seed of many crops. The production of such seed requires that genetic crosses be carefully made each year. If seed from hybrid plants is saved and planted, the resulting plants will not have the vigor of the parent plant.

To produce hybrid corn seed, the tassels that produce pollen at the top

Fig. 6-12. Hybrid corn seed production in which the seed-producing plants have been detasseled.

of the corn plant must be prevented from fertilizing the ears of corn on the same plants, which serve as the female parent of the hybrid cross. The plants are detasseled by hand to prevent them from being self-pollinated. In a seed corn field, rows of corn plants with tassels serve as the male parental line and alternate with rows of detasseled corn of another cultivar that serve as the female parents. Seed is harvested only from the detasseled plants because this seed is certain to be the progeny of the hybrid cross. Needless to say, the process of hand detasseling before pollen shedding requires much expensive hand labor, but only for a short work period. Growers were always worried, however, that weather or other problems would prevent complete detasseling before the pollen was shed.

Then a genetic factor was discovered that caused plants to become male-sterile. Plants from the female parental line no longer had to be detasseled to prevent self-fertilization, which greatly reduced labor costs. The genetic factor involved was governed by genes in the cytoplasm, later found to be those associated with the mitochondria. The so-called **Texas male-sterile (TMS) cytoplasm** is inherited only through the female parent because the mitochondria are contributed to the zygote by the egg, as described in Chapter 4. In hybrid corn crosses, the male parent contained a genetic "restoring factor" that overcomes the male-sterile genetic trait, so the hybrid progeny would produce normal pollen. In 1970, nearly 80% of all hybrid field corn produced in the United States contained TMS cytoplasm. This genetic uniformity was the first step on the road to disaster. It represented a change in the susceptibility of the host—one of the four important factors

Fig. 6-13. A corn plant with Texas-male-sterile cytoplasm is very susceptible to infection by *Cochliobolus heterostrophus* Race T (left). Corn with normal cytoplasm is much less susceptible to southern corn leaf blight (right).

necessary for an epidemic.

In the meantime, a genetic change occurred in the pathogen population, as well. A new race of *C. heterostrophus* was found that was particularly virulent on corn with TMS cytoplasm. The new race was named **Race T** to differentiate it from the common **Race O** that caused only a minor leaf spot disease. Very soon it became apparent that Race T was much more aggressive. New generations of inoculum could be produced in as little as 51 hours after infection. In addition, the fungus infected leaves, husks, and even destroyed the developing ears. In the southern states, epidemic development was so rapid that brown paths could sometimes be seen in fields following the prevailing air movement. Despite economic constraints, airplane application of fungicides was begun to try to save some of the corn crop. In many southern states, entire fields were destroyed, and losses of 80–100% were common.

The new race was specific to the TMS corn, causing much less severe disease on corn with normal cytoplasm. It is now known that *C. heterostrophus* Race T produces a toxin that specifically affects the mitochondria of TMS-cytoplasm corn. This is known as a host-specific toxin because it does not poison cells of normal-cytoplasm corn or other plant species and was the first example of the effects of cytoplasmic genes on disease susceptibility. It led to further research into these important genetic factors.

The third component of the disease triangle now enters the picture and, luckily, the environmental factors were on the side of the farmers in the more northern states. *C. heterostrophus* Race T grows most quickly in hot and moist conditions, so disease losses were much less in the cooler states. It was also not a particularly wet year, which helped reduce infection.

Fig. 6-14. The 1970 southern corn leaf blight epidemic developed very rapidly. The heavily infected corn leaf on the right first became infected about 10 days before the corn plant on the left, which has only a few leaf lesions.

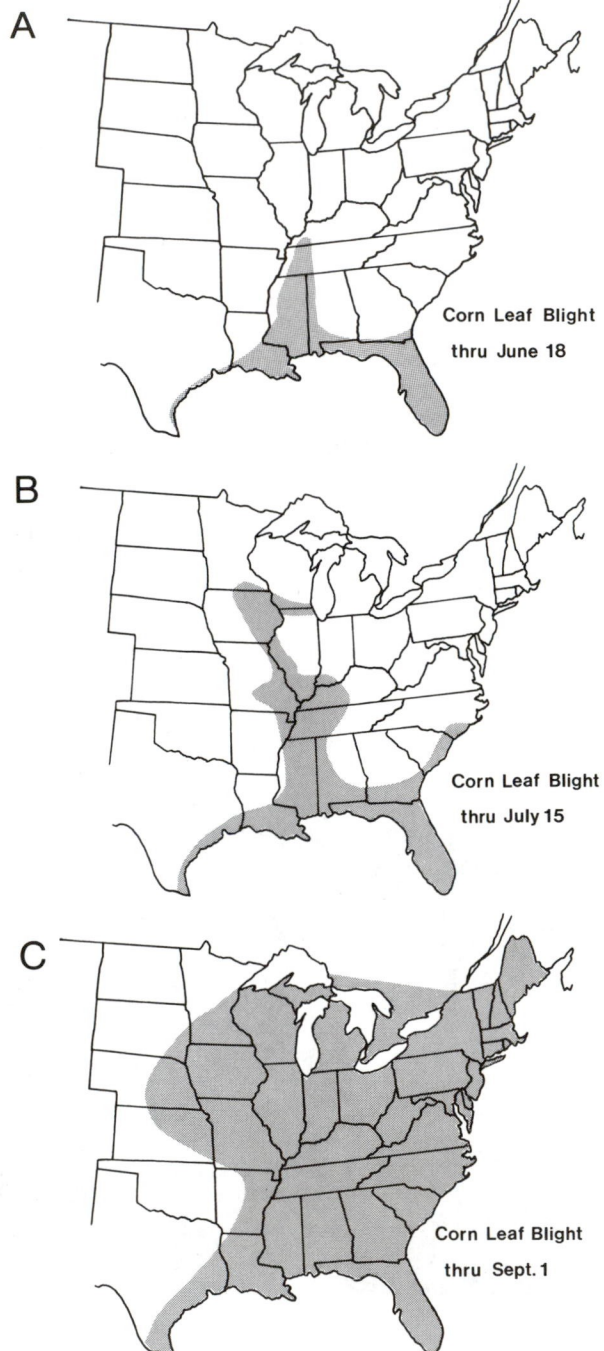

Fig. 6-15. Progress of southern corn leaf blight in North America in 1970.

Corn prices doubled late in 1970, so northern farmers who lost only 10–20% of their crops came out ahead financially. By the following year, 1971, the connection between TMS cytoplasm and susceptibility to *C. hetero-strophus* Race T was clear, and seed producers quickly returned to normal cytoplasm for winter seed production in the South and Hawaii. Farmers planted normal-cytoplasm seed where it was available, blended it with TMS corn, or temporarily switched to other crops where it was not.

In the scramble to take advantage of the male-sterile genetic factor for hybrid seed production, was there no warning of the possible dangers that lay ahead? Reports of increased susceptibility of TMS corn to *C. heterostrophus* were made as early as 1961 in the Philippines, where the climate is hot and humid, but the information was not widely read and the implications were not realized. Corn breeders and seed producers observed the increased susceptibility of the TMS corn in 1969, but that was too late to prevent the use of the TMS seed the following growing season. In the meantime, inoculum was increasing, resulting in the explosive epidemic of 1970. Plant breeders, working with plant pathologists, continually test new cultivars to determine whether they are particularly susceptible to pathogens, but it is difficult to anticipate sudden genetic changes in parasite populations. Similar situations have occurred, one of the most famous being the severe losses in the oat cultivar, Victoria, which contained a resistance gene against crown rust. The cultivar turned out to be particularly susceptible to a minor foliar fungus, named *Helmintho-sporium victoriae* after its victim, which produces a potent toxin specific to cultivars of oats with Victoria parentage. It is often said that success is the bane of a plant breeder because the widespread use of a popular cultivar greatly increases the chances for such a disaster.

But what happened in the hybrid corn industry? Breeders quickly switched

Fig. 6-16. In the southern United States, the 1970 southern corn leaf blight epidemic completely destroyed the corn crop on many farms.

away from the TMS cytoplasm, and many teenagers once again had summer jobs for manual detasseling. Plant pathologists and agronomists became acutely aware once again of the dangers of genetic uniformity, and now several different male sterile genetic factors are used in hybrid seed production, together with hand detasseling.

We have now provided a framework of epidemiological principles for disease management. These principles will be used in the following chapters to illustrate the means by which humans are learning to coexist with pathogens and still obtain a bountiful food supply. For the past 100 years, and especially since World War II, chemical pesticides, the subject of the next chapter, have become an important part of crop protection. In the past 10–20 years, humans have begun to understand some of the ecological effects associated with the use of these chemicals. This understanding is still another result of the study of the ecological principles that govern agricultural ecosystems.

Selected Readings

Fry, W. E. 1982. Principles of Plant Disease Management. Academic Press, New York.

Horsfall, J. G., and Cowling, E. B. 1978. Some epidemics man has known. Pages 17-32 in: Plant Disease: An Advanced Treatise, Vol. 2. J. G. Horsfall and E. B. Cowling, eds. Academic Press, New York.

Klinkowski, M. 1970. Catastrophic plant diseases. Annual Review of Phytopathology 8:37-60.

Tatum, L. Å. 1971. The southern corn leaf blight epidemic. Science 171:1113-1116.

Thurston, H. D. 1990. Plant disease management practices of traditional farmers. Plant Disease 74:96-102.

Yarwood, C. E. 1970. Man-made plant diseases. Science 166:218-220.

Pesticides

Few topics provoke a more emotional response than the use of pesticides in agriculture and on landscapes. Some people believe that pesticide use should be banned completely because of negative environmental and health effects that are not justifiable in times of food surpluses. Others believe that only "synthetic" or "chemical" pesticides should be prohibited, whereas "natural" ones can be used safely. Still others claim that pesticides are a necessary input to modern agriculture without which our food supplies would become expensive and unreliable. Some people who work in agriculture resent government regulations, which they find poorly defined and overly restrictive. Others feel that the government has a responsibility to protect the health of its citizens through pesticide regulation.

An appropriate response to such a complex issue requires that one become informed about its various components. The controversial aspects of the pesticide issue can be divided into two categories: costs and risks. The manufacture, distribution, and application of pesticides cost money, but crop losses and plant damage by pests result in greater production costs and higher food prices. Some environmental and health costs related to pesticide use can be assessed directly in the cost of pesticide regulation, but ecological effects, contaminated groundwater, and chronic health problems are more difficult to diagnose and describe in economic terms.

Evaluating risk has become a dominating concern of modern life. Although a risk-free existence is impossible, the threat of lawsuits may prevent the use of products with great benefits because of a very slight potential for harm. Certainly the chance of being injured or killed in an automobile is relatively high, yet people ride in cars frequently because of the transportation benefits they provide. These same people might not be willing to accept exposure to extremely low concentrations of certain chemicals in their food or water. Is this response appropriate? What are the relative risks? How does one evaluate such risks?

The purpose of this chapter is not to present a "correct" answer to these questions but to give some background information about pesticide use and regulation that can be used to develop an intelligent opinion about this complex issue. Information is presented about pesticides used to control plant diseases, including their historical development and current status, the regulation of pesticide use, and applicator training and certification. Finally, the chapter discusses attempts to reduce pesticide use through a better understanding of the agroecosystem, using programs of integrated pest management. Because these topics can be presented only superficially in a single chapter, detailed references follow at the end.

Pesticides for Plant Disease Management

The general term **pesticide** refers to something that will kill a pest. In agriculture, a number of more specific terms are used: **herbicides** kill weeds; **fungicides** kill fungi; **insecticides** kill insects; **nematicides** kill nematodes; and **antibiotics** are used primarily against bacteria. Fungicides are the most common type of pesticide used for the prevention of plant diseases.

Pesticides other than fungicides may sometimes be used in plant disease management. Insecticides are applied to some high-value crops to kill the insects that transmit plant pathogens, although, in most cases, this is not an effective management strategy. This can be shown with the familiar example of fire blight. It is impossible to kill all insects that may be carrying the bacteria to the flowers of the apple and pear trees. In addition, insects are necessary for pollination to produce fruit. Insecticides used to kill sucking insects such as aphids that transmit viruses and other plant pathogens may be somewhat effective; this subject is discussed in Chapter 12.

Herbicides are occasionally used to kill parasitic plants and weeds that harbor pathogen-transmitting insects, viruses, and other pathogens, but these, too, account for only a tiny portion of the pesticides used to protect plants. The use of antibiotics for bacterial plant pathogens has been limited by efficacy problems, expense, and the rapid development of resistant strains of bacteria, as discussed in Chapter 4.

Soil fumigants are used in special circumstances for soilborne pathogens, but these are limited to crops with a value that can justify the considerable expense of their use. Nematicides, most of which are highly toxic chemicals that were originally developed as insecticides, are applied to soil to kill nematodes (tiny soil roundworms), but once again the cost of applications must be justified by crop value. The use of nematicides and soil fumigants

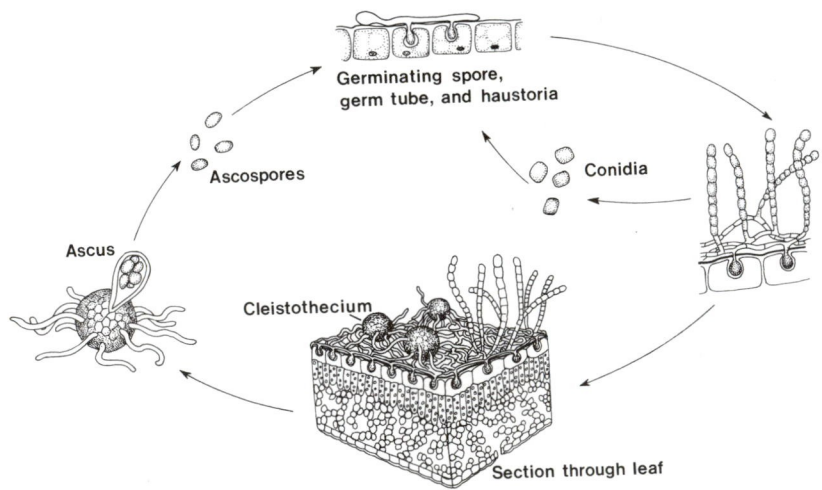

Fig. 7-1. Generalized life cycle of powdery mildews.

is discussed further in Chapter 8. Thus, fungicides remain the predominantly used pesticide for control of plant diseases.

History of Fungicides

Agriculture was practiced for many centuries before any pesticides were used, so the history of pesticide use begins relatively recently. Before we knew that microorganisms caused disease and understood something about their life cycles, farmers were helpless to protect crops from pathogens except by the general means discussed in the previous chapter. The devastation of the potato fields of Europe and the coffee plantations in Ceylon have already been described.

Elemental **sulfur**, applied as a dust, was an early insecticide and control for powdery mildew on plants, but it was not generally applied for plant diseases. **Powdery mildews** are Ascomycetes that grow profusely on the surfaces of plants. These white fungi absorb nutrients from plant cells through haustoria (hyphal projections) from the surface hyphae. Powdery mildews are foliar pathogens that do not permeate throughout the tissues of the host plant as most parasitic fungi do. They remain exposed to the environment and to the toxic effects of sulfur if it is dusted on leaves. Attempts were made to control coffee rust and potato late blight with sulfur, but it was not particularly effective, partially because sulfur is difficult to apply to completely cover foliage, and, like all surface-acting fungicides, it must kill the pathogen before infection, so application to already-infected leaves has no curing effects. Hyphae of most parasites quickly become well protected deep in the leaf tissues. Sulfur must also be applied with caution because it can be phytotoxic and burn foliage, especially when applied in hot weather.

Fig. 7-2. Powdery mildews are named for the powdery white growth of mycelium and conidia (asexual spores) on plant surfaces.

Discovery of the First Foliar Fungicide

In 1885 the first effective fungicide was discovered in France during another terrible plant disease epidemic. The French vineyards were being attacked by a downy mildew fungus, *Plasmopara viticola*. Where had this new plague come from? Once again we find the familiar problem of imported pests. The new downy mildew problem was really the third introduced problem among the grapevines in Europe, a sequence of disasters with which the viticulturists had to contend.

The first new disease that had appeared in the French vineyards was caused by a powdery mildew fungus, probably imported from America. These common plant parasites are generally highly host-specific. For instance, the powdery mildew commonly seen on lilacs cannot infect grapes and vice versa. Shortly after infection, chains of white conidia, the asexual spores, are produced, so the fungus can spread quickly. Later, overwintering ascospores are produced in dark, basketball-like, closed fruiting bodies called **cleistothecia**. Unlike those of most pathogenic fungi, the conidia of powdery mildews are capable of germinating at high relative humidities without free water. Disease development can be quite extensive even under relatively dry conditions, but because the fungus remains on the leaf surface, sulfur gave some disease control. By 1854, French wine production had been reduced by about 80%, but extensive applications of sulfur allowed production to rise once again.

To select grape cultivars resistant to powdery mildew, American grapes were imported; however, a second devastating pest was inadvertently brought along. The exact time of its arrival is not known, and its spread was probably slower than that of the powdery mildew epidemic. The new pest was an

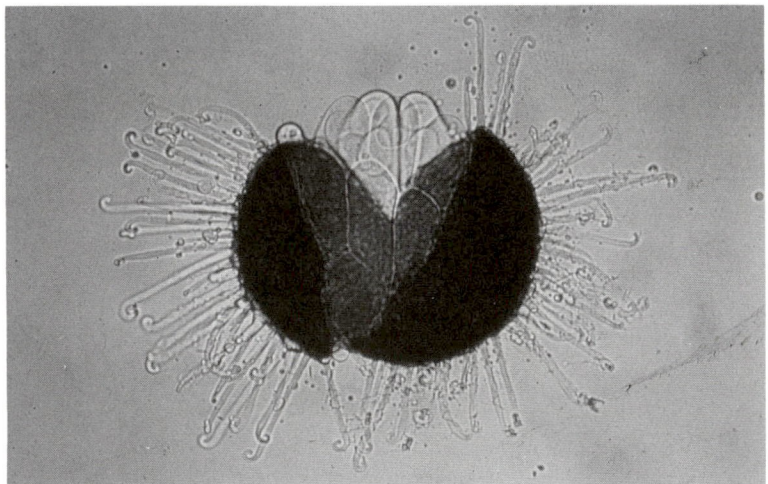

Fig. 7-3. Cleistothecium, the sexual fruiting body of a powdery mildew, broken open to reveal the tips of two asci containing ascospores.

aphidlike insect of the genus *Phylloxera* that largely lived and fed underground on the grape roots. It has not been an important problem in America
because American grapes have coevolved over many years with this native
pest. However, the insect was new to the European grapes, and disaster
struck. Most of the French vineyards were replanted with *Phylloxera*-
resistant rootstocks from America to which French grape scions (or shoots)
were grafted. The grafted plants produced the desirable French grapes while
the roots tolerated the insects.

During the importation of these American rootstocks, the **downy mildew**
fungus arrived as the third, and most devastating, grape plague. As discussed
in Chapter 1, downy mildews survive as thick-walled sexual oospores in
plant debris and soil. Sporangia are produced from the oospores and are
splashed onto the foliage. Zoospores produced in the sporangia swim out
to cause new infections. Shortly after infection, new sporangia are produced
through stomata for air dispersal to other plants, causing extensive necrosis
of both leaves and grapes. Despite the similarity in the names "powdery"
and "downy" mildew, the biological differences between these two pathogens
are great (Table 7-1).

Grapevines suffered from these overlapping pests, beginning with powdery
mildew in 1845, continuing with *Phylloxera* around 1860, and finally
succumbing to downy mildew in 1878. The monetary loss to downy mildew
amounted to nearly $50 billion, with European wine production in the
1880s and 1890s far less than it had been in the 1840s. From this agricultural
and economic disaster, began the chemical pesticide era. The story goes
that a French professor of botany, **Alexis Millardet**, was walking down
a lane observing the grapes infected with downy mildew when he noticed
that some grapes were covered with a bluish-white wash. He also noticed
that the leaves of these plants were healthy, whereas the neighboring plants
were badly diseased. When he questioned the farmer to whom the grapes
belonged, he was told that the grapes along the road had been sprayed
with a mixture of lime and copper sulfate ($CuSO_4$) to discourage pilferers.
This was the accidental discovery of what is now called **Bordeaux mixture**,
named for the area of France where the discovery was made. The copper
ions of the mixture were toxic to the fungus, and the lime reduced the
phytotoxicity of plain copper sulfate. Bordeaux mixture is effective against
most fungi and bacteria, is quite inexpensive, and even today, over 100
years later, is the most widely used fungicide in the world.

Table 7-1. Comparison of Powdery and Downy Mildews

Characteristic	Powdery Mildews	Downy Mildews
Fungal group	Ascomycetes	Oomycetes
Sexual spores	Ascospores in asci in cleistothecium	Oospores
Asexual spores	Chains of white conidia	Zoospores produced in sporangium
Mycelium	White, septate, superficial; absorbs nutrients from plant cells by haustoria	Nonseptate; penetrates deep into plant tissue

Other Early Fungicides

Early attempts to kill seedborne pathogens involved soaking seeds in nearly any strong-smelling or repellent substance such as horse urine. Germination of the seed was often harmed by exposure to such harsh chemicals, however, and many had no apparent effect on the pathogens despite their strong smell. The use of copper for seed treatment was an accidental discovery. In the early 1800s, **M. Prevost** of France found that grain soaked in water in copper containers produced plants free of smut disease. Further investigations showed that even the low concentrations of copper ions that dissolved from the container into the water killed the smut fungus spores when they germinated. His discovery spread slowly but had become commonplace in Europe by the mid-1800s. Application of chemicals to the foliage of plants was limited by the phytotoxicity (plant damage) that many caused and by the lack of effective spraying and dusting equipment.

As humans discovered toxic compounds, they began to use them to poison the pests that infested their animals, their crops, and themselves. Most of the earliest pesticides were inorganic compounds of various toxic heavy metal elements. The scientific definitions of the words **organic** and **inorganic** differentiate between molecules that contain carbon (organic) and those that do not (inorganic). The word *organic* is also popularly used to describe

Fig. 7-4. Alexis Millardet.

something that is produced without synthetic components, which has led to some very misleading concepts.

The elements copper (Cu), arsenic (As), mercury (Hg), and sulfur (S) are toxic to all living organisms. When Bordeaux mixture is applied to leaves, toxic copper ions are absorbed by the vulnerable germinating spores. Only tiny amounts of copper are sufficient to kill the spores, whereas plant and human tissues are relatively protected from the toxic effects by the epidermis that covers each. Arsenic and mercury compounds were commonly applied as pesticides until the 1960s. In fact, until quite recently, arsenic could be purchased from many drugstores for use as an insecticide for household insects and was also commonly used in agriculture. Because the toxic components of such pesticides are chemical elements, they cannot be degraded after application. Residues that wash onto the soil sometimes accumulate in plant products or soils. In some areas, such as citrus groves in Florida, repeated applications of inorganic copper fungicides actually led to toxic levels of copper in the soil that caused toxicity symptoms in the citrus trees. Mercury fungicides were essentially banned in the United States because mercury can accumulate in food chains. A few limited uses, such as for control of snow mold on golf course greens and tees, are allowed in some areas.

Bordeaux mixture was the first fungicide that allowed farmers to reliably protect foliage from invasion by fungi. Such inorganic fungicides rely primarily on copper and sulfur, and in some cases mercury, to kill fungi. They are **broad-spectrum fungicides**, effective against all fungi when properly applied. Inorganic fungicides remained important in the United States until the mid-1940s, when organic compounds began to dominate the fungicide picture.

Fig. 7-5. Dusting potatoes in Waterville, NY, 1911.

Organic Fungicides

The **organic fungicides** are complex molecules that include a number of different elements, primarily carbon, hydrogen, oxygen, nitrogen, and sulfur. In some cases the exact **mode of action**, or the means by which the compound poisons the fungus, is not clear, but they are generally toxic to many fungi and much less toxic to humans and plants. In contrast to the inorganic fungicides, these compounds are degraded by sunlight and soil microorganisms when they are washed into the soil, so that residues do not accumulate.

Because we still do not fully understand the modes of action of many fungicides, the synthesis of new fungicides from theoretical models has not progressed to the sophisticated state found in the pharmaceutical industry. Most fungicides are still discovered by massive screenings of numerous compounds to find possible new ones. Because so many types of organic compounds have already been screened for fungicidal activity, the success rate of finding new compounds is very low. A recent estimate suggested that a new product that will survive the tests necessary for government approval requires the screening of more than 13,500 compounds. This represents great expense in the development of new fungicides. Since 1975, only four new fungicides that account for more than 5% of the fungicide sales for food crop use have been introduced.

The early-developed fungicides, both inorganic and organic, are strictly **protective**. They do not enter plant tissue but remain on the surface and can kill the fungus only as it attempts to germinate and penetrate the leaf. If a fungicide is necessary to protect a crop, the need must be anticipated, and sprays must be applied before infection. Once infection has occurred, the mycelium of the parasite continues to colonize the leaf tissue even if fungicide is applied to the leaf surface. Large volumes of fungicides must be applied for full coverage of the foliage, and repeated applications are

Table 7-2. Chronological Development of Plant Disease Control Agents[a]

Date	Agent
1802	Lime sulfur
1885	Bordeaux mixture
1913	Organomercurials
1934	Dithiocarbamates
1935–1940	Chloronitrobenzenes, o-phenylphenol
1940–1950	Biphenyl, hexachlorobenzene, dichlone, glyodin, streptomycin, cycloheximide
1950–1960	Captan, organo-tins, dodine, anilazine, dicloran
1960–1970	Chlorothalonil, polyoxin, organophosphates, chloroneb, kasugamycin, hydroxypyrimidines, carboximides, benzimidazoles
1970–1980	Sterol biosynthesis inhibitors, phenylamides, dicarboximides
1980–1988	Additional compounds of the classes developed in the 1970–1980 period

[a]Source: Delp, C. J., ed. 1988. Fungicide Resistance in North America. American Phytopathological Society, St. Paul, MN. Table 1.

necessary as the fungicide is degraded or washes off or as plant growth occurs. Large amounts of applied fungicides are essentially wasted and may create environmental hazards at "nontarget" sites such as soil and water.

Systemic Fungicides

Since the late 1960s, systemic fungicides have played a larger role in the chemical picture. **Systemic fungicides** are absorbed into plant tissue and can have some after-infection activity or curative action. Less chemical needs to be applied because it is absorbed into the tissue and does not wash off after application. Some systemic chemicals are redistributed to new growth, so the fungicide does not have to be reapplied as frequently. Systemic fungicides are less likely to reach the environment outside the plant. In addition, compounds specific enough to kill or inhibit fungi without causing toxic effects to plant tissues are often quite specific in their fungal toxicity. For example, some systemics (such as phenylamides) are effective against Oomycetes, whereas others are useful only for certain members of other fungal groups. Whereas broad-spectrum fungicides are toxic to many fungi, including important saprophytic soil inhabitants, the more narrow-spectrum systemic fungicides that do reach the soil may cause less ecological disruption.

The use of systemic fungicides has some important disadvantages. Many do not move well throughout the plant after application, and those that do are transported primarily upward from the roots and must therefore

Fig. 7-6. Modern pesticide application equipment.

be applied to the soil or roots. In addition, although specificity may be an advantage, it may also have some important disadvantages. For example, one systemic fungicide, metalaxyl, is effective against Oomycetes such as *Phytophthora infestans* (the potato late blight fungus) and need not be applied as often as protective fungicides. However, if this is the only fungicide applied, another important potato disease may increase; this is early blight, caused by *Alternaria solani*, which is not an Oomycete. Metalaxyl is not effective against *A. solani*, so the unprotected foliage is susceptible to infection. Previously, many farmers did not deliberately apply fungicides for early blight because the broad-spectrum fungicides were effective against both pathogens, but the importance of *A. solani* became more apparent when the Oomycete-specific systemic fungicide was used. For ecological reasons, one might wish to specifically attack the primary pathogen rather than using a chemical that suppresses all fungi, but, in terms of crop protection, this same specificity may be a disadvantage if other pathogens remain unsuppressed and require further control.

Resistance Problems

A second important problem arises from the specificity of fungicides that attack fungi at a specific physiological site. The early fungicides were very broad-spectrum because their modes of action usually involved the inhibition of major enzyme systems or included a range of toxic effects that interfered with numerous metabolic processes. In such a situation, it is highly unlikely that resistant fungal populations might arise. This was true not only for the inorganic compounds but also for most of the early organic ones. Thus, for many years plant pathologists had the luxury of using these chemicals repeatedly without resistance problems.

This was not the case for entomologists (scientists who study insects) because many organic insecticides have a mode of action that involves inhibition of a specific nerve enzyme, cholinesterase. When insecticides are applied repeatedly, resistant insect populations occur, and the sprays become ineffective. The repeated use of DDT against malarial mosquitoes and its subsequent loss of efficacy against resistant mosquito populations is a famous example.

Such a situation has occurred in plant pathology with the advent of the **site-specific** (mostly systemic) **fungicides** and for the same reason. In 1960, only one fungicide-resistant fungus genus had been discovered, but the number has increased dramatically since then (Fig. 7-7). The modes of action of these compounds are much more specific, so even small genetic changes in the fungus population can lead to widespread resistance.

One of the important challenges in pesticide science is to prolong the effectiveness of these chemicals. Chemical companies face the dilemma that increased sales lead to profits, but repeated use of some compounds makes them ineffective. In most cases, resistant individuals are already present in a pathogen population at the time a chemical is first applied. With repeated application, sensitive individuals are killed while the resistant types continue to reproduce. In rapidly reproducing pathogen populations, this can occur quite quickly, and eventually most of the pathogen population becomes resistant to the pesticide.

In some cases, resistance is probably conferred by a single gene and may develop quite quickly with repeated use. A resistant strain can be selected within one or a few seasons of use, and an apparently sudden "chemical failure" can result. That is, the chemical is applied with little or no decrease in the amount of disease. However, for other systemic fungicides, the genetic response of fungi suggests that several genes are involved and that resistance can be expected to develop more slowly and gradually. Such fungicides are at a lower risk of failure with repeated use and may remain effective if high concentrations of the chemical are applied.

Much of this research is in its preliminary stages, and important questions are still to be answered. For instance, is it best to alternate sprays of a systemic chemical with sprays of a broad-spectrum nonsystemic compound, or should they be mixed at each application? Should systemic fungicides with different modes of action be mixed or used alternately? Also, once resistance is detected in the fungus population, should applications of the fungicide to which the fungus is resistant be applied at higher concentrations, used less frequently, or abandoned altogether?

An important question that remains to be answered for most fungicides concerns the competitive ability of the resistant strains. Once the use of a fungicide has ceased, how long will the resistant strains predominate in the fungus population? If they are competitive, they may remain common

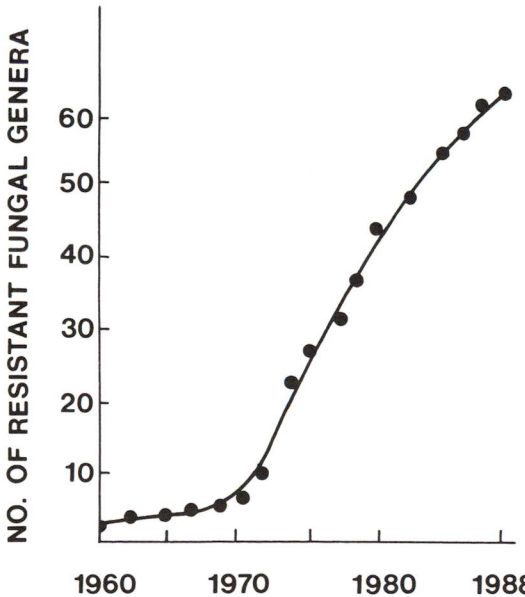

Fig. 7-7. Chronological increase in the number of fungicide-resistant fungal genera. Genera with multiple resistance are repeated.

for a number of years, so the fungicide remains essentially useless. On the other hand, if the resistant population is less competitive than the common strains, the resistant strains may be reduced quite quickly, and successful reintroduction of the fungicide may periodically be possible.

Another area of fungicide research involves a detailed understanding of how the fungicide kills a fungus and how the fungus becomes resistant to the chemical. If the mode of action of two different chemicals is the same, then the genes involved in the resistance may be identical, and a fungus strain resistant to one chemical may also be resistant to a different chemical with the same mode of action. Sometimes the mode of action can be predicted from the chemical structure of the compounds, but such **cross-resistance** is not always predictable. Alternating or mixing two different fungicides does not reduce the development of resistance if the genes that govern resistance to the chemicals are the same.

Because the discovery and marketing of a new fungicide is so expensive and because most major groups of organic compounds have already been screened for their activity as fungicides, it is important that site-specific fungicides be used judiciously. Some systemic fungicides are marketed today only as mixtures so that decisions about mixing or alternating are not left to applicators. Some scientists feel that the use of these compounds should be even more closely controlled, with application by "prescription" only. Their use is still controlled primarily by the applicator, who receives advice from chemical salespeople and private and public sector agricultural advisors.

Fungicide Use

A large number of fungicides are available for the reduction of plant diseases. Some are protective and some systemic, each having its relative advantages and disadvantages and all involving some expense. Systemics tend to be more expensive than the broad-spectrum chemicals, but

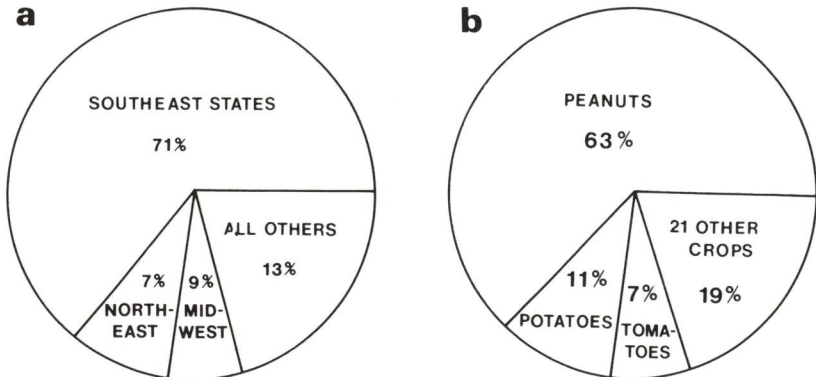

Fig. 7-8. Use of the widely used fungicide, chlorothalonil, in the United States in 1988. **a,** By region; **b,** by crop.

application of a systemic fungicide often requires smaller amounts of the active ingredient at longer time intervals. Even though fungicide costs are relatively low compared to those of most insecticides and herbicides, many crops do not receive regular foliar fungicide applications. Perhaps this is surprising, but the value of many field or agronomic crops does not justify the considerable cost of labor and equipment to apply these chemicals. The following 1984 data compare the use of fungicides with the use of other types of pesticides in the United States: fungicides, 7%; insecticides, 30%; herbicides, 63%.

Only a small percentage of the total acreage of crops receives repeated fungicide applications. Fungicides are applied primarily for diseases caused by polycyclic pathogens. It should already be clear that potatoes require fungicides against *P. infestans* until greater genetic resistance can be developed. Citrus, apples, other fruits, and some vegetables are sprayed regularly to meet consumer demands for unblemished fruit and vegetables. More fungicides are applied in the eastern half of the United States because rainfall is greater and high humidity favors fungal infections. Fungicides are commonly applied to grain crops in Europe because of its humid environment and limited land area. In contrast to Europe, regular application of foliar fungicides to grain crops in the United Sates is uncommon because most U.S. grain is grown in a drier climate and lower yields are acceptable because more land is available for grain production.

Application of a fungicide is an economic decision. The cost of application (chemical, equipment, and labor) must be offset by the increased value of the crop when disease is reduced. In some crops, such as potatoes and peanuts, fungicide applications increase yields by reducing foliar disease, whereas for apples and tomatoes, fungicides improve crop quality by reducing spots and blemishes. Fungicides are not usually applied to low-value crops and are not generally effective against most soilborne pathogens. Because fungicide use is concentrated on a few crops in certain areas, a significant reduction can be made in the total amount of fungicide used by eliminating even a few sprays per season on chemically intensive crops. We return to this idea at the end of the chapter.

Regulation of Pesticide Use

The year 1985 marked the centenary of the discovery of Bordeaux mixture, but the predominant agricultural chemical era really dates from the 1940s, when a relatively inexpensive variety of organic pesticides became available. For a short period, perhaps about 20 years, pesticides were seen as miraculous tools in the fight against pests. The environmental side effects and problems with pest resistance to the chemicals soon began to tarnish their image.

What do we expect from a pesticide? It should be inexpensive and effective. Although toxic to pests, it should not endanger humans and other nontarget organisms. It should be persistent enough so that farmers do not spend too much time on reapplication, but it should not persist in the environment and accumulate in organisms in food chains. While preferably specific to the pest, repeated use should not result in resistance problems.

These conflicting goals for acceptable pesticides are the source of controversies concerning their use. To balance these conflicts, complex bureaucratic legislation has developed to regulate pesticide use. In 1947, the **Federal Insecticide, Fungicide, and Rodenticide Act (FIFRA)** was passed. This heavily amended act and components of the 1954 **Federal Food, Drug, and Cosmetic Act** govern the use of pesticides in the United States.

Pesticide Registration

All pesticides must receive registration approval through the Environmental Protection Agency (EPA). For registration to be granted, the manufacturer must provide extensive information about the chemical to the government. The information required has greatly increased over the years so that total costs now amount to over $28 million from discovery to registration of a new pesticide.

In addition to efficacy and formulation data, chemical companies must provide toxicological information on long- and short-term effects on a variety of organisms including plants, animals, and microorganisms. They must also conduct studies for **mutagenic** (causing gene damage), **carcinogenic** (cancer-inducing), **oncogenic** (tumor-inducing), **fetotoxic** (toxic to a fetus), and **teratogenic** (causing birth defects) effects of the chemical. A 1972 amendment to the pesticide regulations requires that studies must also include economic, environmental, and social effects of the proposed pesticide. This includes studies of the fate of the chemical in the air, soil, and water environment.

Fig. 7-9. Fumigating soil for nematode control. This photograph demonstrates an accepted method of pesticide application in 1962. Note that no protective clothing or respirators were worn.

Final **registration** approval includes the acceptance of a **pesticide label** that is required on all containers. A pesticide label is a legal document that must contain the following items: 1) product name and ingredients statement, 2) information about toxicity and treatment for poisoning, 3) storage and disposal directions, 4) precautionary statements in application, 5) directions for use, and 6) crops for which application of the pesticide has been approved. Any use contrary to information on the label is illegal. This includes changes in the amount or frequency of application; how and where the chemical is applied, stored, or disposed of; and application to crops not specified on the label.

Because registration and label approval is so costly, most chemical companies attempt to get the broadest registration possible so that future sales will offset initial costs. Patent protection for the discovery of a new compound lasts only 17 years. Since registration approval takes part of this time, chemical companies are left with a relatively short time in which to exclusively market enough of the chemical to earn a profit.

Registration of pesticides for food crops is often quite specific because studies must be made on possible pesticide residues in each harvested product. Registration for nonfood crop use is often granted under a broader approval because testing on the hundreds of species of plants used as ornamental and landscape plants would be prohibitive. Many specialty crops, grown on small acreages and having a specific pesticide need, are neglected because registration costs are too great to be covered by eventual

---

Table 7-3. Toxicological Data Requirements for Pesticide Registration[a]

1. Acute studies
   a. Acute oral toxicity (LD_{50}) in rat
   b. Acute dermal toxicity (LD_{50}) in rabbit (preferred species)
   c. Acute inhalation toxicity (LC_{50}) in rat (for volatile substances)
   d. Primary eye irritation in rabbit
   e. Primary dermal irritation in rabbit
   f. Dermal sensitization in guinea pig
   g. Acute delayed neurotoxicity in hen (for cholinesterase inhibitors)
2. Subchronic studies
   a. 90-day feeding studies in one rodent and one nonrodent species (rat and dog)
   b. 21-day repeated-dose dermal tests (rat, rabbit, or guinea pig)
   c. 90-day dermal toxicity (rat, rabbit, or guinea pig)
   d. 90-day inhalation tests in rat
   e. 90-day neurotoxicity in hen and one mammal
3. Chronic and long-term studies
   a. Chronic feeding tests in one rodent and one nonrodent species (rat and dog)
   b. Oncogenicity studies in rat and mouse
4. Teratogenicity tests in two species (rat and rabbit), and two-generation reproduction studies in rat or mouse
5. Mutagenicity studies (in vitro) to evalute: gene mutations, structural chromosomal aberrations, and other genotoxic effects that may be indicated by the test substance
6. General metabolism studies in rat
7. Dermal penetration studies
8. Domestic animal safety determinations

[a]Source: Federal Register. 1984. Data requirements for pesticide registration; final rule. 49:42856 -42905.

sales income. Chemical companies are generally not interested in developing new products for crops that are grown on less than half a million acres. This is a problem similar to that of "orphan drugs" for certain rare diseases. University and Cooperative Extension scientists provide test data for pesticide use on "minor crops" through the **Cooperative State Research Service Inter-Regional Project 4** sponsored by the **U.S. Department of Agriculture**.

Relative Toxicity

For most people, an important concern about pesticides is their **toxicity**, or ability to poison. Since the chemical is capable of poisoning the pest, one must wonder about its toxic effects on humans. This is determined by testing its ability to kill organisms at various concentrations. **Relative mammalian toxicity** is usually determined with rats or mice. When the results of such tests are graphed, a point can be determined at which a certain concentration of the chemical kills 50% of the test population. This is called the LD_{50}, or the **lethal dose** for half of the organisms exposed. This point is chosen because individuals exist who are particularly sensitive

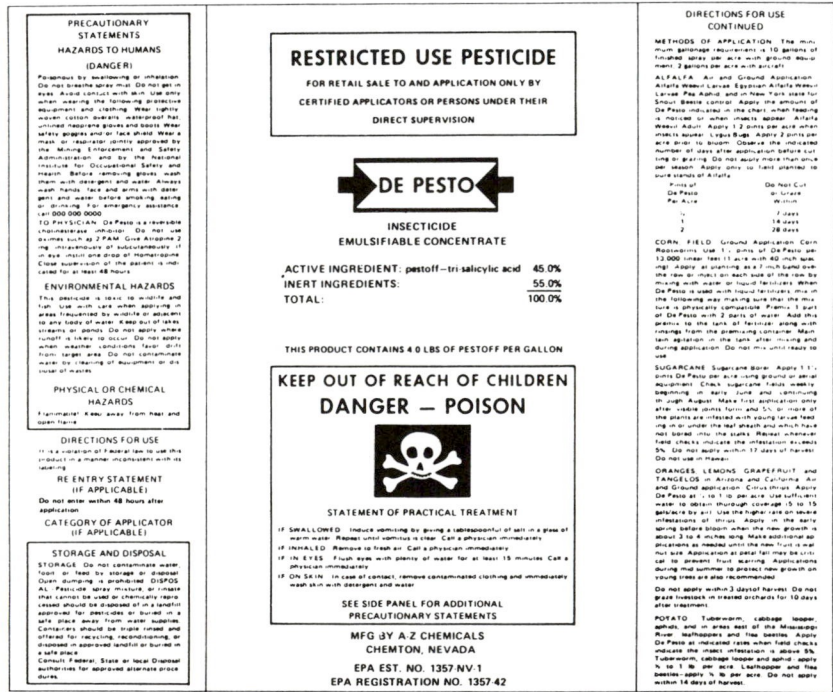

Fig. 7-10. Mock pesticide label. A pesticide label is a legal document that must include, among other information, the active ingredient, precautionary statements, storage and disposal directions, and directions for use.

or resistant to toxic effects, so a middle value is a more accurate representation of the relative toxicity of the chemical. An LD_{50} is usually expressed as milligrams (mg) of chemical per kilogram (kg) of body weight of the test organism. Thus, a chemical with an LD_{50} of 10 mg/kg is more toxic than one with an LD_{50} of 1,000 mg/kg. Insecticides are generally more toxic than fungicides. In fact, most fungicides have a relatively low mammalian toxicity. An LD_{50} represents toxicity for **acute poisoning**, that is, from a single oral dose of the chemical. However, studies have shown that some chemicals are actually absorbed more easily through the skin, and, from a practical standpoint, poisoning through dermal exposure during mixing and application is more likely than from swallowed doses in normal circumstances. Inhalation of pesticides is another important means of poisoning, since lung tissue can rapidly absorb many chemicals. The relative toxicity for inhaled chemicals is expressed as the **lethal concentration** (LC_{50}) in milligrams per liter (mg/L) of air. Although acute relative toxicity is an important factor in pesticide risk, chronic exposure to low levels of

Fig. 7-11. The recommended protective clothing for pesticide applicators includes gloves, water-repellent clothing, boots, goggles, hat, and respirator.

pesticides is more common and much harder to evaluate. An LD_{50} also offers no information about other health risks that may occur from exposure to pesticides over a period of many years, such as increased risk of cancer.

The protection of pesticide applicators and other workers exposed to various industrial chemicals has changed markedly in recent years. Gloves, boots, masks, respirators, and moisture-repellent protective suits are commonly used. Application of some chemicals requires protective clothing as specified on the pesticide label, but prudent applicators prefer to protect themselves from chronic exposure to pesticides whenever possible. Some applicators routinely undergo blood tests for analysis of cholinesterase inhibition. Exposure to carbamate and organophosphate pesticides causes **blood cholinesterase** levels to fall. Routine blood testing can detect pesticide exposure before toxicity symptoms may be noticed.

Careful applicators and people who do not work with pesticides are only infrequently exposed to pesticides, and then at low levels. This kind of exposure is called chronic rather than acute, and the effects are much more difficult to assess. The human body is capable of detoxifying many poisons, including alcohol and nicotine, and may or may not be harmed by low-level exposures. Negative effects from chronic exposure are determined by dosage, frequency, age, health, and other biological factors. Because of these and other complications, relative toxicity is based on LD_{50} values derived from acute oral doses of the chemical even though chronic toxicity is probably the more appropriate concern. Pesticide labels use specific **signal words** to indicate relative toxicity (Table 7-4).

Application of pesticides became more regulated in 1971. To better protect the environment, the general public, and the applicators themselves, certain

Table 7-4. Toxicity Categories and Signal Words Required on Pesticide Labels[a]

| Category[b] | Signal Word Required on Label | Category of Acute Toxicity[c] | | | Probable Oral Lethal Dose for 150-lb Man |
| | | LD_{50} | | LC_{50} Inhalation (mg/L) | |
		Oral (mg/kg)	Dermal (mg/kg)		
I Highly toxic	DANGER—skull and crossbones	0–50	0–200	0–0.2	A few drops to a teaspoonful
II Moderately toxic	POISON WARNING	50–500	200–2,000	0.2–2	Over one tea-spoonful to one ounce
III Slightly toxic	CAUTION	500–5,000	2,000–20,000	2.0–20	Over one ounce to one pint or one pound
IV Relatively nontoxic	CAUTION	>5,000	>20,000	>20	Over one pint or one pound

[a] Source: Northeastern Regional Pesticide Coordinators.
[b] Most fungicides (79%) are classified in toxicity categories III and IV. Most insecticides (71%) belong to categories I and II.
[c] LD_{50} = lethal dose for 50% of a population, in milligrams of chemical per kilogram of body weight; LC_{50} = lethal concentration for 50% of a population when inhaled.

pesticides were placed on a "restricted use" list. Only certified applicators who have received training and passed a written test and who receive continuing education in pesticide application are now allowed to purchase and apply chemicals on the restricted use list. The federal government sets general standards for certification, and individual states administer the licensing, sometimes adding their own stricter standards.

Pesticide Residues in Food

Pesticide residues in food concern many people. EPA registration requires that a **tolerance**, or level of pesticide acceptable at the time of harvest, be set for each food crop use listed on a pesticide label. The tolerance is governed by the Federal Food, Drug, and Cosmetic Act under two sections. **Section 408** sets tolerances on raw products, based on a balance between the risks to humans and the benefits for food production. Tolerances are also governed by **Section 409**, which applies to products to be processed. Different tolerances are needed because processing may concentrate pesticide residues. For instance, fungicide residues on tomatoes may become concentrated during the production of tomato paste. Section 409 includes the famous **Delaney Clause**, which specifies that no food additive, which includes pesticide residues, may be used if it "induces cancer" in animals at any concentration. The current EPA interpretation of this clause is that any substance found to be oncogenic, causing benign or malignant tumors, is excluded under the Delaney Clause. No risk-benefit analysis is allowed in the decision concerning Section 409 tolerances.

For food crops, the application of a pesticide near harvesttime is of particular importance. The EPA tolerance standards reflect several important measures such as the **acceptable daily intake** of a chemical and the **no-observable-effect level**. These figures are determined by physiological studies of the effects of the chemical on test animals and by studies that determine how much of a particular food is normally ingested. For instance, the tolerance for a pesticide in a common food such as wheat or peanuts might be lower than the tolerance for more exotic or unusual foods. The tolerance also reflects a safety factor of 100–1,000 times the no-observable-effect level.

Days to Harvest

After a tolerance is set, studies are conducted to determine how long it takes for the residue left in a food crop to reach the tolerance level after an application. This time reflects normal weathering and degradation of the chemical by wind, rain, sunlight, and microorganisms. An additional safety factor is included so that the time between the final application and harvest should result in no residue at all or residues well below the legal tolerance. The pesticide label indicates this time as the **days to harvest** of the final application. For some pesticides, applications can be made within a day or two of harvest because they degrade rapidly. Other chemicals, such as some systemic insecticides, remain in plant tissues for longer periods and cannot be applied within 90 days of harvest.

EPA inspectors may inspect any harvested crop and seize it if residue levels exceed the tolerance. Pesticide applicators are also required to keep detailed written records of all pesticide use. Of course, as with all laws, enforcement is expensive, and safety really lies with growers who cooperate with the restrictions. Training meetings and additional education for pesticide applicators emphasize these regulations.

Fungicide Residues and Cancer

In 1987, the Board of Agriculture of the National Research Council (NRC) published a report in response to an EPA request for evaluation of tolerances for pesticides and human health risks. The EPA was serving two different legislative mandates—one with a risk-benefit consideration and the other (the Delaney Clause) that specified zero risk. Although the long-term effects of the report's conclusions are not yet clear, the short-term effects have nearly eliminated the use of some major organic protectant fungicides.

The report concluded that 80–90% of the total dietary oncogenic risk from pesticide residues was due to only 10 registered compounds and that 60% of this risk was due to fungicides. Four of the fungicides belonged to a chemical group called the **ethylenebisdithiocarbamates**, the **EBDC**s. This group alone accounts for 28 million pounds of active ingredient applied per year to U.S. food crops.

A number of controversial topics are involved in the subsequent decision by the EPA on how to proceed, based on the results of the NRC report. Currently there are different standards for raw and processed foods. The EPA has chosen to use the broader definition of "causing cancer" in the Delaney Clause by including oncogens rather than strictly carcinogens. Most of the data are derived from animal studies in which extremely high doses were used. The application of the results of such tests to humans who are exposed to only very low levels, possibly intermittently but probably over long periods, is arguable.

Scientists are looking for a reliable and rapid way to assess such long-term effects. The **Ames test**, developed by Dr. Bruce Ames of the Biochemistry Department of the University of California at Berkeley, has been widely used to rapidly detect mutations (genetic changes) in bacteria exposed to a chemical. Some scientists argue that the relation between **mutagenicity** and **carcinogenicity** is not clear, whereas others argue that mutagenesis is an important indicator of potential carcinogenicity. Ames's research has demonstrated that many chemicals can cause cancer when fed to animals at extremely high doses, including many compounds found in foods we eat every day. Some substances probably present little or no apparent risk. There is also no difference in cancer risk from "natural" compounds or "synthetic" ones, since the risk from a compound is related to its chemical structure, not its origin. According to the Delaney Clause, chemicals that cause cancer when fed to animals at extremely high doses are prohibited as food additives even when there may be no evidence that low levels of exposure may pose increased health risks. Many common foods contain low levels of natural products that are toxic or carcinogens but that are not banned because the chemicals were not added by human beings. These include ergot, aflatoxin, and other mycotoxins discussed in Chapter 9. Thus, the Delaney Clause is a simplistic and inappropriate solution to an exceedingly complex and important medical problem.

The calculated risks determined by the NRC assume that the entire crop receives pesticide application and that all residues are at the maximum legal tolerance, which is not likely in most cases. However, since so much of the apparent dietary risk was due to a few compounds applied to only about 15 different foods, the report recommended stricter tolerances to reduce the risks. Because most of the fungicides were registered some time ago, a new amendment to the Federal Insecticide, Fungicide, and Rodenticide Act signed by President Reagan in 1988 requires that pesticides originally registered before 1984 must be updated to meet the new requirements by 1997 or their registrations will be cancelled. Pressure from various groups has caused the EPA to temporarily suspend a number of uses of EBDCs until further evaluations can be made. Reregistration of a pesticide costs between $500,000 and $1 million, so large numbers of pesticide uses are being withdrawn because manufacturers do not expect that future sales will offset reregistration costs.

Table 7-5. Estimated Oncogenic Risk from Fungicides in Major Foods[a,b]

	Estimated Risk	
Crop	Number	Percentage of Total Risk from Pesticides
Tomatoes	8.23×10^{-4}	14.1
Oranges	3.72×10^{-4}	6.3
Apples	3.18×10^{-4}	5.4
Peaches	2.86×10^{-4}	4.9
Lettuce	1.81×10^{-4}	0.1
Potatoes	1.29×10^{-4}	2.2
Beans	1.17×10^{-4}	2.0
Grapes	1.08×10^{-4}	1.8
Wheat	6.65×10^{-5}	1.1
Celery	6.04×10^{-5}	1.1
Total risk from fungicides		39.0
Total risk from herbicides, insecticides, and fungicides		42.0

[a] Source: National Research Council. 1987. Regulating Pesticides in Food: The Delaney Paradox. ©National Academy Press, Washington, DC. Used by permission.
[b] Note: These *worst-case risk estimates* are derived using data and methods of the Environmental Protection Agency (EPA). They assume residues are at the tolerance level, although actual residues may be different. These numbers are the totals of the committee's upper-bound estimates of dietary oncogenic risk for oncogenic fungicides with tolerances on these crops. As more accurate data are received by the EPA, crops may move on or off the list.

The Risk of Cancer

To gain some perspective on the NRC report, one must consider that the estimated risk of a U.S. citizen developing cancer is considered to be about one in four, or 0.25. Cancer studies suggest that multiple factors, including diet, stress, lifestyle, use of tobacco, exposure to sunlight, and genetic predisposition, contribute to this risk. The EPA has temporarily adopted a one-in-a-million level, or 1×10^{-6}, as a "negligible risk" level. That is, if a pesticide tolerance is set at a level that increases cancer risk by a negligible amount, the cancer risk for an average person would rise from 0.25 to 0.250001. For registrations requesting only a Section 408 tolerance (for products that are not to be processed), a slightly higher risk may be accepted if a strong case can be made for its use. Whether the EPA is acting appropriately still remains to be decided in the courts, and the standards just described may change in the near future. Fungicide manufacturers must now decide whether their product can meet the new requirements and whether future sales will offset the further investment that obtaining the new data will require. Captafol and dodine have already been withdrawn from use. Food uses for folpet have been suspended, and certain uses of captan and EBDCs have been withdrawn.

Major emphasis is now being put on reevaluation of cancer risk from pesticide residues. This reevaluation is based on studies of actual human exposure to the residues. Since EBDC fungicides are only on the surfaces of crops, they can easily be removed by washing and peeling. In addition, **market basket surveys** of actual residues on produce purchased by consumers indicate that residues are far below legal tolerances and often undetectable. In other words, the safety factors built into the calculation of days to harvest have been very successful. These new data have been used to recalculate estimated cancer risks based on these new alternative assumptions about the amount of pesticide residues on agricultural products (Tables 6 and

Table 7-6. Comparison of Estimated Cancer Risks from Selected Pesticides Under Alternative Exposure Assumptions[a]

		Excess Cancers per Million		
		Food and Drug Administration (FDA) Estimates[c]		
Chemical	National Research Council Estimates[b]	6–11 Months Old Children	14–16 Years Males	60–65 Years Females
Acephate	37.3	0.01725	0.02139	0.03243
Linuron	1,520	0.328	0.0984	0.1312
Captan	474	0.04462	0.02024	0.05612
Permethrin	421	2.13	0.9	1.215
Chlorothalonil	237	<0.0024	<0.0024	0.0024
Parathion	14.7	0.01116	0.00126	0.00288
Folpet	324	0.0273	0.01015	0.0336

[a] Source: Archibald, S. O., and Winter, C. K. 1989. Pesticide residues and cancer risks. California Agriculture (Nov.-Dec.):6-9. Table 2. Used by permission.
[b] Worst-risk assumptions (see Table 7-5).
[c] Exposure assumptions based on FDA's total diet study.

7). In 14,492 food products sampled by the U.S. Food and Drug Administration in 1989, pesticide residues were undetectable in 57% and exceeded
legal tolerances in less than 1%. Even so, the controversy about food safety
and pesticide residues will undoubtedly continue for some time.

Reducing Fungicide Use

What would be the disease consequences of the loss of these fungicides?
In many cases, no adequate substitutes exist, since the fungicides in question
are by far the most commonly used broad-spectrum protectant materials.
Most of the systemic compounds do not appear to be oncogenic, but they
are commonly mixed or alternated with protectant materials to prevent
resistance problems. Elimination of the need for fungicides is the long-
term goal of plant pathologists, but that time is not yet here. In humid
environments, which means essentially the eastern half of the United States,
crops such as apples, potatoes, beans, onions, carrots, tomatoes, oranges,
raspberries, and grapes are difficult, and often impossible, to grow without
fungicides.

Application of pesticides costs money and has certain associated risks.
In the late 1960s and early 1970s, research emphasis shifted to development

Table 7-7. Comparison of Estimated Cancer Risks from Selected Pesticides Under Alternative Residue Assumptions[a]

	Excess Cancers per Million	
Commodity, Pesticide	National Research Council Tolerance	Food and Drug Administration (FDA) Residue Data[b]
Tomatoes		
Acephate	14	0.0017
Azinphos-methyl	0.00015	1.5×10^{-8}
Captafol	191	0.0033
Captan	29	0.0026
Chlordimeform	479	0
Chlorothalonil	61	0.23
Folpet	45	0.00017
o-Phenylphenol	8	0
Parathion	0.92	0.0002
Permethrin	31	0.088
Total	859	0.33
Lettuce		
Acephate	16	0.025
Captan	55	0.011
Folpet	42	0.054
Parathion	0.43	0.00025
Permethrin	143.4	0.8
Pronamide	3.8	0.0044
Total	261	0.89

[a] Source: Archibald, S. O., and Winter, C. K. 1989. Pesticide residues and cancer risks. California
Agriculture (Nov.-Dec.):6-9. Table 2. Used by permission.
[b] Authors' recalculations are based on residue data from the FDA Los Angeles Laboratory
1982–1986.

of ways to reduce pesticide use. Part of this change in emphasis was due to pressure from the environmental movement (partly due to concern about the rapidly increasing costs associated with pesticide production and fuel for application, both of which are fossil fuel products), and some of the change came from the continuing studies of the ecology of agriculture. Pesticides are not the miracle cure for agricultural pests that many had hoped for. New information from two main areas—economics and pest life cycles—was critical to the reduction of pesticide application.

Integrated Pest Management

Detailed studies of the relationship between the cost of pesticide application and the resulting increased value of the protected crop demonstrated that application at set intervals might actually cost more than the value gained at harvesttime. Thus, the concept of **economic damage thresholds** became a guiding principle of the new pest management. The cost of the loss must meet or exceed the cost of control to justify a pesticide application. The economic threshold concept was dramatically successful for many insect pests. Rather than spraying weekly, farmers were advised to wait until the insect pest population reached a predetermined threshold. This meant that the pest population was monitored, and insecticide was applied only when necessary. Insect monitoring has become much more sophisticated. Life cycle studies have made the timing of sprays more precise and even more effective. For many crops, insecticide applications have been reduced while insect damage has been maintained at acceptable levels.

Similar studies have been applied to plant diseases. One of the major limitations in the use of economic thresholds for many plant diseases is that plant parasites damage plants very differently than most insect pests. Although it may be possible to monitor insect populations and apply an insecticide when thresholds are reached, similar activity with plant pathogens may lead to massive losses because damage has already occurred by the time disease symptoms appear. To reduce fungicide use, detailed information is needed about pathogen life cycles, damage, and yield losses due to disease. Life cycle information can reveal when the pathogen is likely to cause infection, so fungicide timing can be more precise. Economic studies demonstrate how much disease can be tolerated before fungicide applications will offset the losses.

Besides allowing a more accurate economic and biological justification of pesticide application, these new studies emphasize a more complex approach to pest management. Rather than relying on pesticides alone, various cultural practices, genetic resistance, and other management techniques are used to reduce the need for pesticides. This has been called **integrated pest management**, or IPM, and such programs are common in agriculture as well as in other pest problem areas. Plant pathologists have always practiced IPM because so many diseases cannot be controlled by pesticides alone but must be accompanied by sanitation, crop rotation, and other cultural practices that reduce pathogen populations. Unfortunately, IPM programs will always be limited in their ability to reduce fungicide application, because most fungicides function in a protective manner, so it is not usually possible to wait until disease reaches a certain threshold before applications begin.

Control of Apple Scab

Fungicide use has been significantly reduced in some crops, however. The two examples that follow demonstrate two very different disease management situations. The first example is the scab disease of apples and crab apples. While heavy infection by the Ascomycete, *Venturia inaequalis*, can cause severe distortions and scabby lesions as well as fruit drop, the intensive application of fungicides results from market demands for unblemished fruit. Small corky scabs on apples cause no harm to the people who eat them but greatly reduce the market quality and therefore the profit.

Fig. 7-12. Leaf symptoms of apple scab (*Venturia inaequalis*) on crab apple.

V. inaequalis overwinters as mycelium in fallen apple leaves on the orchard floor. During the spring, the **initial inoculum**, i.e., the ascospores, mature and are forcibly discharged from the sexual fruiting bodies, called **pseudothecia**, that are produced in the fallen leaves. Some ascospores land on newly expanding apple leaves, germinate, and grow into mycelium between the cuticle and upper epidermal cells. Soon the **secondary inoculum**, asexual conidia, are water-dispersed to other leaves and developing fruits, causing new infections. The infections on leaves and fruit result in superficial, corky scabs, giving the disease its name.

In the past, apple orchards received routine fungicide application on a regular schedule from leaf bud expansion in the spring and through the growing season to protect against this polycyclic pathogen. Leaves must

Fig. 7-13. Symptoms of apple scab disease on apple fruits.

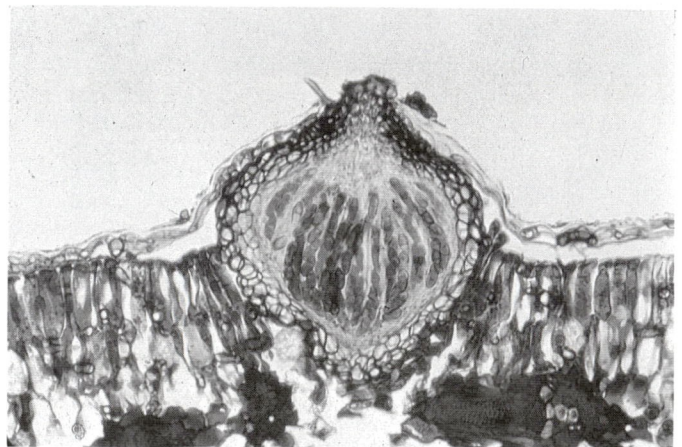

Fig. 7-14. Pseudothecium (sexual fruiting body) of *Venturia inaequalis*. Ascospores ejected from pseudothecia produced on fallen apple leaves are the primary inoculum that begins apple scab disease on newly expanding apple leaves in the spring.

be protected from infection because early scab lesions provide secondary inoculum for later infection of developing fruits. Infection by *V. inaequalis* can be divided into primary infections by ascospores coming from the orchard floor and secondary infections from conidia on new leaves. Plant pathologists determined that if fungicides were used to protect plants during the time of primary infections from ascospores, fungicide applications later in the season could be reduced because secondary inoculum would be very limited. Cooperative Extension personnel carefully monitor the pseudothecia in fallen leaves and produce spring **ascospore reports** to warn growers of the onset, peak, and end of the "ascospore season," during which primary infections can occur. If an orchard is relatively scab-free when ascospores have all been released, usually about two weeks after flower petal-fall, the apple trees require much less fungicide protection for the rest of the season.

Besides knowing when primary inoculum is present, it is also helpful to know how successful infection is likely to be. This is called disease severity. It can be predicted by determining the average temperature and the number of hours that leaves remain wet so that spore germination can occur. An American plant pathologist, **W. D. Mills**, provided this important information through tedious studies that have been compiled in the famous **Mill's Tables**. Growers can place specialized equipment in their orchards that detects and records temperature and leaf wetness in the orchards. From the data and the tables, one can determine whether an **infection period** has occurred. Commercial apple scab predictors are now available that monitor the environmental conditions and then analyze the data with a computer program that can warn growers when fungicide applications are necessary.

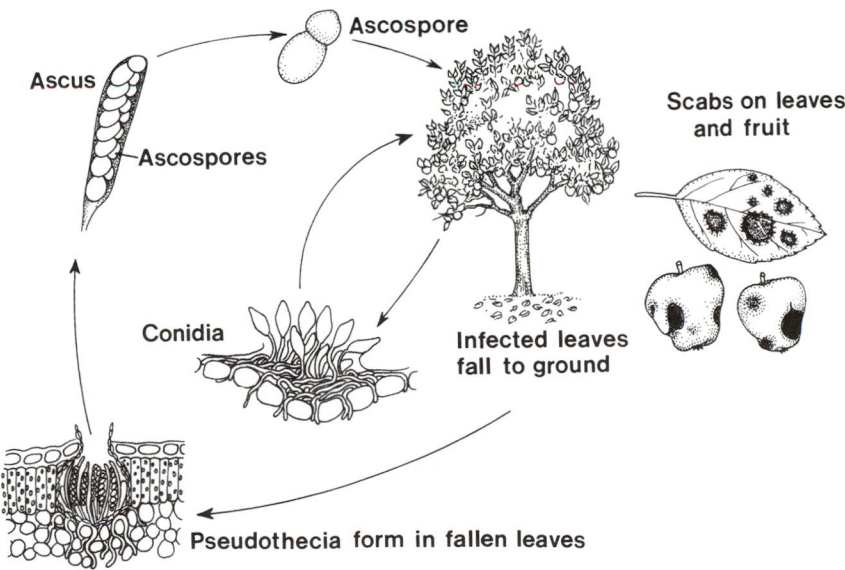

Fig. 7-15. Disease cycle of apple scab (*Venturia inaequalis*).

Mill's infection periods have become even more useful with the introduction of some new fungicides. Some fungicides now available have a certain amount of curative or "kick-back" action to kill the fungus after infection has occurred. Some fungicides must be applied within 24 hours of infection to be effective, while others can eliminate *V. inaequalis* even several days after infection. The superficial nature of *V. inaequalis* infections makes this fungus more susceptible to after-infection control than most fungal parasites. Thus, if the Mill's table predicts that a severe infection period has occurred and a grower suspects that the recent rain has probably washed away much of the residual protective fungicide, it might pay to use a fungicide with some after-infection activity. These fungicides are generally more expensive than protective ones and are also more likely to select resistant strains of the scab fungus, but their judicious use can eliminate primary infections and thus reduce the need for repeat applications later in the season.

Some attempts have been made to reduce the initial inoculum coming from the fallen leaves. Theoretically, total removal of fallen leaves eliminates the fungus and is a recommended practice for homeowners with a few apple or crab apple trees. In commercial orchards, fungicides with after-infection activity are sometimes applied before leaves fall to reduce overwintering mycelium and prevent formation of ascospores. In relatively scab-free commercial orchards, fungicide protection during ascospore production has reduced the season-long need for scab fungicides, although some summer applications are necessary for other apple diseases. New scab-

Table 7-8. Apple Scab Temperature and Wet-Period Guide, Based on Mill's Tables[a]

Average Temperature (°F)	Hours of Moist Foliage Necessary for Leaf Infection
78	13
77	11
76	9
61–75	9
60	9
57–59	10
56	11
55	11
52–54	12
51	13
50	14
49	15
48	15
47	17
46	19
45	20
44	22
43	25
42	30
33–41	More than 2 days

[a]Source: New England Cooperative Extension.

resistant apple cultivars, such as Liberty and Freedom, are now available. **Genetic resistance** is the most effective way to reduce fungicide applications, but replanting is expensive and consumers are often slow to change their preferences in fruit. The resistant cultivars are too new to judge whether the resistance will be durable or whether *V. inaequalis* will be able to overcome it.

Another Success Story — Peanuts

A second example of successful reduction in fungicide applications involves a disease that affects yield and quality only indirectly. In the southeastern United States, peanuts are grown in a warm and humid climate that favors foliar infections. Two important fungi that reduce yields by causing leaf spots are *Cercospora arachidicola* and *C. personatum*. Peanuts are in the legume family, which includes peas, beans, alfalfa, and clover and is characterized by podlike fruits that open naturally along two sides. After peanut plants flower and are pollinated, the flower stalk grows into the ground, and the familiar peanuts mature underground where they are protected from infection by *Cercospora* species. Aboveground, however,

Fig. 7-16. Peanut leaf spot disease caused by *Cercospora* species. **Top,** symptoms; **bottom,** field plots with and without fungicide treatments, showing healthy plants and extensive defoliation in unsprayed plots.

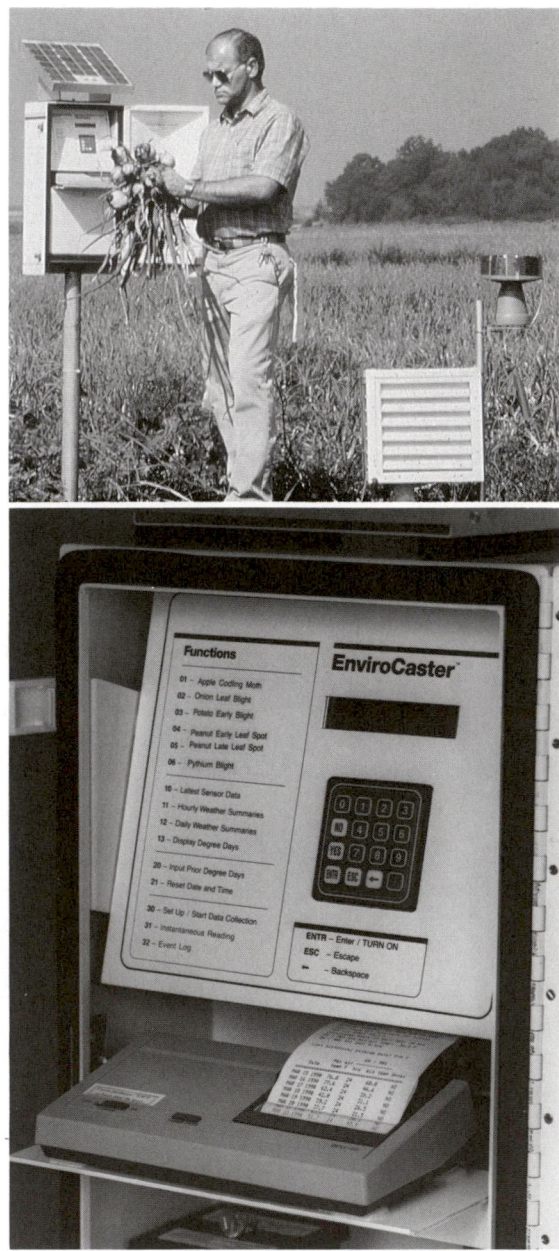

Fig. 7-17. Computerized environmental monitoring equipment (**top**) and close-up view (**bottom**). Environmental data such as air and soil temperatures, rainfall, relative humidity, and leaf wetness are determined and stored. Models predict disease outbreaks based on environmental data. Predictions and management recommendations can be accessed by crop managers.

foliar disease reduces the photosynthetic capacity of the leaves and thus the yield of peanuts.

Growth models of peanut plants have determined how much disease can be tolerated before yield reduction reaches the threshold that justifies fungicide costs. Because leaf spots do not lead to rapid, total destruction of a plant, as we have seen with late blight of potatoes, periodic monitoring of the peanut plants can determine when fungicide sprays are needed to protect new foliage. As with apple scab, infection periods of these leaf spot fungi have been determined, so fungicides can be applied with more precision.

An important technical problem that impedes accurate decisions about reapplication of fungicides is that the amount of fungicide residues that remains on foliage cannot be detected rapidly and inexpensively. However, when integrated with other disease management strategies, such as irrigation timing and destruction of contaminated crop debris, the growth model studies have allowed fungicide applications to be reduced on a chemically intensive crop without sacrificing yield.

Each time a pesticide application is eliminated, costs are reduced and so are possible health risks for applicators and consumers. Pesticides will continue to be necessary to protect certain crops, but they must be applied in the safest way possible. State and federal governments regulate storage, application, and disposal of these chemicals. They also regulate the applicators and the allowable residues in agricultural products. One more important area of concern about pesticide use lies in the environmental hazards they may cause. Two significant topics that need investigation are the nontarget effects of these chemicals and their eventual fate in the environment. Since most pesticides applied to reduce plant diseases find their way to the soil as runoff during application, wash-off from plants during rain and irrigation, or residues left on plant debris at the end of the growing season, we need to consider what happens to pesticides in the soil. So far, our pesticide discussion has focused on aboveground issues. In the next chapter we consider soilborne pathogens and the soil environment of plants.

Selected Readings

National Research Council. 1987. Regulating Pesticides in Food: The Delaney Paradox. National Academy Press, Washington, DC.

Delp, C. J., ed. 1988. Fungicide Resistance in North America. American Phytopathological Society, St. Paul, MN.

Marco, G. J., Hollingworth, R. M., and Durbam, W., eds. 1987. Silent Spring Revisited. American Chemical Society, Washington, DC.

Ottoboni, M. A. 1988. The Dose Makes the Poison. Vincente Books, Berkeley, CA.

Ware, G. W. 1989. The Pesticide Book, 3rd ed. Thomson Publications, Fresno, CA.

Winter, C. K. 1990. Chemicals in the Human Food Chain. Van Norstrand Reinhold, New York.

Soil, The Rhizosphere, and Soilborne Pathogens

An unseen world teaming with life exists wherever there is soil. A single pound of soil contains about eight billion organisms. Although mostly ignored by those of us living aboveground, the world below the surface of the soil is inhabited by incredibly diverse living organisms that are significant to plant pathology. Some of these organisms are beneficial and some are detrimental, but all are members of the community to which underground plant parts belong.

While the underground environment of a plant is no less significant to plant health than that above ground, far less attention is given to the world below our feet. There are several reasons for this. Unfortunately, it is difficult to study the soil environment without disrupting it. In addition, the interactions between soil organisms are exceedingly complex and difficult to study in isolation without creating simplistic or misleading situations. A third factor is the small size of many of the soil organisms (algae, fungi, bacteria, actinomycetes, mites, springtails, nematodes, to name only a few), that reproduce quickly and require painstaking observation before accurate conclusions may be drawn.

The underground parts of plants, consisting mostly of roots, are constantly exposed to the myriad organisms in the soil as well as to the variable physical environment, which includes moisture, temperature, pH, and a wide range of chemical factors. It is impossible to fully understand underground plant growth without including the influences of the soil environment, and we cannot fully understand the soil community without consideration of the tremendous influence of plant roots.

Origins of Soil

The soils that dominate the landscape in so many parts of the world were created slowly over many millions of years. On the rocky surfaces of the land masses, erosion by wind, rain, glacial ice, freezing, and heating broke up the underlying minerals of the larger rocks into smaller pieces. Primitive land plants contributed to the breakdown of the rock as they colonized the cracks and depressions. The mineral particles became mixed with decaying organic material from microbes, animals, and plants that had once been alive. Thus, a soil is influenced from the beginning by the inorganic underlying bedrock and rocky material carried by glaciers and water as well as by the organic matter contributed by the organisms living on the surface.

If you slice straight down into most soils, you will find a typical profile consisting of a relatively thin organic layer at the top that contains mostly dead and decaying organic matter (the **A horizon**), below which is a layer containing a mixture of decayed organic material and small pieces of minerals (the **B horizon**). The final layer in a soil profile, just above the bedrock, consists of small pieces of the mineral components but little or no organic matter (the **C horizon**).

The size and relative proportion of the mineral particles determine the soil type. Mineral soil particles vary in size and range from **sand** (0.002–2 millimeters [mm]) to **silt** (0.002–0.02 mm) to the smallest, called **clay** (<0.002 mm). Where sand predominates, the soil contains relatively large spaces between the particles, allowing rapid drainage after a rain. In clay soils, the smallest particles predominate; the spaces between particles are tiny; and drainage may be slow. Some of the best agricultural soils are **loams**, in which a balance exists between sand, silt, and clay. This results in a variation in the size of spaces between the soil particles.

Water is present in soils in a thin layer on the surface of the soil particles and also as free water in the spaces between soil particles. Because water molecules have a tendency to cling together, the water in the tiny spaces

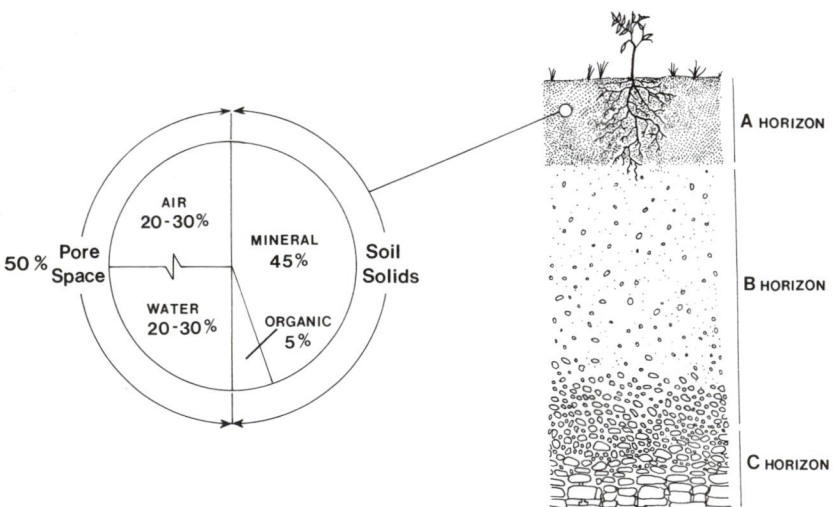

Fig. 8-1. Composition of soils. **Left,** silt loam soils, when in good condition for plant growth, contain approximately half pore space and half soil solids. In a well-drained soil, pore space is filled half with water and half with air. **Right,** the three primary horizons, or soil layers, in a typical soil. The A horizon contains relatively high amounts of organic matter and plant roots. The B horizon contains less organic matter. The C horizon is composed primarily of inorganic particles of the underlying bedrock.

between soil particles is drawn downward by gravity but also held in the pore space by the affinity of water molecules. This same phenomenon can be seen when water drips from a faucet. The drop stretches and enlarges slowly until finally the forces of gravity pull it down.

In a silt loam soil, approximately half the soil consists of **pore spaces** between soil particles. When soil moisture is optimal, half of the pore spaces are filled with air and half are filled with water. Thus, a silt loam soil can maintain soil moisture but also allow good drainage after rain. The air spaces are critical because oxygen (O_2) is available to soil organisms, including plant roots. When drainage is blocked or very slow, soils become **anaerobic** (without oxygen), which causes plant roots and other **aerobic** organisms (those requiring O_2) to die.

The Carbon Cycle

The mineral soils described above contain 1–10% decaying **organic matter**. The decay of organisms after death is an important part of the **carbon cycle**. This cycle refers to the continual processing of carbon atoms into organic molecules by photosynthesis and the release of these same carbon atoms back to carbon dioxide (CO_2) by respiration. These processes are essential to life because the chemical bonds of an organic (carbon-containing) molecule contain energy that can be released in a controlled manner during respiration and used in the life processes of organisms. In addition, the carbon in organic molecules is available to organisms for building the organic

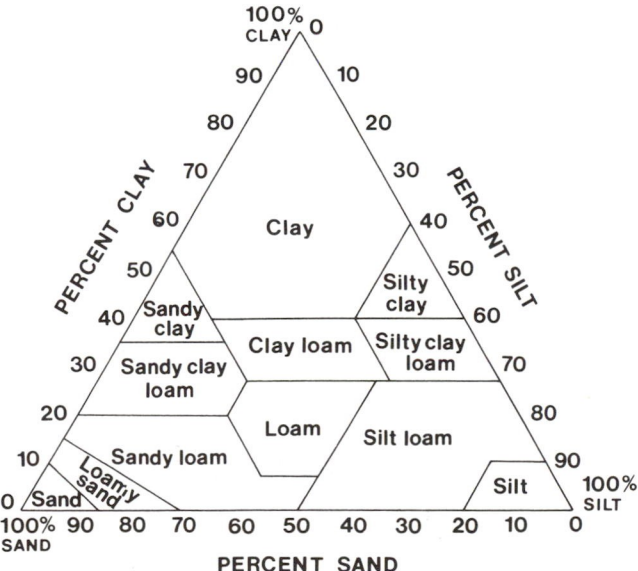

Fig. 8-2. Soil classification. The class name of a soil is based on the particle size distribution. Silt loam soils are excellent for agricultural uses.

molecules necessary for growth—proteins, nucleic acids, carbohydrates, lipids, etc.

The original source of all this energy is the sun. **Photosynthesis** is the process by which the energy of the sun is captured and stored in the chemical bonds of organic molecules produced from CO_2 and water (H_2O). In plants, these compounds are simple sugars, which are often combined and stored as starch. In eukaryotic cells, photosynthesis takes place in specialized organelles (**chloroplasts**) that trap sunlight and "fix" carbon atoms from CO_2 into organic compounds, releasing O_2.

Photosynthesis is the only source of atmospheric O_2, so this process not only accomplishes **carbon fixation** but also provides the O_2 necessary for **aerobic respiration**. Until photosynthesis evolved, all life on earth was anaerobic, functioning without free O_2. Aerobic respiration is more efficient than **anaerobic fermentation** because it derives more energy from the same organic molecules. Thus, the dominant organisms on the earth today are aerobic.

The presence of O_2 in the atmosphere also led to the formation of the **ozone** (O_3) layer in the upper atmosphere, which shields the earth from much of the ultraviolet light coming from the sun. By reducing mutagenic ultraviolet radiation, photosynthesis had a second major effect—allowing life forms to leave the protection of deep water to colonize the land.

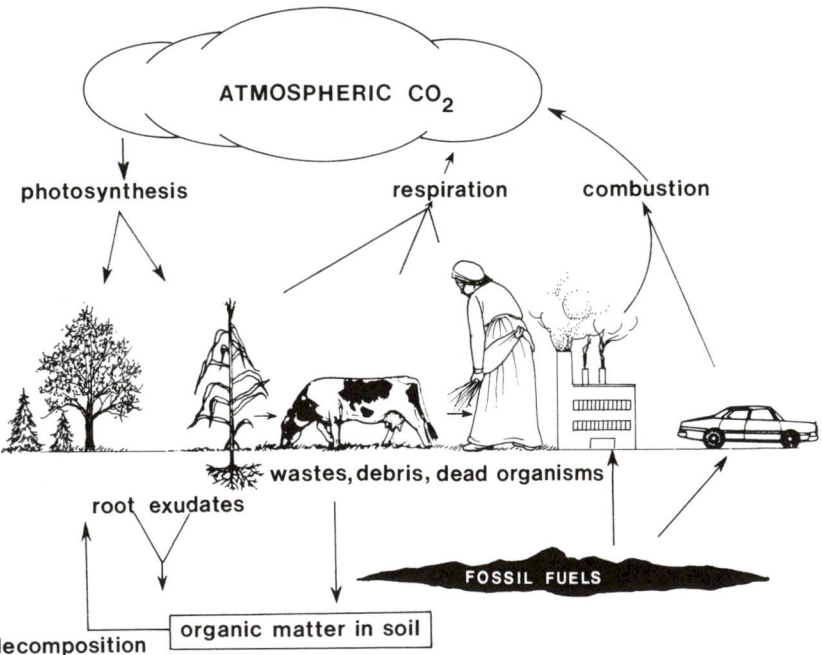

Fig. 8-3. The carbon cycle. Photosynthesis "fixes" inorganic carbon into organic forms. Respiration and combustion return organic carbon to its inorganic form. The carbon cycle becomes unbalanced if one of the factors is disturbed.

The sun's energy is trapped in organic molecules by chloroplasts during photosynthesis. Plants obtain energy from these organic molecules through the process of respiration that takes place in other specialized organelles, the **mitochondria**. Carbon dioxide is released during respiration. When humans or other animals eat plants, the complex compounds are broken down during digestion into simpler compounds. Then, through respiration, humans and animals are able to use the energy of the chemical bonds to run the metabolic processes necessary for their existence, again accompanied by the release of CO_2. Upon death, all types of organisms decay, primarily due to bacteria and fungi, which use the organic molecules of the decaying organisms as a carbon (energy) source. All organic molecules are eventually degraded to CO_2 and H_2O as part of the continuous carbon cycle.

The carbon cycle demonstrates our total dependence on photosynthesis for providing energy in the form of organic molecules. It also demonstrates the eventual return of carbon atoms in organic matter to carbon in the form of CO_2 through the decay processes and respiration. Human beings have become an important component in the carbon cycle recently, due to our rapid degradation of many organic molecules through the burning of **fossil fuels** (whose origin is biological) and other industrial processes. We have also destroyed large areas of photosynthetic activity through **deforestation** for agriculture, fuel, and lumber. As a result, the carbon cycle, which for millions of years seems to have maintained a balance between the carbon "fixed" in organic molecules and the CO_2 in the atmosphere and dissolved in the oceans, is now unbalanced. The level of CO_2 in the atmosphere is slowly rising.

Scientists vary in their opinions about the effect this will have. Some believe that the increased concentration of CO_2 will increase photosynthesis and plant productivity. More are concerned that increased CO_2 will trap heat from the sun and increase atmospheric temperatures. This phenomenon is called the **greenhouse effect** because it is similar to the trapping of heat when sunlight passes through glass. Recent evidence suggests that the temperature of the earth's atmosphere has increased over recent years, which may have major ecological effects, among them effects on temperature zones for agricultural production and on ocean levels (as polar caps melt). It is not yet clear whether the apparent warming is due to the increased levels of atmospheric CO_2 or is part of a natural cycle that also results in periodic ice ages during cooler periods.

Agricultural Soils

Although the carbon cycle is more appropriately applied to the whole world, we can look at a small portion of the cycle in an agricultural field. Before agriculture came to be practiced, plants grew naturally so that organic matter was returned to the soil each year as plants decayed. The organic matter in the soil has several important functions: 1) retaining moisture, 2) being a source of certain elements essential for plant growth, and 3) absorbing and retaining certain chemicals and nutrients.

In agriculture, humans are interested in harvesting organic matter produced through photosynthesis in the form of crops, eliminating the contribution of this material to the organic portion of the soil. Many crops are grown in weed-free rows, leaving large portions of the soil uncovered by vegetation and reducing the amount of organic matter that may be contributed to the soil. Significant amounts of the topsoil are simply blown or washed away where vegetation is not present to hold the soil. At the same time, organic matter already present in the soil continues to be degraded by soil microorganisms. Through harvest, wind and water erosion, and the continuous degradation of the remaining organic material, the organic matter in the soil slowly declines.

For many centuries, humans have tried to maintain the level of organic matter in soil by adding manures, compost, and other organic material. Even so, many soils have not been successfully maintained and have been abandoned. Productive agricultural soils are not a permanent feature of the landscape but a fragile resource that was created over millions of years and that has badly deteriorated in the brief time human beings have been busy with intense cultivation.

Many soils have very shallow profiles and very low levels of organic matter. Such soils deteriorate even more quickly and are much more vulnerable to destruction. These fragile soils are found in many tropical areas, particularly where rain forests have been cleared. They can be cultivated productively for only short periods. We will return to the very important problem of conservation and maintenance of arable lands in Chapter 14.

Plant Roots

Just as the aboveground parts of land plants are adapted to withstand the variabilities of the aerial environment, belowground parts of plants are generally well adapted to the soil environment. Environmental variations in soil moisture and temperature are usually more gradual and less extreme than weather aboveground, but the range is still considerable. In addition, plant parts that exist in the soil must withstand physical stresses from rocks and smaller mineral fragments. They also exist in an intimate relationship with animals, plants, and microorganisms of the soil, with each organism influencing and being influenced by the others.

All vascular plants produce roots that provide anchorage and support for the aerial plant parts. Roots are responsible for absorption and conduction of minerals and water from the soil. Roots also serve a storage function in many plants, in some cases resulting in large, starchy structures that we harvest and eat, such as carrots and sweet potatoes. At seed germination, the primary root of a vascular plant must develop quickly to begin absorbing water for the growing seedling. Even though many seeds contain enough stored food to maintain the seedling until it can begin to photosynthesize, moisture is needed immediately to keep the young tissues from drying out and also to provide the water molecules used in converting complex stored foods such as starch into smaller molecules for respiration in the growing seedling.

Flowering plants, the angiosperms, are divided into two large groups based on, among other features, the morphology of the seedling. If the seedling possesses one cotyledon (seedling leaf), as do grasses, corn, lilies, and onions, they are called **monocotyledons**. "Monocots" usually have leaves with parallel veins. Monocot seedlings have a short-lived primary root that is quickly replaced with a fibrous mass of roots that arise from the base of the stem near the soil line. **Dicotyledons**, the second group of angiosperms, possess two cotyledons and usually have broad leaves with netlike veins and a taproot that continues to grow, producing many lateral branches. Many familiar plants including maple trees, chrysanthemums, peanuts, and carrots are "dicots."

Although roots vary tremendously in structure and anatomy, the roots of vascular plants share several important features for efficient absorption and protection from the soil environment. Roots continually grow through the soil from the tip by means of an **apical meristem**, an area of undifferentiated cells. These cells show no specialization for a particular function but serve as the source of new cells for the growing root. (Plants also have an apical meristem at the apex of aerial plant parts, as discussed previously.) The apical meristem of the root is protected by the **root cap**, a small group of cells that continually slough away as the root pushes past mineral fragments and other obstructions in the soil. Just behind the apical meristem, the cells continually divide to produce new cells. These cells contribute to the **zone of elongation**, the area in which elongation of cells produces all increase in root length.

The new cells differentiate, becoming specialized for their various

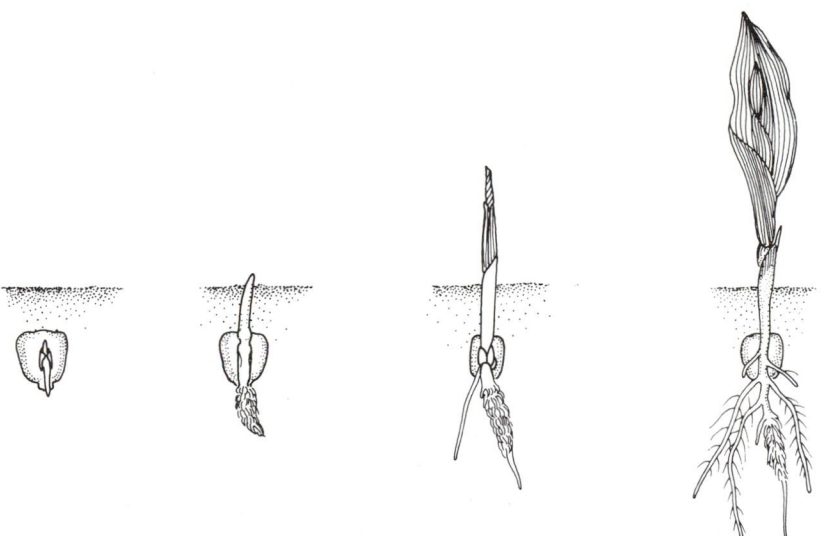

Fig. 8-4. Germination of a monocotyledon (corn). The seedling possesses one cotyledon and adventitious roots. The leaves have parallel veins.

functions. The outermost layer of the root is composed of **epidermal cells**. Many epidermal cells have protrusions called **root hairs** that are very small but greatly increase the absorptive surface area of roots. The **cortex** parenchyma cells just inside the epidermis store nutrients. Water is easily

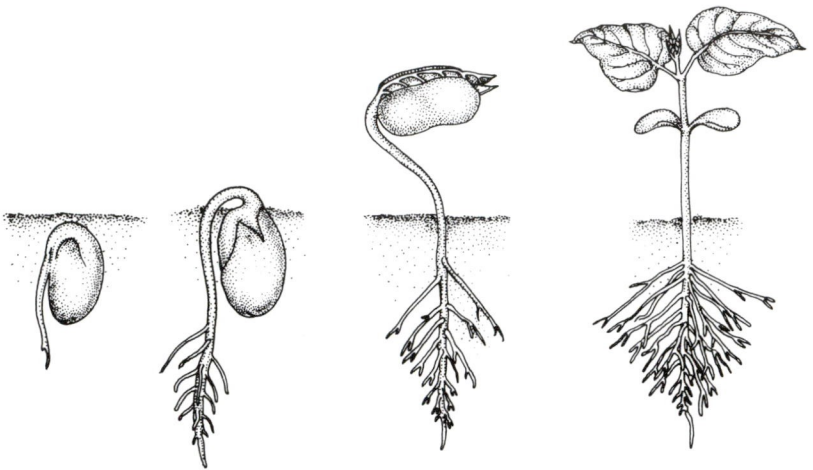

Fig. 8-5. Germination of a dicotyledon (bean). The seedling possesses two cotyledons (seed leaves) and a taproot that develops lateral branches. The leaves have netlike veins.

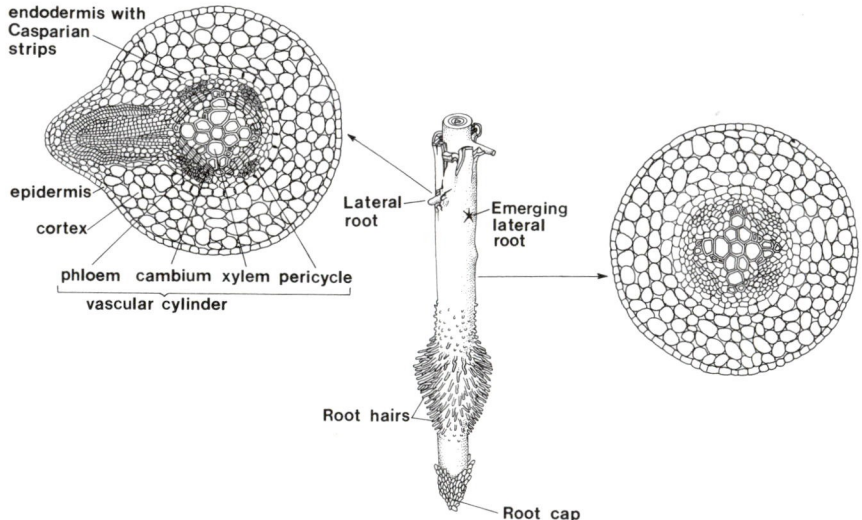

Fig. 8-6. Portion of a dicot root, with cross sections of the main root (right) and of an area where a lateral root is emerging (left). Note that lateral roots emerge through broken cells of the cortex. The growing tip is protected by a root cap. The surface area for water and mineral absorption is increased by root hairs.

absorbed, apparently moving mostly between the epidermal and cortical cells until the **endodermis** is reached. Absorbed minerals move into the cytoplasm of the epidermal cells and then pass through the cortical cells via the cytoplasmic channels (the **plasmodesmata**) that connect the cells.

The endodermis serves as a barrier to increase control of water and mineral absorption. Impervious **Casparian strips** exist between the cells so that all materials absorbed by the roots must pass through the cytoplasm of the endodermal cells before entering the **vascular cylinder** in the center of the root. Water is absorbed by **osmosis**, a process in which water molecules move from areas of higher water concentration to areas of lower water concentration until an equilibrium is reached. Water moves almost continuously into the cytoplasm of the endodermal cells and then into the vascular cylinder called the **stele**. When plants are exposed to very salty water, osmosis no longer moves water into the plants, which explains why most plants must receive fresh water to grow properly. The movement of minerals is more complex and is discussed in Chapter 13.

The stele, inside the endodermis, is composed of several kinds of cells. The **pericycle** is composed of parenchyma cells, which play an important role in secondary growth, discussed in Chapter 11, and also serve as the origin of **lateral, or secondary, roots**. The **xylem** tissue, which conducts minerals and water upward, is in the center of the cylinder and may have a starlike shape, with a varying number of arms depending on the plant species. **Phloem** tissues, found between the arms of xylem, bring organic compounds produced during photosynthesis down from the plant shoot; these compounds are used in root respiration and are stored as excess food. As the root continues to grow, lateral roots arise in the pericycle and exit by breaking through the cortex and epidermis. As they mature, a vascular connection is made between both the xylem and phloem tissues of the lateral root and the primary root.

Water and Mineral Absorption

All absorption of minerals and water takes place in the relatively short-lived and fragile root segment containing recently differentiated tissue. Root hairs are destroyed quite quickly, and soon secondary tissues begin to develop. For continued absorption, plant roots must continue to grow. Root development must be very extensive below a large plant because a balance is necessary between the photosynthetic area aboveground and the absorptive area belowground. Because the absorptive areas of plant roots are particularly fragile, they are easily lost during transplanting. Unless care is taken to minimize disturbance of the soil around the root system, plants may suffer "transplant shock" and grow slowly for some time after transplanting.

Plants require much more water than animals because they are continuously losing water through evaporation from aboveground plant parts during transpiration. Moist cells must be exposed to the air to absorb CO_2 for photosynthesis. This results in tremendous water loss for the plant. Plants control water loss with the protective waxy cuticle, releasing most water through the stomata that open and close by means of guard cells. Although this gives the plant some control, stomata must be open for extended periods to absorb enough CO_2 for photosynthesis. Thus, unlike animals who need to drink only periodically, plants continuously absorb and transpire water, losing over 90% of the absorbed water to evaporation in order to take in enough CO_2 for photosynthesis.

Roots require many of the same factors for growth and life as aerial plant parts. Soil moisture is absorbed continuously. This moisture is necessary, but if water fills the air spaces between soil particles, the root cells are deprived of oxygen. Where soils consist of heavy clay with poor drainage and small air spaces, crops are sometimes planted in raised beds to increase drainage and improve soil aeration. Plants grown in poorly drained soils deficient in O_2 grow slowly or wilt and die because root development is reduced. The photosynthetic area aboveground must remain in balance with the absorptive capacity of the roots, so plants are often stunted and may exhibit symptoms of nutrient deficiency. House plants grown in pots without drainage holes are often killed because the air spaces in the soil fill with water; the roots are then deprived of oxygen and die.

Root Exudates and the Rhizosphere

Plant roots clearly need O_2, water, minerals, and suitable environmental factors such as temperature and pH for continual growth to support the

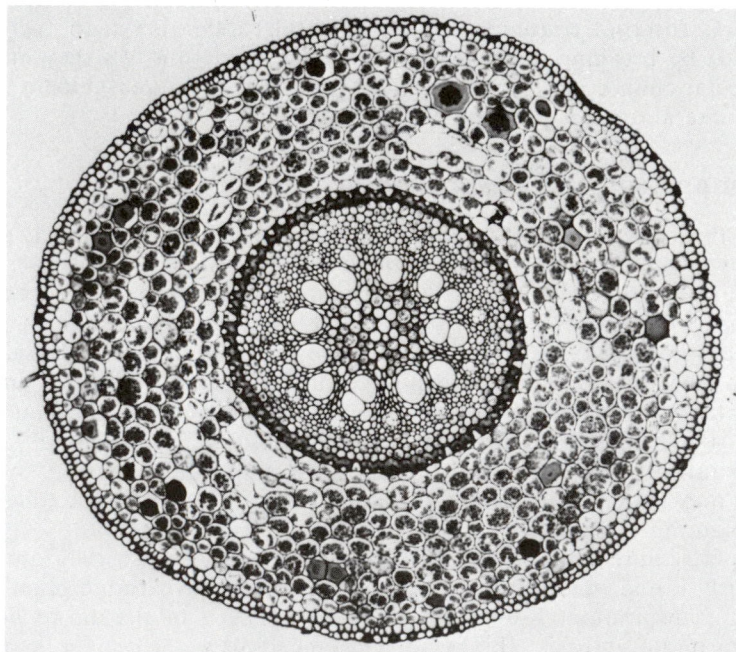

Fig. 8-7. Plant root (of the genus *Smilax*) in cross section.

aboveground shoot. The physical environment of the soil directly affects plant roots. However, the presence of plant roots also affects the soil environment. Roots are a food source for many herbivores and omnivores, including small mammals and insects, that chew and burrow into the moist vegetation. Dead roots are also an important food source for these animals, as well as for the decay microorganisms.

Many chemical compounds, called exudates, leak from plant roots during their growth. These **root exudates** include various nutrients such as amino acids, sugars, organic acids, and various growth factors. These may serve as nutrients, inhibitors, and stimulants for various soil organisms. Several important sources of these compounds occur on the growing root: 1) cells continuously sloughed off from the root cap, 2) nutrients that leak out of cells in the area of elongation behind the meristem, and 3) contents of cells that are torn open at the point of exit of lateral roots originating in the pericycle. These compounds can also be lost in considerable quantities from root wounds caused by insects, nematodes, cultivation, and rocks. Root exudates create a special environment in the thin "halo" surrounding the actively absorbing roots and extending out some distance from the root surface.

This halo, influenced strongly by the presence of root exudates, is called the **rhizosphere**. It plays a significant role in the activity of soil

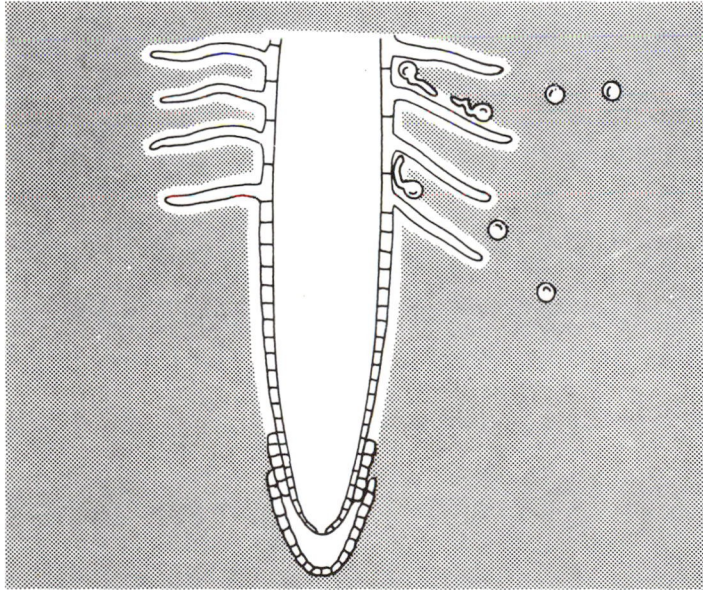

Fig. 8-8. The rhizosphere. Exudates from plant roots contribute to a "sphere of influence" near the surface of plant roots. Many fungal spores remain dormant until influenced by the nutrient-rich rhizosphere. This phenomenon is called "fungistasis."

microorganisms near root surfaces. Very different populations, activities, and species of organisms are found in the rhizosphere than exist in soil just outside the rhizosphere. Significant thermal, pH, and gaseous differences also exist. Plants affect the microenvironment of the soil by contributing roots, other underground plant parts, and dead aboveground plant parts to the soil. In addition, they contribute root exudates to the soil microenvironment.

Soil Microbes and Plant Pathogens

What organisms live with plant roots below the soil surface? We cannot begin to list them all, but some important groups include burrowing mammals (moles, ground squirrels, mice), numerous insects (some of which feed on plant roots), other arthropods (millipedes, centipedes, spiders, mites, and springtails), mollusks (snails, slugs), and other small invertebrates such as nematodes. Of course, the most numerous "neighbors" of plant roots are microorganisms, particularly fungi and bacteria.

Soil microorganisms are quite limited in their ability to move significant distances in the soil due to their small size. Thus, most soil microbes essentially play a waiting game until some organic matter becomes available. The arrival of a growing root tip and its nutrient-rich rhizosphere causes a burst of biological activity. Intense competition springs up between the organisms. Those that reproduce quickly and perhaps produce inhibitory compounds against other organisms have an advantage. Certain organisms are better adapted for competition and survival in the rhizosphere and will predominate there. Because nutrient-rich environments may be available only for a short time, microbes that can exist for long periods in a dormant state are better able to survive until more organic matter becomes available.

Many soilborne pathogens also play a waiting game, trying to survive in the soil until a plant root arrives. In many cases, the pathogens are in a dormant state, especially fungi that commonly exist in the soil as thick-walled **spores** or **sclerotia**. Structures that are capable of renewed growth and development are called by the general term **propagule.** When a plant root grows in close proximity to dormant propagules, the propagules receive a stimulus to begin active metabolism and growth.

How does this occur? Many of these propagules exist in a state of **exogenous dormancy**. This term means that the spores are not dormant because of an internal control but because they require an outside nutrient source to overcome dormancy. This exogenous dormancy is called **fungistasis**. Nutrients are present in the soil, but the highly competitive populations of saprophytic bacteria and fungi rapidly metabolize these nutrients. Thus, it is only in the nutrient-rich rhizosphere that sufficient nutrients are present to overcome fungistasis. Dormant propagules usually become active only when a potential host is present. In most cases, the response of the dormant propagules is nonspecific; that is, the propagules become active in the presence of nearly any plant root.

Some species of pathogens produce propagules that require a specific factor to stimulate growth and that can survive in a dormant state for long periods. This is a common feature of obligate parasites, which die if they become active when no host is present. All dormant propagules risk the possibility of ingestion, parasitization, or degradation by other soil organisms. Their populations decrease each year if no host plant roots become available for colonization and reproduction.

Many soilborne pathogens can exist also in a saprophytic state in the soil. Highly competitive soil organisms can maintain their populations over long periods of time and are termed **soil inhabitants**. Soil inhabitants may successfully exist in soil by one or several means: 1) rapid growth to take advantage of available nutrient sources, 2) production of various substances antagonistic to other organisms, and 3) production of survival structures during adverse conditions. Humans exploit some of the antagonistic substances for use as "antibiotics," most of which have been isolated from various soil microorganisms, primarily fungi and actinomycetes.

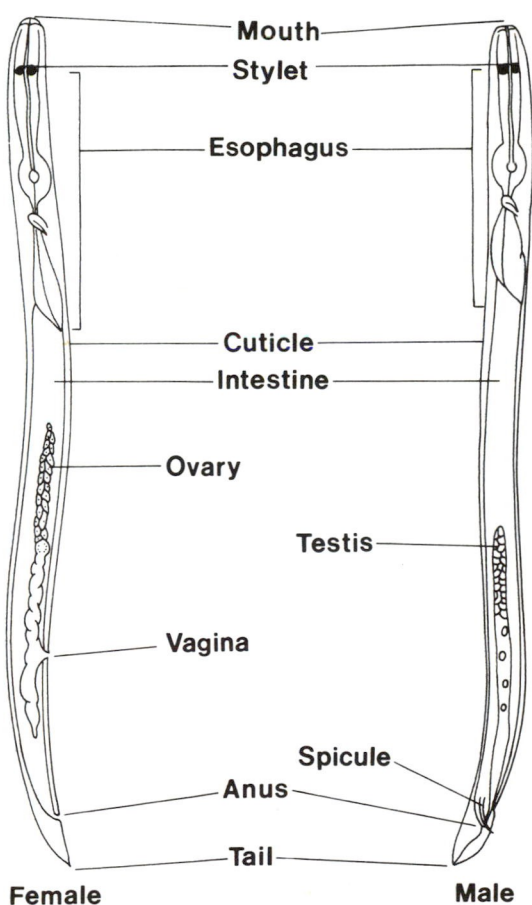

Fig. 8-9. Generalized nematode anatomy.

Parasites seem to sacrifice some of their competitive ability as they become specialized. We have already seen in Chapter 6 that pathogens are generally more temperature-sensitive than saprophytic organisms. Most pathogen populations decline quite dramatically in soils if no host plant is available for several years because they are far less competitive than the soil inhabitants. Most soilborne pathogens are categorized as **soil invaders**, which indicates their ability as facultative saprophytes and also their relatively short existence in the soil environment. Unfortunately, a few very important soilborne pathogens are very competitive facultative saprophytes and maintain their population in soil for long periods of time. *Rhizoctonia* and *Pythium* are two important pathogenic genera of fungi that are soil inhabitants.

Biological Control

An understanding of the rhizosphere and the biology of soilborne pathogens has important practical applications. Soilborne pathogens are well-equipped for survival in soil as saprophytes or as propagules subject to fungistasis. These adaptations allow the pathogens to survive despite the absence of a host plant. The presence of the rhizosphere most commonly results in renewed pathogen activity. In an agricultural situation, it would be most desirable to plant a crop having roots that would stimulate pathogen activity but that would not serve as a host. The roots would thus induce the propagules to expend their food reserves and starve. Pathogens that are facultative saprophytes also decline if no host is provided for several years, due to competition with resident soil microorganisms. Many soil pathogens parasitize botanically related plants, so common crop rotations recommend a planting cycle that varies the type of crop, such as potatoes followed by corn and a legume crop. If such rotations are followed, most pathogen populations from one type of crop are reduced by starvation or competition in subsequent crops.

Crop rotations are commonly assumed to be for the purpose of reducing excessive absorption of nutrients, which causes "tired" soils. However, soilborne pathogens commonly cause root rots and vascular wilts that result in aboveground symptoms such as stunting, yellowing, and wilting that can easily be confused with nutrient deficiencies. The misleading conclusion might be drawn that the soil was depleted of nutrients. When a crop is planted that is not susceptible to the predominant pathogens present, however, normal plant growth can be expected.

Crop rotation is our most effective means of preventing soilborne pathogen populations from reaching damaging levels. It is actually a means of biological control because it depends on the competitive soil organisms to reduce pathogen populations. Damage due to soilborne pathogens can be devastating, and pesticide applications to protect underground plant parts are expensive and often ineffective. Genetic resistance is limited by its specificity, and resistant cultivars are available for only a few potential soilborne pathogens.

In the past, soilborne pathogens have forced farmers to stop growing certain crops and have caused good farmlands to be abandoned. Crop rotation could make that unnecessary. However, in modern times, many forces reduce the ability of farmers to take advantage of crop rotation as a biological control for soilborne pathogens. Modern agriculture often requires expensive, specialized planting and harvesting equipment that must be used frequently to be economically justified. Also, many crops vary greatly in their value, so farmers face difficult decisions about planting low-value crops in some years. Fallow years are now almost economically impossible.

Besides economic constraints, there are biological complexities. Crops vary in their susceptibility to pathogens, so rotation with one crop may result in a decline in one kind of pathogen but an increase in another. Weed species are often hosts to some important pathogens, so poor weed control can negate an otherwise appropriate rotation plan. Rotation of perennial crops, of course, is impossible. Soilborne pathogens of tree fruits, grapes, asparagus, and other perennial crops increase each year and can eventually make it necessary to replant with other kinds of plants. For high-value crops, such as apples, strawberries, and tobacco, soil fumigation is often necessary to reduce pathogen populations before replanting is possible.

For 100 years, plant pathologists have been isolating antagonists from the soil. These organisms appeared, in the laboratory, to be good candidates

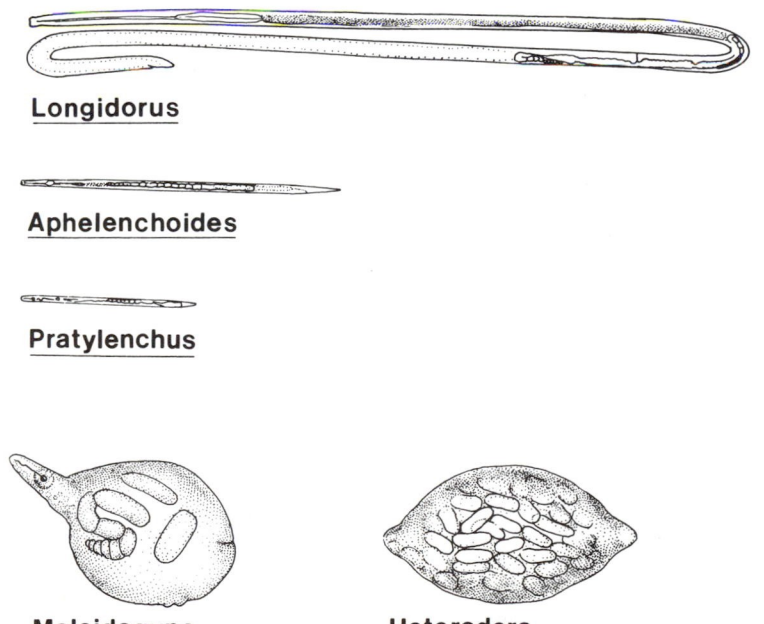

Longidorus

Aphelenchoides

Pratylenchus

Meloidogyne   **Heterodera**

Fig. 8-10. Morphology and relative size of some important plant-parasitic nematodes.

for biological control, but when they were added to soil, the results were disappointing. However, the study of the rhizosphere and its effects on soil organisms has led to a better understanding of how to effectively foster biological control in the rhizosphere to protect roots in the soil without massive inundations of competitive or antagonistic organisms. Bacteria and fungi recently isolated from the rhizosphere, called **rhizosphere-competent**, seem to hold more promise as agents for biological control.

Commercial rhizosphere-competent biological controls are now available as seed treatments to protect against soil pathogens. In many cases, they are not only as effective as chemical seed treatments but hold the possibility of extended protection because they are living organisms. As such, they may multiply in the rhizosphere and continue to protect the growing root surfaces. The bacteria *Bacillus subtilis* and *Pseudomonas* species and fungi in the genus *Trichoderma* are currently receiving the greatest attention as biological control agents.

The success of these biological controls confirms some important principles derived from rhizosphere research. It is not necessary to introduce huge quantities of antagonists into the soil when only the rhizosphere requires protection. Rhizosphere competence is critical to the success of the biological control agent. Also, organisms that compete with pathogens should be established on the seed or root surface from the beginning of planting.

It may seem wasteful and even dangerous for a plant to leak so much of its photosynthetic products into the soil environment and thus stimulate the activity of pathogens. Rhizosphere studies have begun to answer this puzzle. The nutrient-rich environment produced by plant roots allows microorganisms to mineralize organic matter so that important elements become more available to the plant. The active rhizosphere community also protects against the activities of root-attacking pathogens. Some rhizosphere bacteria seem not only to be protective but actually to be able to increase the yield and growth of plants. These **growth-promoting pseudomonads** appear to compete with pathogenic organisms for iron while not reducing its availability to the plant. This phenomenon indicates that pathogenic organisms have been reducing crop yields in the absence of obvious disease symptoms more than we had imagined. Yield increases of 20–30% and more have been obtained in several crops that were **bacterized** before planting.

Nematodes

Microbial populations can effectively protect against infection by bacteria and fungi, but they do little to reduce attack by one important pathogen group, the **nematodes**. Nematodes are tiny, translucent roundworms that are found in all soils and bodies of water. In fact, they represent the largest number of multicellular organisms in nature. Nematodes may be parasites of animals, humans, or plants or may feed on bacteria or organic matter in the soil. Well-known human parasites include "pinworms" and worms responsible for elephantiasis and trichinosis. Pets and farm animals are frequently "wormed" by medication to kill parasitic nematodes. Most nematodes are free-living aquatic organisms. In soils, they live in water in soil pores and the narrow layer of water on soil particles. Because they must remain in water, they move most easily through sandy soils where particles are bigger.

Nematodes have a simple life cycle: they begin as an egg and mature into an adult after four molts. In between, the nematodes exist as "juveniles" of increasing size. In most cases, nematode juveniles look essentially the same as but smaller than adults. Adults are sexually mature. Although **sexual reproduction** between males and females is common in nematodes, **hermaphrodites** (possessing both male and female sex organs) and **parthenogenic** (capable of reproducing without males) individuals are found in many species.

The nematodes that parasitize plants are **obligate parasites**. Second-stage juveniles hatch from eggs and move to host plants to feed. Some feed from the surface of plants (**ectoparasites**), while others penetrate the plant tissues (**endoparasites**). Plant-parasitic nematodes possess a sharp, hollow **stylet** that allows the nematode to pierce plant cells and ingest their contents.

Historically, the first nematodes discovered to damage plants were the **seed-gall** nematodes that parasitize many grains and grasses. Nematodes swim from the soil to the developing grain head in a layer of water on the plant surface and burrow into the developing flowers. At harvesttime, seed galls are harvested along with healthy grains. Until the life cycle of this nematode was understood, farmers planted the shrivelled, blackened, nematode-filled galls along with the healthy seeds, which returned the

Fig. 8-11. Foliar nematodes (*Aphelenchoides* species) swim in water to aboveground plant surfaces, damaging many ornamental plants such as chrysanthemum. Feeding damage produces necrosis, generally from the bottom of the plant up.

nematodes to the soil, where new plants would become infected. Proponents of the spontaneous generation theory used nematode-infected grains as evidence for their theory before the true life cycle was understood. Nematodes of the genus *Aphelenchoides* also swim in water on plant surfaces, feeding on and damaging foliar tissue and other aboveground plant parts of many ornamental plants such as chrysanthemums and begonias.

Two important kinds of plant nematodes never live in the soil at all. They parasitize trees and are carried to new host plants by hitching a ride on insect vectors. One of these nematodes causes an important tropical disease called **red ring of coconut**. Another, the **pinewood nematode**, infects and kills pine trees. Pines in the United States are not usually killed by the presence of these nematodes. Scotch (or Scots) pine, widely planted in the United States in landscapes and on Christmas tree farms, is highly susceptible to pine wilt disease. Infected trees often die rapidly. Dead trees can be replaced with Norway or blue spruce, Douglas fir, cedar, or hemlock, all of which are virtually immune to the disease. Japanese red and black pines are very susceptible, and huge pine forests in Japan have been killed since the introduction of this pathogen from the United States. Restrictions on the export of U.S. lumber to Europe and Asia because of the pine wood nematode have caused economic hardship to the lumber industry.

Most plant-parasitic nematodes remain in the soil and attack roots and other belowground plant parts. As previously mentioned, they may be ectoparasites or endoparasites. An important genus of endoparasites, *Pratylenchus*, the **lesion nematode**, penetrates and feeds on root cells, leaving behind dead and dying cells that become brown. The lesions are often invaded by other microorganisms. The dark streaks of decaying cells are

Fig. 8-12. Red ring disease of coconut, caused by the insect-vectored nematode *Rhadinaphelenchus cocophilus*.

a common symptom of these nematodes, which are commonly found on hundreds of host plant species.

Two important genera of plant-parasitic nematodes are endoparasitic

Fig. 8-13. Typical symptoms of nematode damage in the field include stunted plants, often in a somewhat circular pattern. The carrots in this Massachusetts field are heavily infected by the lesion nematode, *Pratylenchus penetrans*.

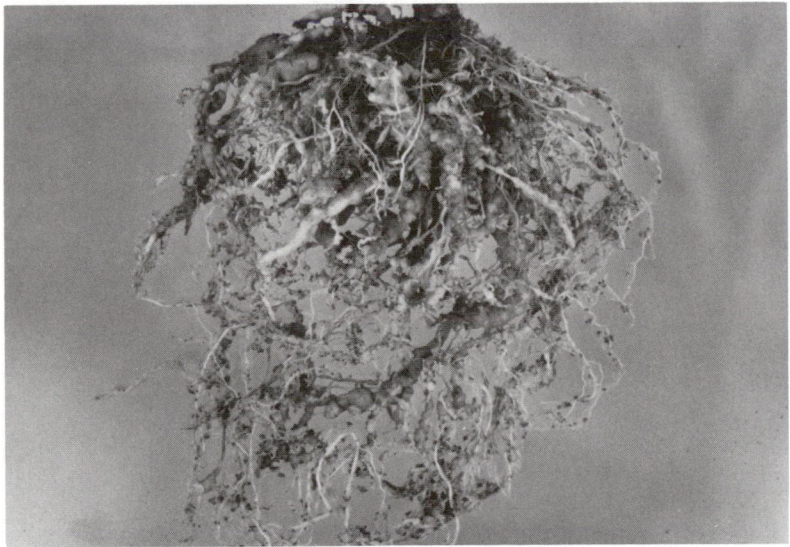

Fig. 8-14. Root galls on a tomato plant. Each gall contains at least one female root-knot nematode (*Meloidogyne* species).

but do not cause root necrosis. They feed in a sedentary manner, and the hormones in their saliva actually modify the roots to accommodate them in long-term feeding sites. Typically, multinucleate **giant cells** form near the vascular system. **Root-knot nematodes** cause the production of giant cells and small "knots," or galls, at feeding sites. Root-knot nematodes are in the genus *Meloidogyne*, the name reflecting the swollen appearance of the mature female at the time she lays a mass of several hundred eggs. Similarly, mature female **cyst nematodes** (of the genera *Heterodera* or *Globodera*) lose their wormlike shape and actually become a protective sac (or cyst) for their eggs.

One important biological difference between these two kinds of nematodes is their host range. Root-knot nematodes parasitize many species of plants, whereas cyst nematodes tend to be very host-specific. They are most damaging in sandy soils in warmer climates, where they reproduce more quickly. Crop rotation is generally not very effective, due to the wide host range of these nematodes, but genetic resistance has been reasonably successful and is available in some important host plants such as tomatoes. Genetic resistance has also been somewhat effective for cyst nematodes. Management by crop rotation is an important method to prevent cyst nematodes from reaching damaging levels. However, the eggs can survive for years in the protective cysts, even though a large percentage of them hatch in the first year.

Many genera of nematodes are represented in the ectoparasitic group that feeds on underground plant parts, particularly the roots. The importance of this group of plant parasites has been appreciated only in recent years, since relatively inexpensive soil fumigants became available following World War II. Ectoparasites often do not cause obvious damage to roots, but

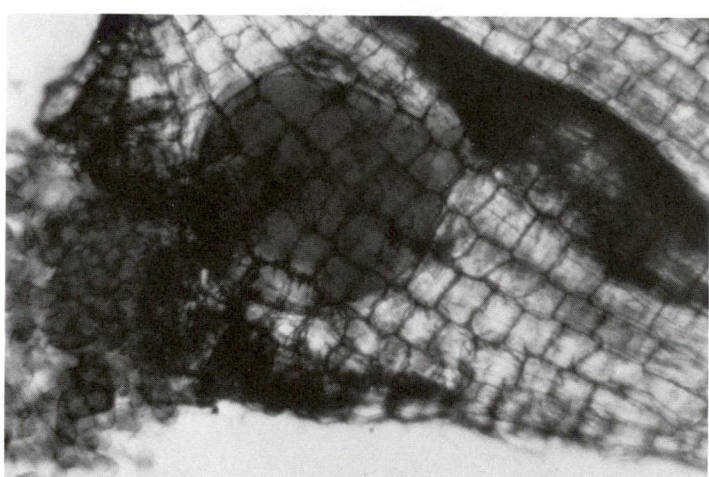

Fig. 8-15. Swollen female root-knot nematode in root gall. Note that her head is placed near the vascular tissue of the root for feeding. A mass of eggs in a gelatinous matrix has emerged from the gall.

their feeding can result in yield losses, reduced plant growth, and increased infection by other soilborne pathogens in wounds made by nematode feeding. As with endoparasitic nematodes, ectoparasites may have sedentary or migratory feeding habits. The sedentary types may cause root malformations and distortions. The migratory types may serve as virus vectors as they move from plant to plant. Genetic resistance to ectoparasitic nematodes has not been particularly successful, and crop rotations are difficult to recommend, again due to the wide host ranges of many species.

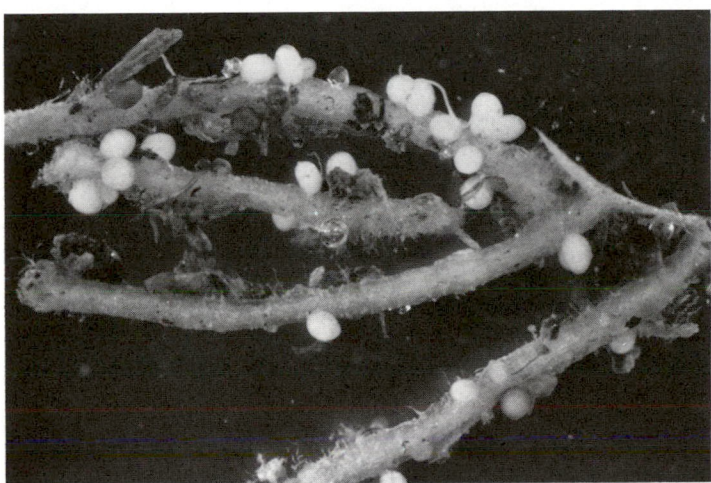

Fig. 8-16. Cyst nematode females (*Heterodera* and *Globodera* species) become swollen and eventually become protective containers, or cysts, for eggs.

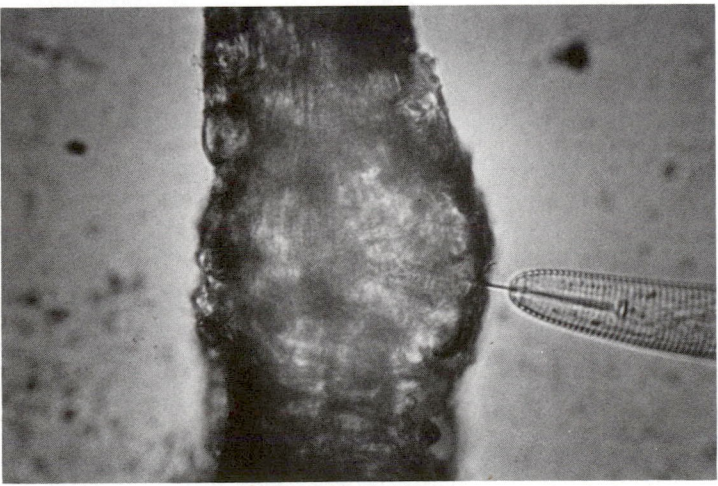

Fig. 8-17. An ectoparasitic nematode feeding on a plant root.

Nematicides and Soil Fumigants

Numerous potential biological control agents have been discovered for nematodes. These include predators and parasites of amazing variety. Some fascinating fungi parasitize nematodes, some by actually trapping them in hyphal nooses that close after a nematode swims in. Other fungi parasitize eggs and cysts. Unfortunately, commercial application of biological controls has not yet been very successful. Use of biological control is limited by its inability to compete with the tremendous killing potential of chemical controls.

Due to the biological variation among the many nematodes that parasitize a particular crop species, plant pathologists and nematologists (scientists who specialize in the study of nematodes) have searched for a more broad-spectrum means of nematode control. **Nematicides** and **soil fumigants** are two types of chemicals effective against all nematodes to rapidly reduce their population. Nematicides, mostly carbamates and organophosphates, are nerve poisons. Most are also insecticides and were first developed for that purpose. They were chosen for soil application because of their high toxicity and high water solubility. Aldicarb (trade name Temik) is a highly toxic nematicide/insecticide that has been found in groundwater in Florida after application to citrus, on Long Island after application to potatoes, and in other places as well. It is now a restricted-use pesticide, and its registration may be withdrawn.

Soil fumigants are volatile chemicals that are applied to soil to kill pests.

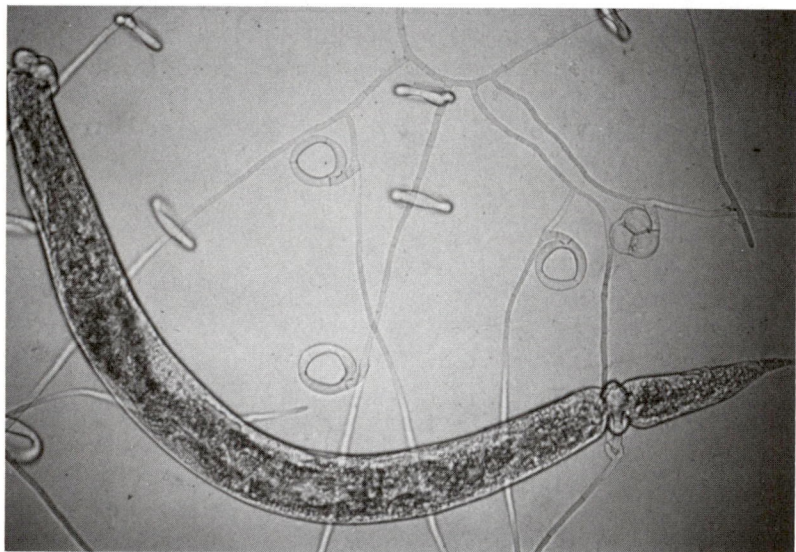

Fig. 8-18. Potential biological control for nematodes. This fungus, *Dactylaria brochopaga*, actually traps nematodes by constricting a hyphal noose. The nematode is then parasitized and killed by the parasite.

Many are **biocides**, which means that they are toxic to all living things, including weeds, crop plants, fungi, bacteria, insects, and nematodes. Because of their broad-spectrum activity, they must be applied before a crop is planted and be allowed to "clear" from the soil to avoid crop damage. Soil fumigation is expensive and time-consuming and is usually done to soils in which high-value crops will be planted. Soil fumigation is especially popular for perennial crops such as grapes, strawberries, and raspberries.

Unfortunately, a number of nematicides and soil fumigants have been found in groundwater. **Groundwater** is found in subsurface accumulations known as **aquifers**. Aquifer contamination by pesticides became an important concern after the first discovery of a nematicide in groundwater in 1979. Since then other pesticides, primarily herbicides, soil fumigants, and other nematicides, have also been discovered in groundwater, generally at low levels measured in parts per billion.

Since groundwater is the source of water for 50% of all U.S. citizens, contamination of this water presents problems. If pesticides and other impurities that threaten human health are found in groundwater, multimillion-dollar water treatment plants must be built to remediate the groundwater, or the site must be abandoned. In addition, piping systems must connect rural homes to a water supply, or individuals will have to resort to bottled water. However, if contamination is extensive, bottled "spring" water—so popular with people seeking an alternative to tap water— may not be safe either. Testing water for numerous organic compounds and their breakdown products is prohibitively expensive in many cases. Also, scientific data about the health effects of minute amounts of

Fig. 8-19. Fumigation of soil with methyl bromide requires that the area be sealed with plastic to prevent escape of the chemical.

contaminants in water is lacking. Recent advances in detection technology allow scientists to measure chemicals present at parts per billion, trillion, and even less. Many scientists do not believe that such low concentrations are biologically significant. Water may also contain very low levels of natural elements such as arsenic, lead, mercury, and other heavy metals. At this time, health standards for pesticide contamination are based more on the technology of detection than on proven health risks.

To understand how to best protect groundwater from further contamination, an accurate view of the problem is necessary. The pesticides that have been detected so far have mostly been chemicals applied to bare soil and/or injected into soil, especially in certain environments. These include areas where soils that are primarily sandy allow water to penetrate quickly, thus increasing the chances of carrying chemicals below the biologically active soil layer. Organic matter in the soil has tremendous binding power and holds pesticides tightly, but sandy soils have lower amounts of organic matter than other soils. When the water table is high, the distance between the soil surface and the aquifer level is short. Areas with shallow water tables, such as Florida, Long Island, and river valleys, have had the greatest contamination problems. The climate has effects, too, because abnormal amounts of rainfall increase the chances of chemicals being washed through the soil and into the groundwater.

We must also consider what happens to pesticides when they are applied to soil. Organic pesticides in soil are no different than other organic compounds. Like organic matter from dead plants, animals, and microbes, pesticides are degraded by microorganisms. How long it takes for a pesticide to be degraded depends primarily on its chemical structure. Some compounds are rapidly biodegradable; others take longer. Most pesticides in use today degrade quite rapidly and have a half life of only 30–60 days in the soil.

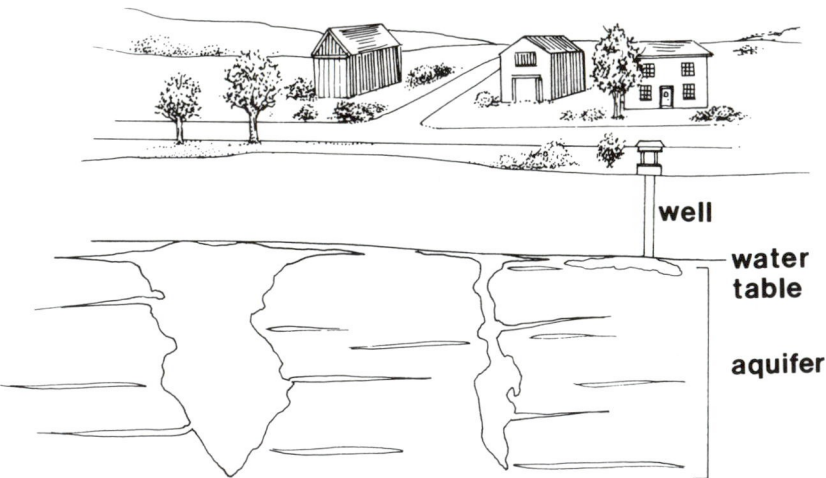

Fig. 8-20. Diagrammatic view of groundwater sources in aquifers.

Since agricultural pesticides are biodegradable, why are they not degraded before they reach the groundwater? Actually, groundwater contamination was a surprising phenomenon to the scientists who were familiar with the tremendous microbial activity in the soil. For many years, huge amounts of pesticides have been applied to agricultural soils without groundwater problems. For example, a single fungicide application to midwestern potato fields involves application of perhaps 500,000 pounds of active ingredient, and potatoes require two to 10 applications per season. Similar applications are made to other crops, yet these fungicides have not been found as groundwater contaminants.

Pesticides, like other organic components of the soil, are degraded before reaching the groundwater except in certain situations. Factors that increase the chances for groundwater contamination include the physical characteristics of the soil, the climate or irrigation patterns, chemical characteristics of the pesticide itself such as solubility and persistence, and the application rates and patterns. Nematicides and soil fumigants are highly water-soluble chemicals injected or applied directly to bare soil. Nematode damage to plants is worse on sandy soils, where groundwater is also most vulnerable. Despite the 1.2 billion tons of pesticides applied annually, only low levels of a few compounds have actually been detected and only in concentrations measured at parts per billion.

This should serve as a warning, however, that care must be taken to stop the contamination while it is still limited. We have the incredible luxury

Fig. 8-21. Newspaper advertisement from Long Island, New York, where groundwater contamination has caused problems for residents.

of a vast resource of clean water in the aquifers beneath us. Federal and state governments are acting to ban the use of certain highly soluble pesticides that are threats to groundwater. They are also restricting use of these pesticides on soils particularly vulnerable to leaching because of rainfall, soil type, and water table depth. Finally, considerable effort is being made to train pesticide applicators in techniques that will reduce the chances of contamination. Pesticide containers must be triple rinsed and, when not recyclable, disposed of properly. Waste and spill water must be contained and treated as hazardous waste if it cannot be reused. Wells receive special protection from back-siphoning and spills.

Pesticide professionals have become well educated in groundwater protection, but pesticide applications are not the only source of potential contaminants. Other agricultural chemicals such as fertilizers have been found at even higher concentrations in groundwater. Householders dump numerous toxic substances such as cleaning compounds, motor oil, gasoline, paints, varnishes, thinners, insecticides, and herbicides into drains that lead to septic tanks or sewage lines. Some simply pour wastes directly onto soils and into streams and ponds. Gas stations, farms, and many homes have underground fuel storage tanks that are subject to leaking. Industries

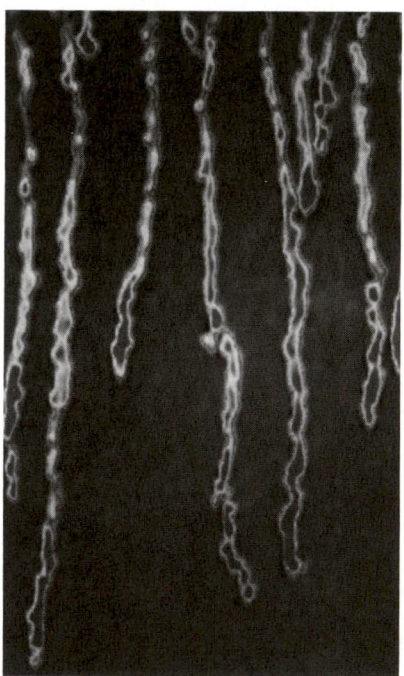

Fig. 8-22. The movement of compounds through soil is complex. Laboratory studies of "preferential paths" contribute to an understanding of such movement. Soil column studies of the light intensity of water moving through soil record and statistically analyze movement under varying conditions.

have deep-well injected wastes. Nuclear wastes are stored deep in the earth. The waste of human beings from sewage sludge, septic tanks, leaking sewage pipes, and landfills seeps into the earth and contaminates water that, until now, has remained clean for millions of years.

The reason groundwater is so clean is due to the incredible "scouring power" of microbial populations in the soil, particularly those associated with the rhizosphere of plant roots. Once contaminants reach the aquifers, they may remain for a long time because deep parts of the earth are cold and there are few decay bacteria to degrade the contaminants. Contamination occurs when we overload the soil ecosystem or send chemicals directly to the aquifers before they can be degraded. Contamination of groundwater reflects an ecological abuse by human beings. In the near future, it may be that no water can safely be pumped into homes untreated. The cost of protecting groundwater is high, but the cost of cleaning it is even higher. We all participate in this pollution, and we must all consider how we can reduce this ecological overload.

Fig. 8-23. Groundwater contamination comes from many sources. Technicians at a Superfund site are testing buried wastes for volatile organic contaminants.

Selected Readings

Bergdahl, D. R. 1983. Impact of pinewood nematode in North America. Journal of Nematology 20:260-265.

Bruehl, G. W. 1987. Soilborne Plant Pathogens. MacMillan, New York.

Chet, I., ed. 1987. Innovative Approaches to Plant Disease Control. Wiley-Interscience, New York.

Cook, R. J., and Baker, K. F. 1983. The Nature and Practice of Biological Control of Plant Pathogens. The American Phytopathological Society, St. Paul, MN.

Dropkin, V. H. 1989. Nematology, 2nd ed. John Wiley and Sons, New York.

Lockwood, J. L. 1986. Soilborne plant pathogens: Concepts and connections. Phytopathology 76:20-27.

Stone, A. 1977. Cyst nematodes—Most successful parasites. New Scientist 276:355-366.

Fungi in Food: Natural Poisons and Gourmet Delicacies

CHAPTER 9

Countryside scenes of people eating foods unadulterated by the contaminating influences of modern agriculture are a popular image for advertisers. We perceive these foods as safe because they have been sampled and eaten over the millennia by foraging humans seeking nourishment. Even so, everyone knows that some of these plants are poisonous and should be avoided, although the specific folklore about the edibility of natural plants is familiar to few people in industrial societies. Most of us simply choose from among the foods displayed in supermarkets and farm stands.

Even plants commonly eaten are not necessarily nontoxic. Leaves and stems of potato plants contain high concentrations of toxic alkaloids. Tubers left in sunlight turn green and accumulate these same alkaloids; therefore, green tubers should not be eaten and potatoes should be stored in the dark. Cassava, or manioc (*Manihot esculenta*), a staple food for more than 300 million people in tropical regions, contains high levels of cyanogenic glucosides. The starchy root of this plant must be ground and washed thoroughly before being eaten to avoid cyanide poisoning. Many natural components of plants that we commonly eat are toxic, and many are carcinogenic. Research by Dr. Bruce Ames at the University of California at Berkeley has demonstrated mutagenic compounds in familiar foods, including raw mushrooms and many vegetables.

Over the years, humans have learned which plants or parts of plants are safe to eat. Our bodies have also developed many detoxification mechanisms that allow us to eat some toxic compounds with little or no harm. Plant breeders who select plants with resistance to insects or pathogens test the resistant cultivars for increased levels of toxic compounds before releasing them for human consumption. Many compounds used by plants to resist these pests may cause toxic reactions in us as well. Such compounds could be present at high levels throughout the plant tissues, resulting in a health risk far greater than low-level pesticide residues on plant surfaces, where they can be washed away.

Ergotism and the Holy Fire

Once the edibility of a plant was established, humans began to select plants that were high yielding and suitable for agriculture. As a result, most people today eat a limited number of plants containing only a few major calorie sources. As this specialization developed, the potential health threat from any particular plant increased greatly because of the disproportionate role these few species play in our diet. One particular risk that is receiving renewed attention in the interpretation of historical records is associated with the cultivation of the cereal rye (*Secale cereale*).

In the Middle Ages, a frightening disease known as **holy fire** or **St. Anthony's fire** was common but unpredictable. Unlike most diseases, which were rampant in the overcrowded cities, holy fire was most common among the rural poor. It did not seem to be contagious and could strike one family or even members of a family without necessarily infecting neighbors. Children and feeble people were more susceptible to the disease. Nursing mothers might see the effects in their babies. A folk cure for these victims

Fig. 9-1. Saint Anthony. Redrawn from a woodcut made in Germany about 1215 A.D.

was to have the nursing mother eat white bread rather than the coarse and inexpensive rye bread.

This scourge of the Middle Ages has an unclear history. Medical diagnostics were more primitive then, and some of the described symptoms could be attributed to several different medical problems, making historical records difficult to interpret. The common symptoms included strange mental aberrations and hallucinations. People described the feeling of burning skin or insects crawling under their skin. Women frequently miscarried when the disease occurred; fertility was generally reduced during outbreaks. Severe cases resulted in gangrene infections due to constriction of blood vessels in the extremities; many victims lost hands and feet and became permanently crippled. Hospitals dedicated to St. Anthony took in such patients, caring for them until their painful and prolonged suffering ceased. In art work from that period, one can see the symbols of St. Anthony on medals and clothing worn by people hoping to ward off the effects of the disease.

Ancient European records of the disease before the Middle Ages are missing, and the Greeks and Romans do not seem to have described the malady. The disease became prominent following the introduction of rye as an agricultural grain after the invasion of eastern Europe by nomadic groups such as the Vandals. Disease outbreaks were sporadic in the Middle Ages, particularly in years in which cold winters put the rye under stress and cool, wet springs prolonged the flowering period.

France was the center of many severe epidemics, primarily because rye was grown as the staple crop of the poor, and the cool, wet climate was conducive to the development of the fungus that caused the mysterious malady. We now know that alkaloids produced by a fungus parasitic on rye were the source of the varied symptoms observed in human beings. Harvested along with the regular rye grains were hard, purplish black, grainlike structures (called **ergots** or **sclerotia**) produced by the fungus, *Claviceps purpurea*. When the grain and the ergots were ground together during milling, the flour became contaminated by the potent toxic alkaloids of the fungus.

It takes only a few ergots mixed in the flour to produce the frightening symptoms of the holy fire. Hundreds of thousands of humans as well as cattle and horses died or became seriously ill in the years before its cause was understood. Not until well into the 18th century was the direct relationship between the disease and the fungal parasite of rye widely recognized. Once the ergots were known to be the source of the misery, they could easily be separated from the grain before milling to prevent the poisoning, but accidental ergot poisonings, especially among domestic animals, have been recorded into the 20th century. Today, government standards prevent the sale of grain containing more than 0.3% ergot by weight for human consumption. A cultural preference for wheat bread over rye bread remains in France and among many people in other parts of Europe and in the United States.

As the rye grain matures, the ergots fall to the soil. These masses of mycelium with a dark rind are survival structures for the fungus similar to those produced by many other fungi. They help protect the fungus from desiccation and ensure its survival through the winter. In the spring, small mushroomlike structures are produced from the ergots on the soil surface. They forcibly eject the microscopic ascospores (sexual spores) of the fungus upward, to be picked up by wind currents in the field of rye. Rye, like many grasses, is wind pollinated. The **stigma** of the rye flower is large and featherlike to help trap the windborne pollen. This same mechanism helps trap the airborne ascospores expelled from reproductive structures of the fungus on the soil surface.

If water and temperature conditions are favorable, the ascospores germinate and infect the ovaries of the flowers. Like many fungi, *C. purpurea* then produces asexual conidia to increase the number of infections in the rye field. This must happen quickly because the flowers remain susceptible to infection for only a short time. To help spread the conidia, the fungus also secretes a sweet, sticky **honeydew** over the infected flower, which attracts flies and other insects. As they feed on the honeydew, conidia become attached to their feet and bodies. The insects move among the rye plants searching for more honeydew, effectively disseminating the conidia. Depending on the weather conditions, this honeydew stage may last for a few or many days.

Rather than producing a rye grain, infected flowers may be sterile, which directly reduces yield. Other infected grains are of reduced quality, while still others are totally replaced by the sclerotium, or ergot, of the fungus. Control of the disease is very successful if the stubble is plowed under after harvest to bury the remaining ergots and the rye crop is rotated to different fields.

In a recent book by historian M. K. Matossian, *Poisons of the Past: Molds, Epidemics, and History*, the author builds quite a strong case for the theory that ergot toxicity played a multifaceted role of in European

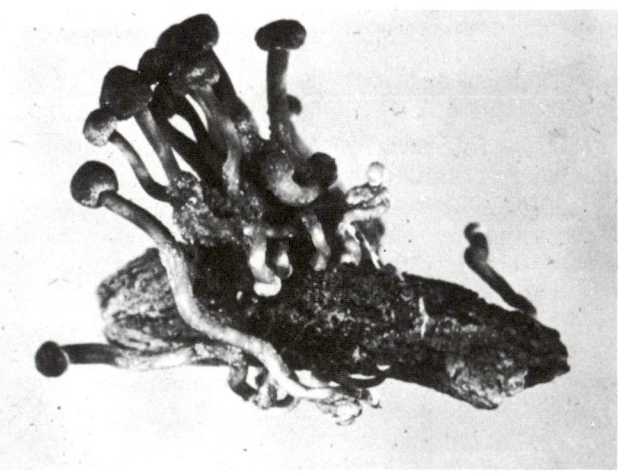

Fig. 9-2. Germinating sclerotium, or ergot, of *Claviceps purpurea*.

history. She challenges a common theory that the introduction of the potato from the New World was responsible for the European population explosion between 1750 and 1850. She presents convincing evidence that the population growth was directly related to the removal of toxic compounds (including ergot alkaloids and other mycotoxins discussed below) from the human diet when the potato replaced rye and other susceptible grains as the preferred food of poor people. Her population studies of individual countries also provide evidence for the role of ergotism in demographic depression, particularly in France, where rye remained the predominant food of the poor long after the potato had become more popular in other countries.

She also presents evidence that ergot poisoning had a role in the mental state of the French peasants, which contributed to the French Revolution in 1789, and in the accusations of witchcraft in Europe and the United States. The hallucinogen lysergic acid diethylamide (LSD) is derived from an alkaloid of ergot. Accusations of witchcraft have been shown to correlate with historical weather data, including cold, wet weather in winter and spring. These accusations are concentrated in areas of France, Central Europe, and the Rhine Valley where rye was a staple cereal.

Midwives have long used ergot to induce abortion and as an aid in childbirth, which explains the ergot's common name of "mothercorn." The powerful alkaloids can be purified and, with the dose carefully measured, used in modern medicine to reduce postpartum bleeding. Other modern pharmaceutical uses of ergot derivatives include the migraine headache medication cafergot and various compounds used for the treatment of mental

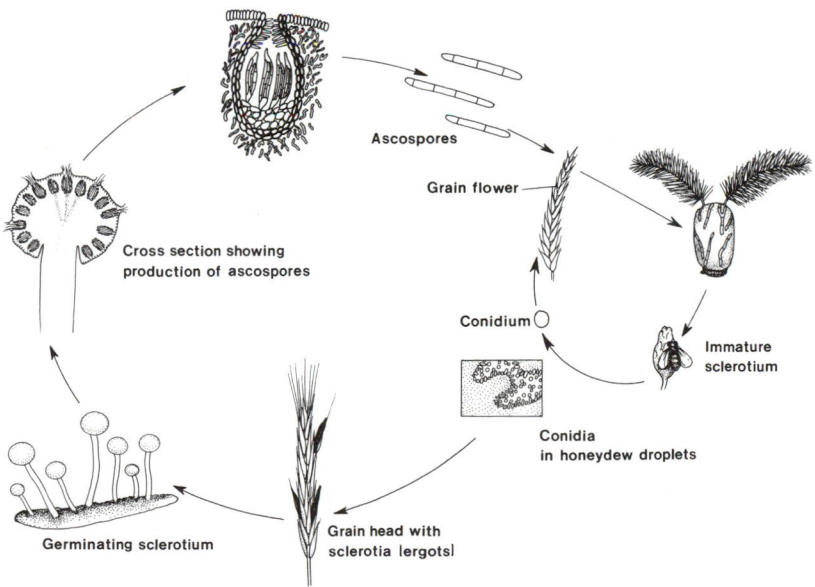

Fig. 9-3. Disease cycle of ergot of grains caused by *Claviceps purpurea*.

problems, blood pressure, and other ailments. Ergot infection of rye is deliberately induced in some areas for these uses.

G. L. Carefoot and E. R. Sprott, in *Famine on the Wind*, also describe the role of ergotism in history. For instance, Peter the Great was halted in his attempt to capture Constantinople (Istanbul) and gain access to warm-water ports when his troops and their horses were poisoned by ergoty grain at the mouth of the Volga River in 1722. Many serious outbreaks of ergotism have been recorded in Russia and probably would have continued if the New World potato had not replaced rye as the major food crop of the poor.

C. purpurea can infect other grain species in addition to rye, including the less susceptible wheat and many wild grasses. Careful harvesting and milling practices make ergot poisoning unlikely in modern times, although animal poisonings are occasionally reported because animal feed is not as carefully monitored. This is most common when animals, such as horses, feed on wild grasses in pastures following prolonged wet springs, when ergot infection is more severe. Since the 1970s, when male-sterile breeding

Fig. 9-4. Ergots in rye (left) and bromegrass (right).

lines of wheat and barley were developed, there has been concern about increased susceptibility of these lines to ergot infection because they have an extended flowering period, leaving the plants open to infection longer. Breeding lines are being carefully evaluated to avoid selection of new grain cultivars in which ergot might be an important problem. Other *Claviceps* species infect other plants, as well, causing problems in corn in Mexico and in sorghum and pearl millet, especially in tropical Africa.

Related fungi have been discovered in pasture grasses, where they have caused serious diseases of grazing animals. Investigations demonstrated that the hyphae of these fungi are found within the tissues of the infected plants. The fungi are **endophytes**; that is, they exist entirely within the host plant. The mycelium infects developing seeds, so new seeds contain the fungus in the next generation. Although this is a disadvantage in pasture grasses because of their toxic effects on animals, these same fungi have been found to deter insect feeding on turfgrasses (grasses used for lawns and golf courses). Endophyte-containing seed is now sold as a biological control against sod webworms and other turfgrass pests in perennial ryegrasses and fescues.

Decay Fungi in Foods

We find rotting, moldy food objectionable because the microorganisms responsible for the rot produce distasteful or even toxic compounds. An example of such a compound is ethyl alcohol, a product of anaerobic fermentation produced by yeasts, which are Ascomycetes, and bacteria. Yet, most human cultures have developed a taste for alcoholic beverages despite their negative health effects, and people deliberately create conditions conducive to enhanced production of alcohol in various products such as corn, rice, rye, potatoes, and grapes. We are not the only species to find the alcoholic

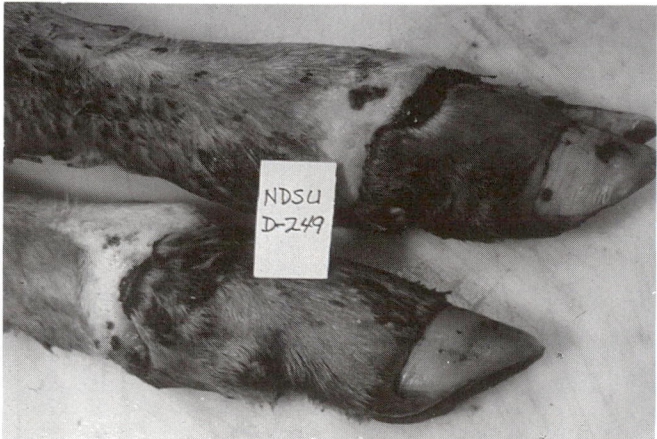

Fig. 9-5. Leg lesions of a calf affected by ergotism. Note the sharp demarcation between living (top) and dead (bottom) tissue.

content of foods attractive; some birds and small mammals ingest fermenting berries and other fruits to the point of intoxication. Even the word *intoxication* reflects the knowledge that the product is not healthful despite its deliberate ingestion.

Many fungi invade moist food products and contribute to their decay. Most do not produce toxic compounds. Even though the fungi make the food distasteful, it may still remain edible. An unappealing but edible example is a potato tuber rotted by *Phytophthora infestans*. Some of the rot fungi, such as the common secondary invaders of the genus *Alternaria*,

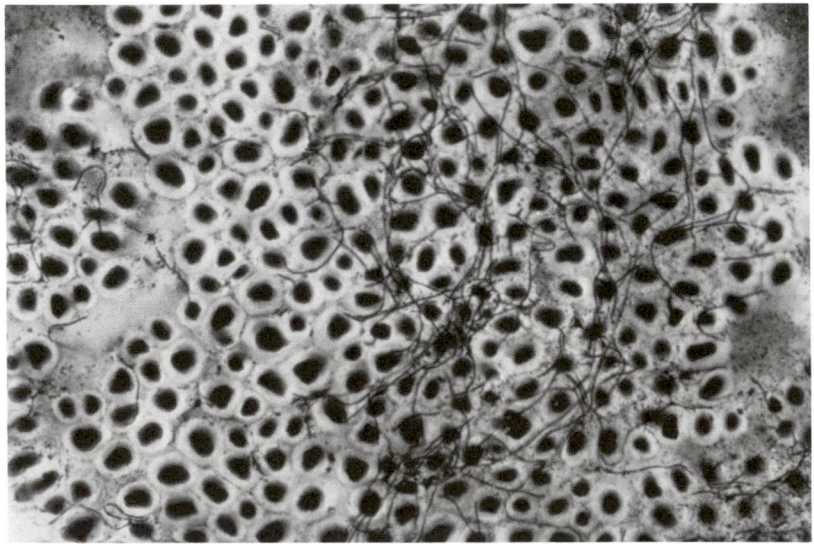

Fig. 9-6. Endophyte mycelium in grass seed. The fungus is related to the ergot fungi and causes toxic effects in animals that eat endophyte-infected plants.

Fig. 9-7. Endophyte-infected turfgrasses are used as a biological control against feeding by certain turfgrass insect pests. The perennial ryegrass cultivar Pennant (right) resists bill bug damage better than ryegrass cultivars without endophytes.

do produce toxic compounds. The most important **mycotoxins**, or poisons produced by fungi, are those that invade grains and seeds. Infections generally occur while the food products are in the field before harvest. Certain conditions increase the chance of invasion, including wounds in the seed coat and high seed moisture levels that allow more extensive invasion by the fungi. Seeds containing high amounts of oil are particularly vulnerable to infection and mycotoxin contamination. Examples include sunflower seed, peanuts, soybeans, cottonseed (which is commonly used as a cattle feed), walnuts, pecans, and coconuts. Grains are also very susceptible to mycotoxin contamination.

Grain Mycotoxins

Hogs, cows, and other animals often reject moldy grain and feed, which suggests that instincts have developed from previous unfortunate experiences. Fungi of the genera *Penicillium, Aspergillus, and Fusarium* produce toxic and carcinogenic compounds that threaten the health of animals and humans who ingest them in food. While animals commonly eat unprocessed grain or feed and refuse badly contaminated products, humans usually mill or otherwise process grain products until we can no longer detect the harmful compounds in our food. We must rely on careful inspection of grain and the rejection of contaminated grain by federal or state inspectors and food processors.

Aspergillus flavus produces several of the most potent mycotoxins, **aflatoxins**, named by combining the parts of the Latin binomial of the fungus. A closely related fungus, *A. parasiticus* also produces aflatoxins. These toxins are tasteless, odorless, and colorless. Environmental conditions affect the amount of aflatoxins produced, but even grain that shows no obvious infection by the fungi in the field or in storage may be contaminated by the poisons. Aflatoxins are some of the most powerful toxins and carcinogens known. Rats fed a diet containing one part per billion (ppb) of aflatoxin B_1 develop liver cancer. Aflatoxins are thought to be 100 times more likely to induce cancer than polychlorinated biphenyls (PCBs) are. They are also mutagenic, toxic to the liver, and immunosuppressive. Besides affecting the body's immune response to pathogens, aflatoxins in the diet also affect the success of vaccinations against other important pathogens.

The historian M. K. Matossian, who has studied the role of ergot and mycotoxins in history, suggests that the tremendously high mortality due to the Black Death, or bubonic plague, in 1348–1350 may have been due to the immunosuppressant effects of mycotoxins in moldy grains. These same mycotoxins may have killed rats, which fed on the grain as well. This increased the number of fleas looking for alternate sources of a blood meal and, in the process, transmitting the bubonic plague bacteria to humans and to domestic animals, which also died in high numbers during that period.

Aflatoxins proliferate in peanuts and corn, particularly when the seeds are not well dried before storage, when water leaks into storage areas, or when storage insects feed on the seeds, making wounds that allow easy invasion by fungi. For instance, aflatoxin levels can increase from 200 to 2,000 ppb in just three days in high-moisture corn. Aflatoxins may remain even after the processing and baking of contaminated products. When dairy cows eat contaminated grain, aflatoxins are metabolized into another highly carcinogenic compound that appears in the milk. Because cottonseed is a significant feed source for dairy cows, the occurrence of aflatoxins in cottonseed is of concern. Treatment of cottonseed with ammonia greatly reduces the level of aflatoxin so that it can be fed to animals. *Aspergillus* species can also cause serious lung infections, known as aspergilloses, with symptoms similar to those of tuberculosis. Peanut growers in the south-

Table 9-1. Major Mycotoxins and Toxin-Producing Fungi from Corn, Cereals, Soybeans, Peanuts, and Other Products, and Some of Their Effects on Animals[a]

Toxin or Syndrome	Fungal Source	Feeds or Foods Affected	Possible Effects on Animals
Aspergillus toxins			
Aflatoxins (primarily B$_1$, B$_2$, G$_1$, and G$_2$). B$_2$, G$_2$, M$_1$, and M$_2$ are metabolites; M$_1$ and M$_2$ are important regional contaminants in milk.	*Aspergillus flavus* and *A. parasiticus*	Cereal grains, peanuts, and other foods	Liver injury; carcinogenic; reduced growth rate; hemorrhagic enteritis; suppression of natural immunity to infection; decreased production of meat, milk, and eggs
Ochratoxins (nephrotoxins)	*A. ochraceus* and *Penicillium viridicatum*	Cereal grains	Toxic to kidneys and liver; abortion; poor feed conversion, reduced growth rate, general unthriftiness; reduced immunity to infection
Sterigmatocystin	*A. nidulans* and *A. versicolor*	Cereal grains	Toxemia; carcinogenic
Tremorgenic toxin	*A. flavus, P. cyclopium,* and *P. palitans*	Cereal grains, soybeans, peanuts, and other foods	Tremors and convulsions
Fusarium toxins			
Zearalenone (estrogenic syndrome)	*Fusarium graminearum* (sexual state *Gibberella zeae*), *F. tricinctum,* and in a minor way, *F. moniliforme*	Cereal grains, corn	Hyperestrogenism, infertility, stunting, and even death
Emetic or feed refusal factor (deoxynivalenol [DON] or vomitoxin)	*F. graminearum* (sexual state *G. zeae*)	Cereal grains, corn	Food refusal by swine, cats, dogs; reduction in weight gain
Other trichothecenes (T-2, monoacetoxyscirpenol [MAS],	*F. tricinctum,* some strains of *F. graminearum, F. equiseti, F. lateritium, F.*	Cereal grains, corn	Severe inflammation of gastrointestinal tract and possible hemorrhage; edema; vomit-

(continued on next page)

Table 9-1 continued

Toxin or Syndrome	Fungal Source	Feeds or Foods Affected	Possible Effects on Animals
diacetoxyscirpenol [DAS])	*poae*, and *F. sporo-trichioides*		ing and diarrhea; infertility; degeneration of bone marrow; death; reduced weight gain; slow growth; sterility
Fumonisins	*F. moniliforme* (possibly other *Fusarium* species, including *F. proliferatum*)	Corn	Fumonisin B_1 causes leucoencephalomalacia in horses ("blind staggers")
Penicillium toxins (primarily luteoskyrin)	*P. islandicum*	Rice	Liver injury; carcinogenic
Patulin	*P. urticae, P. expansum, P. claviforme*, and *A. clavatus*	Cereal grains, apple products	Hemorrhages of lung and brain; edema; toxic to kidneys; possibly carcinogenic
Rubratoxin	*P. rubrum*		Liver damage and hemorrhage
Citrinin	*P. citrinum*	Grains	Kidney damage

[a]Source: Shurtleff, M. C., Kirby, H. W., and Eastburn, D. M. 1990. Mycotoxins and Mycotoxicoses. Report on Plant Diseases 1105. University of Illinois, Urbana. Used with permission.

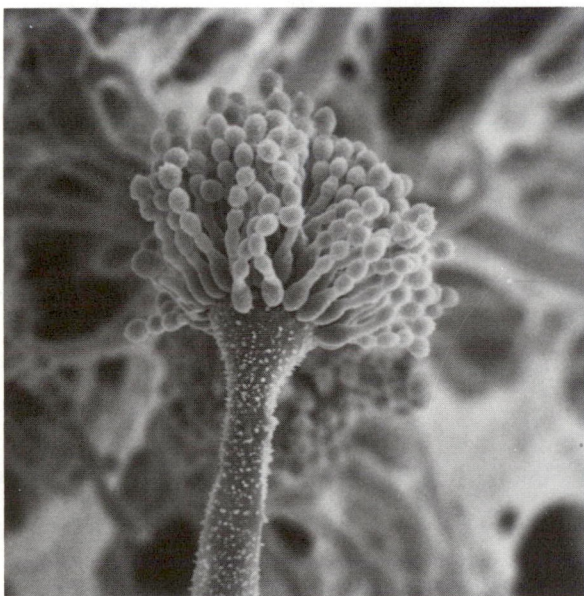

Fig. 9-8. *Aspergillus flavus* (×1,000).

eastern United States have strict aflatoxin testing standards because of the continual risk of contamination of peanuts in the humid Southeast.

Knowledge of the health effects of aflatoxins and other grain mycotoxins is relatively recent. The Food and Drug Administration set the first "action level" (permissible level) for aflatoxins in food products in 1965. Federal regulations on interstate shipment of grain in foods such as cornmeal and peanut butter set the legal limit at 20 ppb. The same threshold is used for feed for dairy cows, but higher levels of aflatoxins may be present in feed for animals raised for meat. The recent development of immunoassays

Fig. 9-9. Sclerotia of *Aspergillus flavus*. **a,** Germinating in soil; **b,** producing conidia.

for rapid detection of aflatoxin in food products has improved our ability to screen food products for safe aflatoxin levels. However, there is some concern about the 20-ppb level. Some scientists believe that the limit should be more restrictive. The Soviet Union has a 5-ppb standard, and Japan is considering a 0-ppb limit. Many countries, including the Soviet Union and some countries in Europe, have rejected U.S. grain that does not meet their stricter aflatoxin standards. Controversy also exists about the statistical methods used for sampling large quantities of grain accurately, since pockets containing high levels of aflatoxins are common in large storages. The

Fig. 9-10. Moldy corn. **Top,** in storage; **bottom,** in ears taken from a field where hail damage occurred.

government sometimes allows grain that is excessively high in aflatoxins to be diluted with other grain to reach the 20-ppb threshold.

The health effects of aflatoxins are well known. They were first dramatically demonstrated in Great Britain in 1960 when 100,000 turkeys and many other birds died of "Turkey X Disease" after eating a shipment of peanut meal from Brazil. The primary area attacked is the liver, and ingestion of high levels of aflatoxins has been demonstrated to cause liver cancer. Further evidence comes from epidemiological studies of high liver cancer rates in Africa, where large amounts of peanuts are eaten, and Asia, where large amounts of soybeans are eaten (although these same areas also have high rates of Hepatitis B virus infection, which is also correlated with the incidence of liver cancer). Aflatoxin contamination of soybeans is a major health concern in Taiwan. It is interesting that the Delaney Clause, which was added to the Food, Drug, and Cosmetic Act to protect food supplies from carcinogenic food additives, does not apply to aflatoxins, which are some of the most potent natural or synthetic carcinogens known.

High aflatoxin levels in crops occur periodically. In 1977, aflatoxin contamination of more than 60% of the corn grown in the southeastern United States resulted in losses of $200 million. The severe drought in the midwestern United States in 1988 resulted in aflatoxin contamination of 5–25% of the corn crop. One third of the 1988 corn crop tested in Iowa and Illinois contained dangerous levels of aflatoxins. Since aflatoxin is not commonly a problem in midwestern corn, milk had to be dumped in more than five states when the contaminated grain was discovered. Since the average American consumes about 160 pounds of corn and corn products each year, this represents a significant public health concern. Increased rates of abortions in hogs in 1989 have also been linked to aflatoxins in feed.

The midwestern corn crop became highly contaminated during the recent drought because of poor husk coverage of the tips of the ears, ear feeding

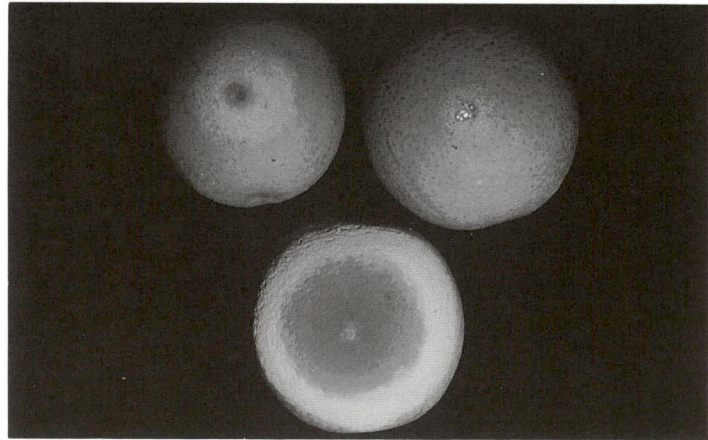

Fig. 9-11. Green mold on sweet oranges caused by *Penicillium* species.

by insect pests that introduced the *Aspergillus* conidia, and cracks in the corn kernels that increased infection by fungi. Speculators bought aflatoxin-contaminated corn and held it until later in the year when desperate cattle raisers would pay higher prices for the contaminated corn because good feed was no longer available.

Aspergillus species are not the only fungi that produce mycotoxins in grains and other foods. *Penicillium* species are familiar causes of mold and decay of leather, books, and other materials stored in basements and damp closets. *P. italicum* and *P. digitatum* cause the common blue and green molds on citrus fruits. A number of *Penicillium* species produce mycotoxins that damage the liver, lungs, brain, and kidneys; some are also carcinogens. Several species produce the mycotoxin patulin, which is often found in cider squeezed from damaged apples already invaded by these fungi. Some scientists have speculated that the level of patulin in apples might increase as growers, bowing to public pressure, stop using Alar, a

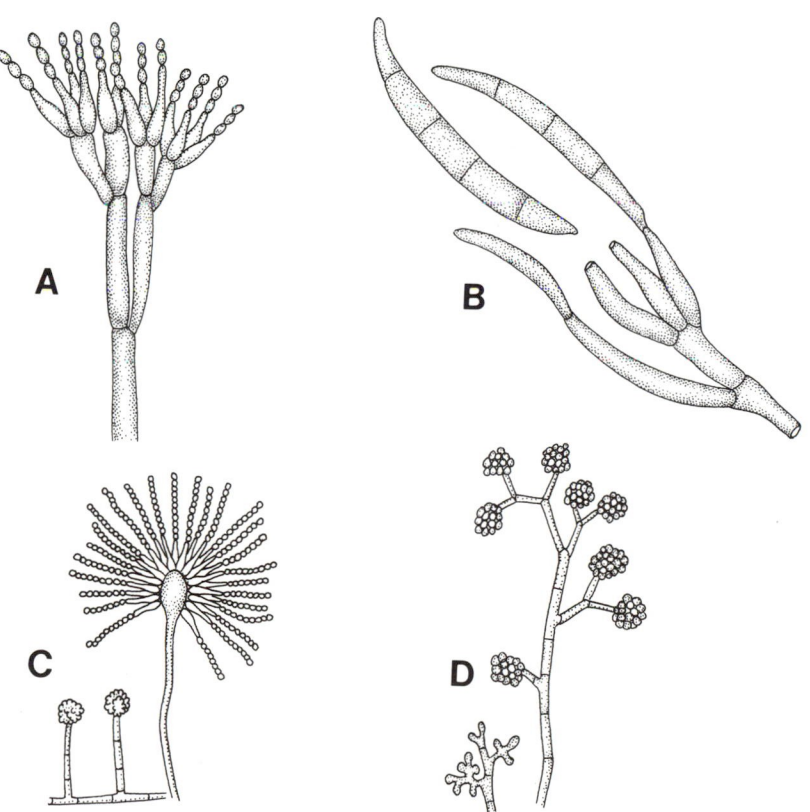

Fig. 9-12. Fungi important in postharvest losses: **A,** *Penicillium;* **B,** *Fusarium;* **C,** *Aspergillus;* **D,** *Botrytis.*

growth regulator that helps apples resist decay by increasing the natural hemicellulose levels in the fruit. *Penicillium* species also invade grains and contaminate them with mycotoxins.

A third important fungal genus, *Fusarium*, also commonly invades grains and may spread under conditions of high moisture in storage. A rule of thumb has generally been that aflatoxin is a southern problem and *Fusarium* toxins are a northern problem. *Fusarium* toxins, mostly **trichothecenes**, are common in corn and other grains in late harvests after wet summers. The mycotoxins produced by these pink to yellow fungi include "vomitoxin" and "refusal factor," both of which cause animals to refuse or avoid contaminated grain. **Zearalenone**, another *Fusarium* toxin, has estrogenic effects in animals that reduce fertility. Fumonisins in corn cause a disease of horses called "blind staggers." **T-2 toxin** was first identified as the cause of alimentary toxic aleukia in Russia in the early 1900s, but it and other fusarial toxins have probably been a health threat in moldy, poorly stored grains for centuries.

By the time we purchase processed grain products that may be contaminated by these mycotoxins, we can no longer detect their presence by the observation of mold, off-taste, or odor changes. We must trust farmers, food manufacturers, processors, and government inspectors to appropriately harvest, store, and test the foods before they are processed. Since many of the infections take place in the field before harvest, management strategies revolve around protection of the ripening grains and seeds. These include good growing conditions, sufficient field drying, and harvesting techniques that reduce injuries to the seed coat. Once harvested, grain must be kept cool, dry, and free of rodents and insects that cause wounds. Organic acids, such as proprionic acid, reduce the growth of fungi that produce mycotoxins, although the treated grain can be used only as animal feed. Most scientists agree that improved accuracy and speed in mycotoxin detection will result in improved healthfulness of the food we eat.

Edible Fungi

Humans eat fungi throughout the world. It is well known that some fungi, such as the "death cap" mushroom (*Amanita phalloides*), are deadly, whereas others, such as the common meadow mushroom (*Agaricus bisporus*), are cultivated for supermarkets by the ton. A number of fungi are deliberately added to food and food products to enhance their flavor with no apparently toxic effects.

The knowledge of the edibility, toxicity, and medicinal value of plants and other foods has slowly accumulated over centuries of trial and error by people throughout the world. Mistakes can be deadly, as when a group of newly arrived Asians died a painful death in California after eating mushrooms that appeared to be the same as some edible types from their homeland. Other fungi are toxic to some individuals only. Mushroom hunters who decide to try a new species, after careful identification by an expert, are cautioned to eat only a small amount at first. Some mushrooms, like the inky caps of the genus *Coprinus*, can cause gastric upsets when ingested with alcohol, such as wine, at a meal.

One very familiar fungus, *Penicillium roquefortii*, produces the blue-green color and distinctive flavor of blue cheese. It apparently originated as an accidental contaminant of cheese ripening in caves in France. The name *Roquefort cheese* is reserved for cheeses from that area of France, just as the name *champagne* can only be used in France for sparkling wines from the province of Champagne. The mycelium and conidia of the fungus colonize the cheese until the blue-green veins are visible throughout. Another species, *P. camemberti*, is used for the famous surface-ripened cheese, camembert. The genus *Penicillium* also contains a number of species that produce powerful antibiotics known as penicillin, useful for medicinal purposes. *P. notatum* and *P. chrysogenum* are the most prominent species in commercial antibiotic production.

Aspergillus is another common fungal genus that is responsible for mold and decay of food and stored materials but that also includes species used in food production. One species of the genus, *A. oryzae*, is added to rice and soybeans and allowed to ferment for the production of sake and soy sauce, respectively.

Gray mold is a common disease of small fruits, flowers, and the older leaves on many kinds of plants. The distinctive brownish gray mold is visible on infected plant materials in humid weather. The fungus is *Botrytis cinerea*, which produces prolific conidia for dissemination and sclerotia for survival in plant debris. This mold is commonly observed on strawberries, raspberries, and aging leaves and flowers of petunias, tulips, marigolds,

Fig. 9-13. Tulip "fire," or Botrytis blight.

and geraniums. It is also a common parasite of grapes. Under ideal conditions of cool night air, humid soils, and sunny, warm afternoons, infected grapes

Fig. 9-14. Strawberries with postharvest development of gray mold fruit rot caused by *Botrytis cinerea*.

Botrytised Select Harvest

JOHANNISBERG RIESLING
1986

Harvest Date: October 23, 1986

Harvest Sugar: 25° Brix

Residual Sugar: 10° Brix

Harvest Acid: 1.15 g/100 ml

Bottle Acid: 1.0 g/100 ml

This is a rare wine. It is sweet, tasting of apricots and honey, with a concentrated flavor developed by *Botrytis cinerea*. This wine is made from grapes that are picked individually after they have dried virtually to raisins. The result is an extreme concentration of flavor and a nectar-like intensity. This wine may be enjoyed now or cellared for up to 20 years.

Edward O'Keefe, Vintner

C H A T E A U

GRAND
TRAVERSE

Riesling

Botrytised Select Harvest

M I C H I G A N
Johannisberg Riesling
1986

Grown, Produced and Bottled by
Chateau Grand Traverse, Ltd., Traverse City, Michigan
Alcohol 12.2% by volume   Contains Sulfites

Fig. 9-15. Label for a wine produced from grapes infected by *Botrytis cinerea*, the "noble rot."

become sweeter as the natural sugars and tannins concentrate and the water content of the grapes decreases. Once infected by this "noble rot," such grapes are used to make special types of wines, particularly dessert wines (golden Sauternes) and other wines, including Riesling, Semillon, and Sauvignon Blanc.

The correct environmental conditions that produce infected grapes without the development of the destructive gray mold disease are found only in certain areas. The fungus is endemic in the vineyards and, therefore, develops naturally in areas conducive to its growth and spread. The infected grapes must be harvested slowly throughout the season as different grapes reach the proper stage for wine making. *Botrytis* gives a distinctive taste to the wines that bear the name of the fungus or the word *botryized* on their labels.

Smuts

Just as it is not possible to judge the danger of a compound based solely on its origin in nature or the laboratory, one cannot necessarily determine the toxicity of fungi and fungal by-products based on their appearance. Another group of fungi has the unappealing common name of "smut." The name originates with the black masses of dusty spores produced by mature smut fungi. These black spores are formed from the darkening and thickening of the individual cells of the mycelium of these Basidiomycetes. The smut fungi vary in their life cycles, but the black smut spores germinate at some point to produce spores that function in the production of a new dikaryotic mycelium capable of invading a host plant. Smuts are parasites of many kinds of plants, but some of the most serious smut diseases affect the cereals.

One type, loose smut, comprises several species that infect various host plants such as wheat, barley, oats, and rye. Harvested grains from infected plants appear healthy but are already invaded by the mycelium of the smut fungus and will result in a new grain head that bursts with black smut spores instead of healthy flowers, contaminating and infecting the flowers of neighboring healthy plants. Systemic fungicides capable of killing the fungus before the grain is planted have greatly reduced the incidence of loose smut in cereals.

Another important type of smut has several common names including bunt, covered smut, and stinking smut. These smuts are in the genus, *Tilletia*, named for M. Tillet, a French biologist who in 1755 demonstrated the contagious nature of the smut disease. He dusted the black spores onto healthy grain because he believed that a poisonous substance was associated with the dust. His work was a key step in the early study of fungi as plant parasites. Stinking smut has a different life cycle from the loose smuts in that the smut spores are released at harvest so that any newly harvested grain is contaminated with the spores. Stinking smut spores also survive in the soil and infect germinating grain seeds. In years of heavy stinking smut infection, dusty clouds of smut spores are released when infected grains are broken open during harvesting. The air can be so filled with spores that farmers have been killed when sparks from harvesting machinery or at grain elevators touched off explosions.

T. caries and *T. foetida* do not produce mycotoxins, but smutty grain has an unappealing "rotting fish" taste and smell. Flour made from smutty grain is also brownish due to the presence of the smut spores. It is said that gingerbread was invented to disguise flour made from smutty grain by covering the bad taste with spicy ginger flavoring and adding molasses to mask the brownish color. *T. controversa* causes dwarf bunt of wheat, which is most common in the Pacific Northwest. Because this disease is not yet present in the People's Republic of China, fear of dwarf bunt contamination has interfered with U.S. grain exports to that country.

A common smut familiar to home gardeners is corn smut. This fungus is more common in sweet corn than in field corn. It is recognized by the silvery white, translucent galls found nearly anywhere on the corn plant, including the tassels, ears, and leaf whorls. It usually damages the corn yield and quality only when it invades the ear. The galls produce huge quantities of black smut spores that overwinter in debris and infect corn produced the next year. Despite its unappealing look, this fungus is edible and highly prized in its earlier stages when the developing mycelium has colonized the gall tissue produced by the corn plant. It has been cultivated in Mexico for many centuries and has recently come to the attention of U.S. gourmets.

The fungus is called by the Latin binomial, *Ustilago zeae* (and commonly also *U. maydis*) and by the common Spanish names *huitlacoche* or *cuitlacoche*, which do not have appealing English translations. U.S. markets are using the common name, **smoky maize mushroom**. Farmers in Pennsylvania and some midwestern states are now cultivating corn smut for gourmet food markets, where the fungus sells for a retail price of about

Fig. 9-16. Release of "stinking smut" spores (*Tilletia caries* and *T. foetida*) at harvesttime. The spores contaminate harvested grain and soil where they land.

$50 per pound on the corn ear, which includes about 3.5 pounds of fungus. It is used like truffles, another type of edible fungus fancied by gourmets. Truffles are fruiting bodies of parasitic Ascomycetes produced on the roots of host trees. They are found by animals that detect their aroma and dig them up, thus dispersing the spores when the truffle is broken open. Humans who desire these fungi train dogs and pigs to find and dig up the highly prized truffles.

Food safety regulation by the federal government is popular with most citizens. Compounds in food that occur naturally in the original plant, animal, fish, or fungus or potential contaminants such as mycotoxins and parasites are regulated, as well as deliberate food additives such as preservatives, colors, flavors, and pesticide residues. The regulation of these potential health hazards is based on their biological activity as toxins, carcinogens, mutagens, and teratogens. The activities are the result of their

Fig. 9-17. Corn smut.

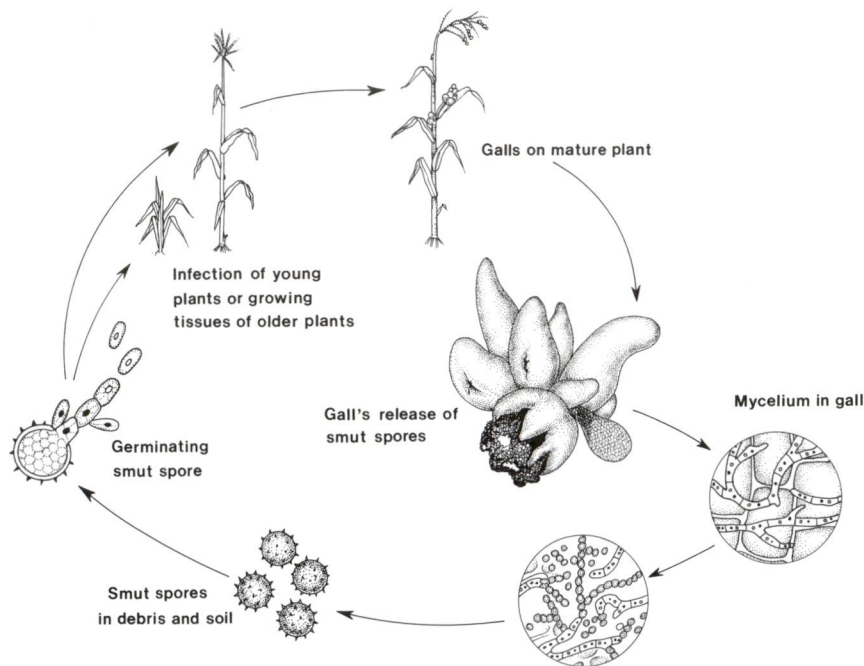

Fig. 9-18. Disease cycle of corn smut, *Ustilago zeae*.

Fig. 9-19. Portion of a mural painted by Guillermo Lourdes in the city of Durango in Mexico, showing the importance of corn in agriculture and in Mesoamerican culture. Young smut galls, which are used as food in parts of Mexico, can be seen in the background. From a painted copy by Rudy Cruz.

molecular structure, not their origin. Some "natural" compounds are distinctly more important health risks than many "synthetic" compounds. Risk assessment must include dose, frequency, and biological activity of the various compounds. This is why many scientists believe that much more study and screening should be done of grains and seeds potentially infected by mycotoxin-producing fungi than of the extremely low levels of pesticide residues found in most foods.

There is also concern that reduced use of some pesticides may actually increase health risks. Food not protected by fungicides is more likely to possess spots and blemishes that are vulnerable to infection by mycotoxin-producing storage fungi. Thus, "organic" produce may not necessarily be more healthy than commercially grown produce. Optimal storage conditions and protection against infection by storage fungi are requisites of a healthy food supply. In temperate climates, cool storage of food supplies is possible during the winter for free. Tropical countries, which have a year-round warm climate, face much higher losses of food to storage fungi and more danger from mycotoxin production. Dry and well-ventilated food storage structures are a high priority in the development of a healthy reserve food

Fig. 9-20. Woman hunting truffles with a pig trained to sniff out and dig up the underground fungi.

supply in some of the most populated parts of the world in Asia, Africa, and Central and South America.

Selected Readings

Anonymous. 1987. Aflatoxin: Everybody's poison. Science of Food and Agriculture (Jan):8-13.

Hoffman, J. A. 1982. Bunt of wheat. Plant Disease 66:979-986.

Lorenz, K. 1979. Ergot on cereal grains. CRC Critical Reviews in Food Science and Nutrition 11:311-354.

Marasas, W. F. O., and Nelson, P. E. 1987. Mycotoxicology. The Pennsylvania State University Press, University Park, PA.

Matossian, M. K. 1989. Poisons of the Past: Molds, Epidemics, and History. Yale University Press, New Haven, CT.

Meronuck, R. A. 1987. The significance of fungi in cereal grains. Plant Disease 71:287-291.

Rusts

CHAPTER 10

> No one can be a statesman who is entirely ignorant of the problems of wheat.
>
> Socrates

Farmers have battled rust fungi since agriculture began. Long before humans understood the role of parasitic fungi in plant diseases and certainly before anyone considered the activities of soilborne pathogens, people had studied and recorded the strange rust-colored eruptions on the leaves of diseased crops. The rust fungi, named for the color of their spores, were believed to be, like the fungus growth on blighted potatoes, an expression rather than the cause of the disease that shrivelled grains and reduced yields.

Robert Hooke, famous for his Cell Theory, spent considerable time studying minute things with the newly invented microscope. He described and drew molds on leather and also published the first illustrations of rust on a rose leaf in 1665. To Hooke, the rust appeared to be tiny plants similar to mosses. Over the next 200 years, many scientists studied the amazing structures produced by rust fungi and began to recognize the clues to their biology and their role as plant parasites.

Historical Records

While some species of rust fungus parasitizes nearly every kind of plant important to humans, the grain rusts are the ones described so often in history. Recent archeological digs in Israel discovered spores of wheat stem rust on grain from at least 3,000 years ago. Wheat (*Triticum aestivum*), an important crop in ancient times, is the most important crop in the modern world. Each year more wheat is grown than rice or corn. About $500 billion is lost to cereal rusts each year, and even more is lost in years with severe epidemics.

Wheat, barley (*Hordeum vulgare*), and oats (*Avena sativum*) have their centers of origin in Asia Minor, and their rusts originated there as well. As the first farmers began to select wheat grains and plant them in plots, the wheat rusts took advantage of the feast. Unlike the pathogen that causes late blight of potato, which can leave a potato crop a rotted mass of vegetation, rust fungi are more delicate parasites. They carefully obtain their nutrients without destroying the entire plant. Yet their presence takes its toll, reducing yield and weakening stems just before harvest. The grain from rust-infected plants is not toxic, however, so farmers had to accept the losses and harvest what was left. Each year they could do little more than pray for a good crop.

Rusts are described in many ancient writings that mention crops and agriculture. Despite some confusion in translation, the Bible seems to mention smuts, rusts, mildews, blights, and blasts. In some cases, the problem may have originated from hail or storms, but in other instances it is very clear that a fungus was observed on the plants, even though it was still considered the result rather than the cause of the plague.

The ancient Greek writings of Aristotle and his student Theophrastus also described rust diseases of crops. In particular, Theophrastus, who is often called the "father of botany," noted that plant species varied in their susceptibility to rust diseases. He also recorded the role of environmental effects such as moisture and air movement in the development of epidemics.

In Roman times, rust diseases continued to plague farmers. The rust-colored spores seen in the early stages and the black spores that appeared later in diseased wheat fields were all too familiar to Roman farmers. A pair of gods—Robigo (female) and Robigus (male)—was honored in the **Robigalia**, a religious ceremony, practiced for over 1,700 years, involving sacrifices of reddish colored animals, such as dogs or cows. The origin of these rituals lay in the belief that humans were punished with rust-

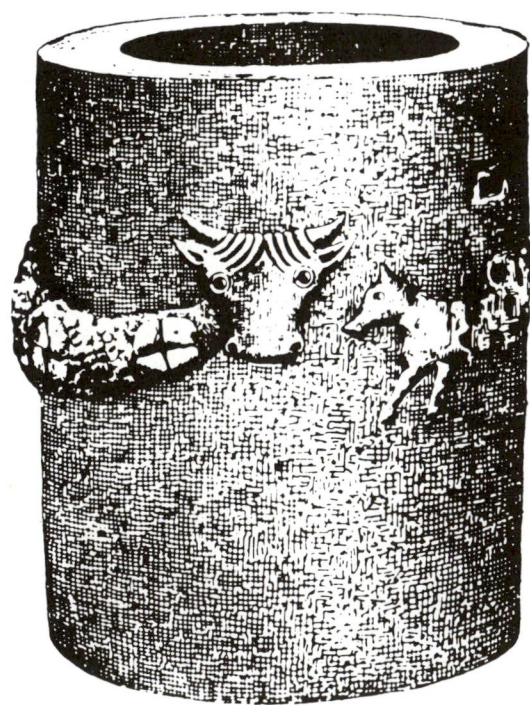

Fig. 10-1. Drawing after the Robigo altar found at Castiglioncelli.

infected crops because a live fox had cruelly been set afire. The red and black colors of wheat rust are repeated in many of the symbols associated with these ceremonies: red dogs and cows, foxes, bloody sacrifices, and fire.

These ceremonies probably arose from folk beliefs that dated back to the beginning of recorded history, perhaps 3,000 years ago, as farmers looked for a way to gain a reprieve from rust epidemics. The Roman calendar contained three agricultural holidays in the spring: the Cerealia (April 12–19), the Robigalia (April 25—about the time when wheat was heading and needed protection from the rust gods), and the Floralia (April 28). The early Christian calendar, which absorbed the traditions of many pagan holidays, included Rogation Days around April 25th as a time for the blessing of crops. Since 1978, Rogation Days have no longer been officially required, but days of prayer for crops are still held in several European countries.

Grain crops have been the focus of agriculture in temperate climates for many centuries. Harvested grain is easily stored and can be ground and baked in an enormous variety of forms. The gluten in wheat allows dough to rise when yeast is added. Grains are also an important trade item that can be shipped around the world. The breaking of bread has provided a social and religious symbol for centuries. The word *companion*, which means a person with whom one eats bread, is used to describe a friend. The Food and Agriculture Organization of the United Nations (FAO) has the motto *Fiat panis,* which translates to "let there be bread."

Wheat and rice are the two important grains used primarily for human consumption rather than as animal feeds. Wheat, the traditional "staff of life," is grown throughout the temperate zone in small plots as well as fields that are measured in square miles rather than acres or hectares. It originated in the Fertile Crescent and spread slowly into Greece and on to Western Europe. Rust epidemics have been a part of wheat production since the beginning. As agriculture has become more modern and fields larger and more genetically uniform, the risk of loss to rusts has increased.

In Italy, which had no natural barriers separating it from Asia Minor, the Romans suffered from rust epidemics whenever weather conditions were conducive. G. L. Carefoot and E. R. Sprott, in their book *Famine on the Wind*, suggest that unusually wet weather in the early Christian era may have contributed to the decline of the Roman Empire. They also note

Fig. 10-2. Symbol of the Food and Agriculture Organization of the United Nations (FAO). Wheat is the grain depicted, and the motto translates to "let there be bread."

that rust was probably not a major problem in Western Europe in the early days of production because grains were carried over the Alps and grown in isolation. As travel between the Middle East and Europe became more common, however, a shrub, the common barberry (*Berberis vulgaris*), was established in Western Europe, and rust epidemics occurred. The barberry plays an important role as an alternate host in wheat stem rust epidemics, but the relationship between barberry and rust took over 2,000 years to become clear.

Barberry and Wheat Rust

Farmers who observe crops first hand year after year are often able to draw empirical conclusions about agricultural problems even without a scientific education. In the case of wheat rust, they looked beyond prayer for a solution to the rust epidemics. Farmers noticed that rust was often worse near common barberry bushes. In 1660, in Rouen, France, the first legislative act concerning a plant disease was enacted to help reduce wheat rust by eradication of barberry plants. This law was made nearly 200 years before the Irish potato famine and more than 200 years before Anton deBary discovered the biological connection between barberries and wheat stem rust.

In the meantime, European colonists had settled in the New World, bringing with them both barberry and wheat plants. Common barberry served as a source of tool handles and yellow dye for the colonists. It also provided a fast-growing hedge to contain animals, and its berries were used in sauces and jellies. The Japanese barberry (*B. thunbergii*), so common in today's landscape plantings, is in the same genus but is not susceptible to the wheat stem rust fungus. Following the legislation in France, laws were enacted in the American colonies of England in the 1700s forbidding

Fig. 10-3. Wheat harvest in the United States.

the planting of barberry near wheat fields. Barberry laws are recorded for Connecticut in 1726, Massachusetts in 1754, and Rhode Island in 1766. Attention to these laws faded as grain production became centered in western territories.

Despite the discovery of the microscope and passage of the Barberry Laws in the 1600s, an understanding of rust epidemics was slow to come. One complicating factor was that a single rust species can produce up to five different spore stages on two unrelated host plants. Scientists began to notice a similarity of spore types among the rust fungi found on various host plants, even though each had been given a separate Latin binomial because the full complexity of the life cycles had not been discerned. Anton deBary, the famous mycologist, had been studying a rust fungus that infects garden beans. He discovered four different types of rust spores and, through careful study, determined the consecutive order of each.

As with many fungi, a dark, thick-walled spore is the overwintering spore. Following the same terminology used with the smuts, which are also Basidiomycetes, these dark survival spores are called **teliospores**. In the spring, four thin-walled **basidiospores**, the products of meiosis, are produced by each teliospore as initial inoculum. Shortly after infection of the bean plants by the basidiospores, small vase-shaped structures, called **spermogonia**, are produced. They protrude through the upper epidermis of the leaf of the infected plant. (Terminology for rust spores and reproductive structures has changed over the years, and disagreement still exists.) Some weeks later, a cuplike structure erupts through the lower epidermis, producing chains of spores that are dispersed by wind. The structure is called an **aecium**, which refers to its blisterlike appearance. Aecia are usually produced in groups and are often called "cluster cups." The spores produced in aecia are **aeciospores**.

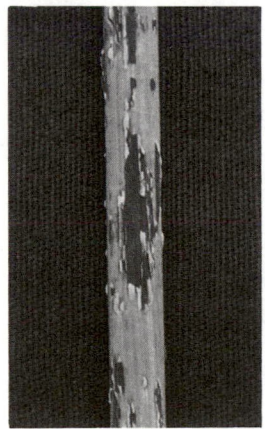

Fig. 10-4. Symptoms of wheat stem rust, *Puccinia graminis* f.sp. *tritici*. The uredial stage produces uredospores, the "repeating stage" that causes epidemics in wheat fields.

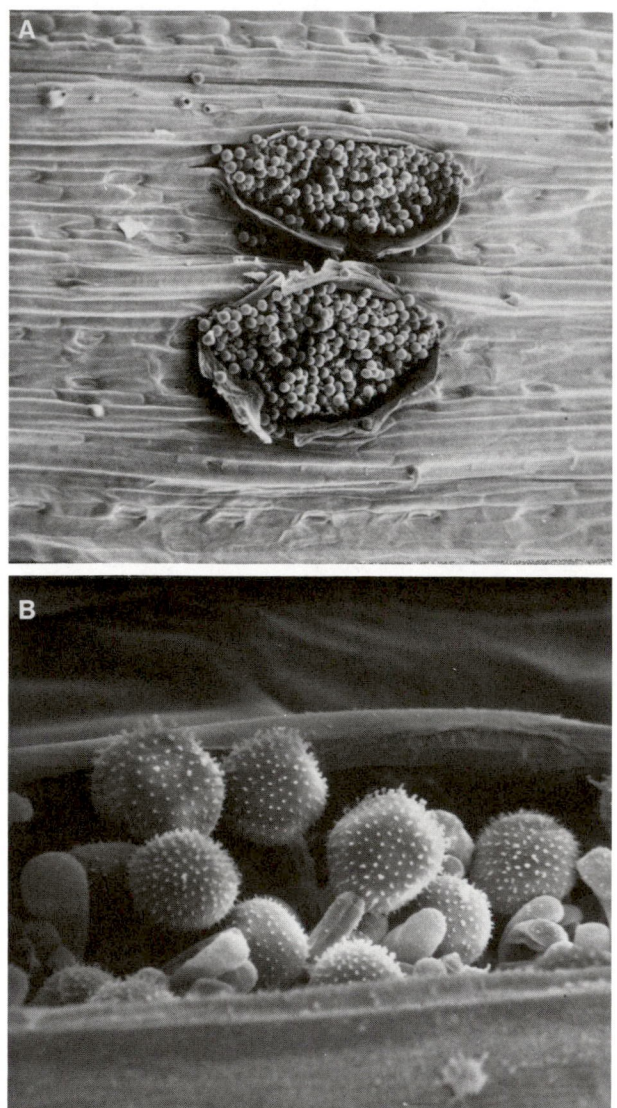

Fig. 10-5. Uredia of uredospores of *Puccinia graminis.* (**A,** ×100; **B,** ×1,000)

During the growing season, another type of spore is produced that is spiny and rusty colored when observed with a microscope. Rather than chains of spores, single "crops" of **uredospores**, each on an individual stem, are produced in the **uredium**. A single uredium may produce 2,000 uredospores per day for two to three weeks. The inoculum potential of these fungi is enormous. It is easy to imagine how quickly an epidemic could develop. Toward the end of the season, the rust fungus produces the dark teliospores for overwintering, completing the life cycle.

After studying the bean rust fungus, deBary turned his attention to the more confusing and economically important disease found on wheat. He harvested the thick-walled black teliospores, allowed them to germinate to produce basidiospores, and inoculated healthy wheat plants as he had done with his garden beans. He was puzzled by the lack of infection.

He then considered the long-time belief by farmers that common barberry somehow played a role in rust epidemics. In 1865, he inoculated barberry leaves with basidiospores that had been produced from the teliospores found on wheat plants. The basidiospores infected the barberry, and the "missing" spore stages were produced, completing the life cycle that he had observed in his bean rust studies. Spermogonia were produced on the upper leaf surfaces of the barberry, followed by the production of aecia and aeciospores on the lower leaf surfaces.

The aeciospores, even though produced on barberry, infect wheat, not barberry. For the first time, the amazing phenomenon known as a **heteroecious** life cycle had been discovered. The parts of this word indicate that the fungus has "different houses." Two unrelated host plants, wheat and barberry, are necessary for the production of all spore stages of the wheat stem rust fungus. This discovery rapidly led to the matching up of various host pairs, and alternation between different hosts was found to be quite common for rust fungi.

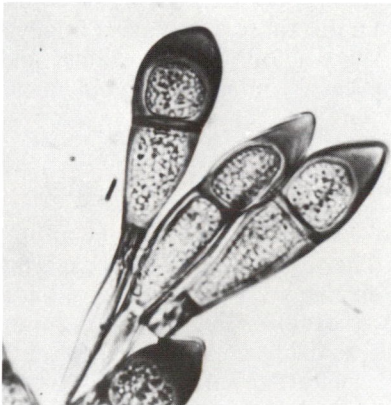

Fig. 10-6. Thick-walled teliospores of *Puccinia graminis* produced on wheat. These spores survive winter conditions. In spring, they produce basidiospores that infect barberry.

Biology of the Rust Life Cycle

A complete understanding of the rust life cycle was not accomplished until J. H. Craigie's work was published in 1927. DeBary had observed the spermogonia but had not understood their function in the life cycle. Craigie, a Canadian scientist working in the Dominion Rust Laboratory in Winnipeg, Manitoba, finally discovered the role of the spermogonia.

Nuclear studies of chromosome numbers have demonstrated the following life cycle. As in other fungi of this group, basidiospores are haploid. After germination, they produce haploid structures called spermogonia, which consist of **spermatia** and **receptive hyphae** of a single mating type. Cross-fertilization between two mating types on different spermogonia is necessary for the life cycle to continue beyond this point. A sticky, sweet substance produced in the spermogonia attracts insects to carry spermatia of one mating type to a receptive hypha of the other mating type. Successful fertilization brings two genetically different nuclei into the same hypha.

As we have seen in previous examples from the fungus world, nuclei may exist in the same hypha without fusing. In a process still not completely clear, a dikaryotic mycelium forms after the nucleus of the spermatium enters the receptive hypha. The mycelium ramifies through the leaf tissue, producing cuplike aecia and dikaryotic aeciospores on the lower leaf surface. Each aeciospore contains one nucleus of each type. No genetic change has taken place. In heteroecious rusts, the dikaryotic aeciospores infect the alternate host, producing more dikaryotic mycelium.

Later, uredia and uredospores, still dikaryotic, are produced; they infect only that alternate species of host plant. Uredospores can cause repeated infection of the same plant or infection of neighboring plants of the same species. These spores are sometimes called the **repeating stage**. This is the spore stage that causes the damaging epidemics so well known to ancient and modern people alike.

The fungus remains dikaryotic until the black teliospores are produced later in the season. Within this spore, karyogamy finally occurs and a diploid nucleus is produced. Meiosis follows karyogamy, and the basidiospores produced in the spring each contain a haploid nucleus, completing the life cycle.

Other Aspects of the Biology of Rust Fungi

It is easy to become so absorbed in observing the complex spore production of rust fungi that one fails to appreciate other aspects of these interesting parasites that produce so many kinds of spores. Rust fungi have been the subject of many kinds of studies. For example, some of the best fungal fossils are rust fungi parasitizing plant tissues. In addition, rust fungi are obligate parasites and very host-specific. Only a very few have been grown on complex artificial nutrient media, so most must be studied in the tissue of their host plant.

Rusts invade plant tissue intercellularly, causing little visible damage in the early stages. They obtain nutrients by absorption via haustoria, and

Fig. 10-7. Rust. **a,** Spermagonia of *Puccinia recondita* (leaf rust) with honeydew that attracts insects. Spermatia contribute nuclei to receptive hyphae, which leads to the development of a dikaryotic mycelium. **b,** Aecial stage of stem rust on leaves of common barberry. Aeciospores infect wheat.

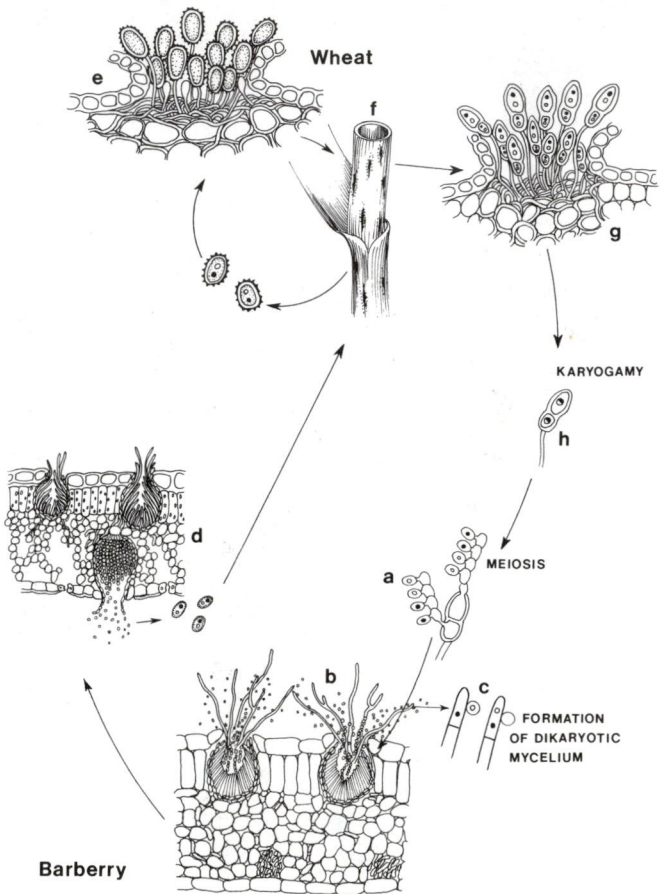

Fig. 10-8. Disease cycle of stem rust of wheat (*Puccinia graminis* f. sp. *tritici*). a, The rust fungus life cycle begins with haploid basidiospores that infect common barberry. b, Barberry infections result in spermogonia composed of spermatia and receptive hyphae. c, Spermatia contribute a nucleus to a receptive hypha from a different mating type, which initiates the formation of a dikaryotic mycelium. d, Aecial cups filled with dikaryotic aeciospores are produced on the lower leaf surface. Aeciospores infect wheat. e, Uredial pustules filled with dikaryotic uredospores develop on wheat stems. Uredospores are capable of infecting wheat repeatedly and are known as the "repeating stage" in field epidemics. f, Pustules of rust spores break through the epidermis of the wheat stem, weakening it. g, Later in the season the same dikaryotic mycelium forms telia filled with dark, thick-walled teliospores that are capable of surviving harsh winter conditions. h, Karyogamy and meiosis occur in teliospores, which results in the formation of haploid basidiospores in the spring.

little necrosis occurs. Because host tissue is not macerated or degraded to any great extent, histological studies can clearly observe hyphal and haustorial development and subsequent spore production. Few fungi, except perhaps the powdery mildews, have been studied genetically in such detail and had the corresponding genetics of their hosts determined. H. H. Flor, it will be remembered, first proposed the **gene-for-gene theory** based on his studies with flax and its rust, *Melampsora lini.*

Many fascinating studies have been conducted to try to understand how rust spores identify and penetrate their hosts. Uredospores penetrate primarily through stomata. The germ tubes appear to locate stomata using both chemical and physical signals. They can locate stomata even on artificial casts of the host plant epidermis. Studies also showed that if the epidermis is stripped away, the lack of physical clues causes germ tubes to wander over the surface of the host plant without penetrating it because they cannot identify their host. Very recent studies using artificial surfaces have suggested that the germ tubes can discern both the distance between epidermal cells and the height of the lip of the stoma guard cells as signals to guide them to an appropriate penetration site.

It is amazing to consider the genetic information necessary for such complex interactions between host and parasite. While these studies are interesting in themselves, they may have very practical purposes. If scientists can understand how a fungus finds a penetration site, they may be able to confuse a germ tube either chemically or through crop breeding procedures that alter epidermal patterns to prevent infection.

Rust fungi that produce all five spore stages are called **macrocyclic**. The heteroecious, macrocyclic life cycle is considered the "primitive" rust life cycle that evolved early. Without going into great detail, one piece of evidence for this theory is that the rusts found on more primitive plants tend to be heteroecious and macrocyclic. Examples include those that alternate between conifer and fern hosts. The host range of rust fungi seems able to remain quite stable. In the last Ice Age, many kinds of plants were eliminated from the British Isles. A number of birches and ferns eventually returned to Great Britain approximately 12,000–14,000 years ago. Because these plants hosted the uredial repeating stage, the rusts have continued to maintain themselves over the years. Despite the long separation from their alternate hosts, the rust fungi were able to reestablish a complete life cycle when firs and larches were planted near the birches and ferns.

Managing Grain Rust Epidemics

As wheat production in North America moved into the western prairies of both the United States and Canada, the common barberry followed. The edible berries of this shrub were carried by birds to new locales, and people, who were probably not wheat farmers, planted the shrub, contributing to its spread. Rust epidemics occurred every year and were particularly severe when the weather was moist and warm. Uredospores are produced about 7–14 days after infection. In 1916, 200 million bushels of wheat were lost in the United States, and an additional 100 million bushels were lost in Canada. The U.S. government decided that wheat production was important not only to feed our nation but for national security. Wheat was needed to supply allies fighting in World War I. An intense federal program of barberry eradication was initiated in 1918 at the urging of E. C. Stakman, a Minnesota plant pathologist, and continued until the program was gradually turned over to individual states during 1975–1980. Over three million barberry bushes were eradicated in Illinois alone. The current cost of the program is about one hundredth of a cent per acre of small grains. The goal of the eradication program is to remove the alternate host of the wheat stem rust fungus, *Puccinia graminis* f.sp. *tritici*.

There are actually three common rusts of wheat. *P. g.* f.sp. *tritici* causes the most important disease, black stem rust of wheat. Grain yield is reduced in both quantity and quality in infected plants. Worldwide, one million metric tons of wheat are lost to stem rust annually. In addition, the numerous pustules that break through the surface of the stem weaken the plant and cause lodging (falling over) of the wheat plants, which can make harvesting difficult or impossible.

P. striiformis, called yellow rust in Europe and stripe rust in the United States, has no known alternate host. It is the most important wheat rust in Europe, where the climate is cooler and more humid; yellow rust accounted for a loss of 70% of the winter wheat crop in the Netherlands in 1956. In the United States, it is mostly a problem in the Pacific Northwest. A third wheat rust, caused by *P. recondita*, has meadow rue as its alternate host. It causes brown leaf rust and can be severe in certain years.

P. graminis exists in over 300 forms that are identical in morphology (appearance) but differ in their host range. The forma specialis that infects wheat, barley, and some wild grasses is called *tritici*. There are six other

Fig. 10-9. Spraying common barberry (*Berberis vulgaris*) with herbicide to eliminate the alternate host of *Puccinia graminis* in Erie County, PA, 1952.

formae speciales of this important species, all of which share the same woody alternate host, common barberry. For example, *P. g.* f.sp. *poae* causes rust on lawns, and *P. g.* f.sp. *secalis* infects rye.

Effects of Barberry Eradication

What biological and epidemiological effects can be expected from the eradication of barberry? If one removes the spore stages found on barberry, the only spores to consider are those produced on wheat, uredospores and teliospores. Teliospores may still be produced, but their basidiospores have no host to infect after barberry is removed. Thus, the only functional spores left in the life cycle are the uredospores, which are destroyed by cold winter weather in the northern United States and Canada.

Unfortunately, wheat stem rust is still with us. Why did barberry eradication fail to eradicate stem rust from the North American continent? The wheat-growing areas in the southern states and Mexico, where winters are not as harsh as they are on the northern plains, serve as a source of uredospores. Each year, uredospores are carried in winds from field to field in a long journey northward. Thus, rust infections are found throughout the **Puccinia Pathway** that stretches across entire grain-growing

Fig. 10-10. The *Puccinia* pathway. Each year uredospores are blown from southern states and Mexico to Canada.

areas from Texas and Mexico north to the wheat lands of Canada.

Was the eradication program useless? Although not a total success, it has greatly improved our ability to manage wheat stem rust. Removal of the barberry delays epidemics for several crucial weeks, since inoculum must make its way northward over many hundreds of miles each year. Since a single barberry bush can be the source of 64 million aeciospores, a significant source of inoculum was removed as the barberries were eradicated. Most importantly, the eradication program broke the sexual cycle of the fungus.

Based on the number of known resistance genes in wheat plants, it has been theoretically estimated that the wheat stem rust fungus has the potential to produce 56,000 races. This potential genetic variation would be overwhelming to any breeding program. Without the normal sexual cycle, however, the major source of genetic change is mutation. Mutation can still be a significant source of variation for a fungus that produces thousands of uredospores in a single uredium. However, in 1918, before the eradication program, 28 different races were detected and seven races predominated, whereas today only about five races are commonly found and only one or two predominate.

Genetic Resistance to Rusts

Single-gene resistance remains our main line of defense against the grain rusts. The long-term goal in breeding programs is a durable, general resistance effective against all races of the rust. This is sometimes referred to as "slow rusting" because it reduces the rate of epidemics. In the meantime, how can single-gene resistance be effective against a polycyclic pathogen that invades such large expanses of cropland? The general use of single-gene resistance has been a matter of much theoretical speculation and practical experimentation.

Some scientists recommended creating a single cultivar with a number of single resistance genes, since it would be unlikely that the fungus could make all of the necessary genetic changes to cause infection. The accumulation of resistance genes in a single cultivar through multiple crosses is called **pyramiding**. Although most scientists agree this would likely be successful, there is always the fear that the use of such cultivars would put selective pressure on the fungus, resulting in a **superrace**. All of the resistance genes used would then be essentially useless. Other scientists proposed the geographical deployment of different resistance genes in bands from north to south, so that arriving rust spores would be met by varying resistance genes throughout the Puccinia Pathway, thus preventing major epidemics. To be successful, this would require total cooperation of growers, seed producers, and state seed associations, and an efficient and continuing international bureaucracy. Such gene deployment has been successfully used for oats against crown rust (*Puccinia coronata*), which has buckthorns (*Rhamnus* species) as its alternate hosts.

In some areas, oats are planted with physical mixtures of seeds of various cultivars that have similar characteristics and vary primarily in a single resistance gene. These mixtures are called **multilines**. The theory behind the use of such mixtures is that the pathogen meets a variety of resistance genes, somewhat like it might find in a natural ecosystem. The other theory behind such mixtures is that no superrace will arise because of something called **stabilizing selection**. Evidence suggests that "extra" genes necessary to attack all the cultivars in the mixture make races less fit and less competitive than common wild type races. Thus, multilines keep disease at a low level and do not select for superraces. Multilines have the added advantage of allowing different resistance genes to be included in the mixtures from season to season.

Although these theories remain speculative and controversial, each year wheat plant pathologists maintain "rust gardens" consisting of a number of wheat cultivars with known resistance genes, to determine which rust races predominate. Resistance genes for the next season are based on predictions derived from the rust garden results. Each year we must hope that the correct seed has been chosen to protect the 50 million acres (20 million hectares) of U.S. wheat.

Because grain crops are commonly grown in extensive plantings, one must always worry about the use of new cultivars. As discussed in Chapter 6, corn with TMS cytoplasm became very popular, and major losses occurred when Race T of *Cochliobolus heterostrophus* increased. A similar example occurred in 1946 in an oat cultivar named Victoria and in closely related cultivars. The cultivar came from South America and, because of its excellent crown rust resistance, quickly became popular on several continents. It succumbed to a new disease caused by a leaf spot fungus then named *Helminthosporium victoriae*. Later the Victoria cultivar was determined to be extremely susceptible to a host-selective toxin produced by the fungus, which is normally a minor leaf spot parasite. The fungus caused such a severe disease that the crown rust resistance was no longer significant.

Fungicide use against grain rusts and other foliar diseases is not yet

Fig. 10-11. A severe outbreak of crown rust of oats (*Puccinia coronata*) near a hedge of its alternate host, buckthorn (*Rhamnus* species).

widespread. Protective fungicides must be used two or three times, which is not usually economical, considering the relatively low value of grain crops. The newer systemic fungicides, which give more long-term and specific control, have a wider application for rust. Their use is more common in humid areas such as Europe, where maximum yields must be derived from limited land areas. In the United States, systemic fungicides are being applied on a small scale, but how widespread the use of fungicides for rust control will become remains to be seen.

Common Home and Garden Rusts

Besides being economically important, many of the 6,000 or so rust species are yellow, orange, brick-red, or brown and interesting to find on plants in natural ecosystems. Especially toward the end of summer, attractive rusts can be found on many wild plants, including willow, clovers, grasses, cinquefoil, and brambles, that rival the fall colors of tree leaves, although on a smaller scale. As with many plant-pathogenic fungi, the economic importance of rusts as pathogens is related to many common agricultural practices, especially genetic uniformity and monoculture.

Rusts are also found on many vegetable and ornamental plants. They do not generally cause severe damage if the watering and sanitary practices described in previous chapters are properly employed. As mentioned earlier, beans are often infected with rusts. Because all spores are produced on a single host, this rust is called **autoecious**. Another autoecious rust is commonly found on brambles such as raspberries and blackberries. Because of the perennial nature of these plants, sanitary practices are more difficult to employ. For the same reason, the autoecious rust *P. asparagi* is a difficult disease problem on asparagus. Sweet corn is commonly infected by *P. sorghi* by the end of summer, and both red uredospores and black teliospores can usually be found on the leaves. The general resistance found in field and sweet corn prevents severe damage by this macrocyclic, heteroecious rust, which has wood sorrel (*Oxalis* species) as its alternate host.

On flowers, roses are commonly infected by autoecious rusts (*Phragmidium* species), and careful sanitation and pruning of overwintering rust cankers are required to reduce disease. A more unusual life cycle is demonstrated by the rust so commonly found on hollyhock. *Puccinia malvacearum* is a rust fungus that produces the minimum spore stages necessary to complete a life cycle—teliospores and basidiospores. Because of its minimal life cycle, hollyhock rust is called **microcyclic**. Geranium, chrysanthemum, and carnation rusts can be important diseases in greenhouses, where the warm, moist air and splashing water can cause rapid spread of the uredospores. Watering of only the soil surface helps prevent the spread of rust and other fungal pathogens.

Other Rusts of Economic Importance

Since 1979, sugarcane rust has been found in all areas of the Western Hemisphere where this crop is grown. It spread rapidly, apparently from transoceanic sources, throughout many of these areas in a single year. Because sugarcane is vegetatively propagated and grown in a moist, warm environment, this newly arrived rust is an important threat. In peppermint production, the rust, *P. menthae*, is an important disease. Deep plowing of crop debris after harvest was recommended until it was found that this practice increased Verticillium wilt problems. Flame sanitation is now practiced in the spring to decrease rust infection. A potentially devastating rust of soybeans, *Phakopsora pachyrhizi*, has not yet been found in the United States but causes losses of 20–30% annually in Taiwan.

Some important rusts infect trees. One commonly seen on apple and crab apple trees is *Gymnosporangium juniperi-virginianae*. It is known as cedar apple rust, a name in which the *cedar* refers to red cedar (juniper) rather than white cedar (arborvitae), which is not involved. The spermogonia and aecia are produced on apple trees. Aeciospores infect junipers, on which a small brown gall develops that is covered with masses of overwintering teliospores. In the spring, starting just as the apple blossoms are ready to open and lasting until about mid-June, orange jellylike protrusions from the gall help the basidiospores disperse on their journey to newly expanded apple leaves and fruits.

This heteroecious rust is different from the macrocyclic rusts already mentioned because it lacks the repeating uredospore stage. Thus, rust infections of apple trees do not serve as sources of spores that can cause an epidemic within the orchard. The aeciospores produced on apple trees can infect only junipers. Rusts of this type can be completely controlled by removal of one host. Junipers are less economically important than apple trees and are removed from areas near commercial apple orchards.

Fungicide protection can be very effective in protecting apple trees for

Fig. 10-12. Hollyhock rust (*Puccinia malvacearum*).

the short period during which basidiospores may arrive from nearby junipers, but it is usually not necessary. Fungicides applied for the more important apple scab disease may be sufficient to protect trees from rusts as well. Many commercial apple and crab apple cultivars have good genetic resistance to rust diseases.

A single juniper gall can produce over 100 billion spores, so severe infec-

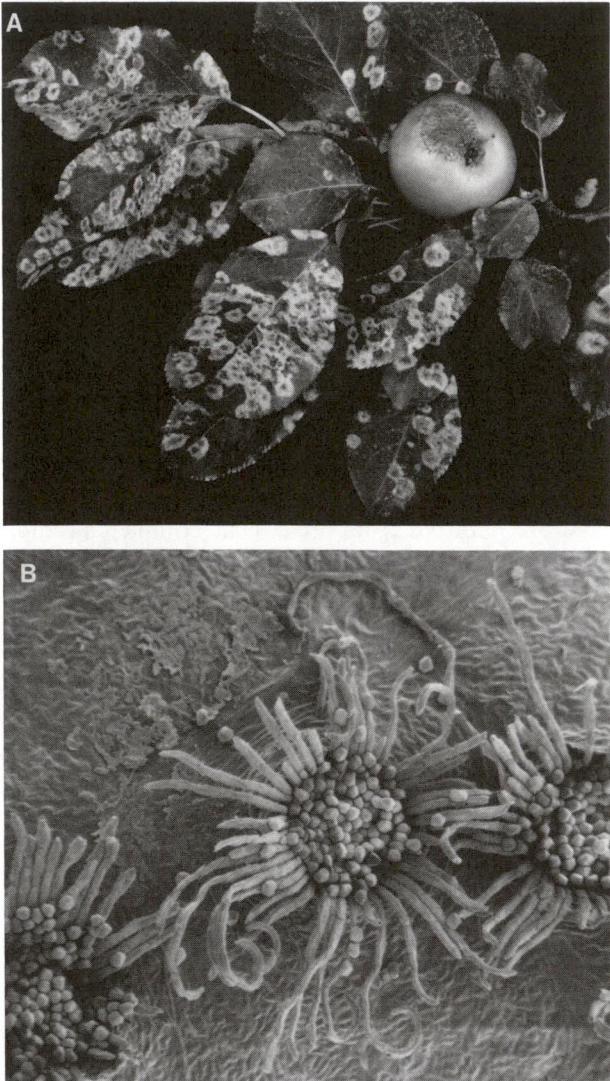

Fig. 10-13. A, Cedar apple rust symptoms on apple. **B,** Aecium with aeciospores produced on the lower leaf surface of an apple leaf infected with *Gymnosporangium juniperi-virginianae*. The aeciospores infect susceptible junipers. (×125)

tions may occur when junipers and crab apples are grown together. In landscaping situations, one should choose between junipers or crab apples unless rust-resistant cultivars are planted. Rust-resistant or immune cultivars of popular juniper and crab apple species are available. As in apple orchards, fungicide applications are effective but are not usually recommended because infections only occur for about a month, when juniper galls are producing spores. Spores from wild junipers in nearby fields may also be an important source of infection, although the spores are viable only for about a mile from their source.

A more serious rust disease destroyed white pine forests of the northeastern United States in the early 1900s. The white pine blister rust fungus, *Cronartium ribicola*, is macrocyclic and heteroecious, with plants of the genus *Ribes* as alternate hosts, as suggested in the specific epithet. This genus includes gooseberries and currants. The rust apparently arrived on white pine seedlings imported from Germany. The widespread devastation it caused made white pine blister rust one of the diseases, along with citrus canker and chestnut blight, that convinced legislators that U.S. borders needed to be protected. This led to the passage of the 1912 Federal Quarantine Act. Unfortunately, the legislation was late and did nothing to stop the pathogens that had already arrived and become widely dispersed.

Knowledge of the rust life cycle suggests a useful control measure. The spermogonia and aecia are found on white pine, whereas the uredial and telial stages are on the *Ribes*. Thus, the repeating stage that causes epidemics among plants of the same species was, luckily, on the host of lesser economic importance. Once the destructive nature of this rust was realized, many

Fig. 10-14. Gall stage of cedar apple rust on juniper.

people were hired to scour the woods cutting and killing wild *Ribes* plants within 1,000 feet of white pines, the distance viable basidiospores can travel. People were also restricted from planting susceptible species of *Ribes* near white pine forests. Eradication programs continue in some areas today. Because rust spores produced on pine can infect only *Ribes*, the fungus rapidly declined with the removal of the alternate host in the eastern United States.

Individual infected pine trees were usually doomed, as the mycelium of

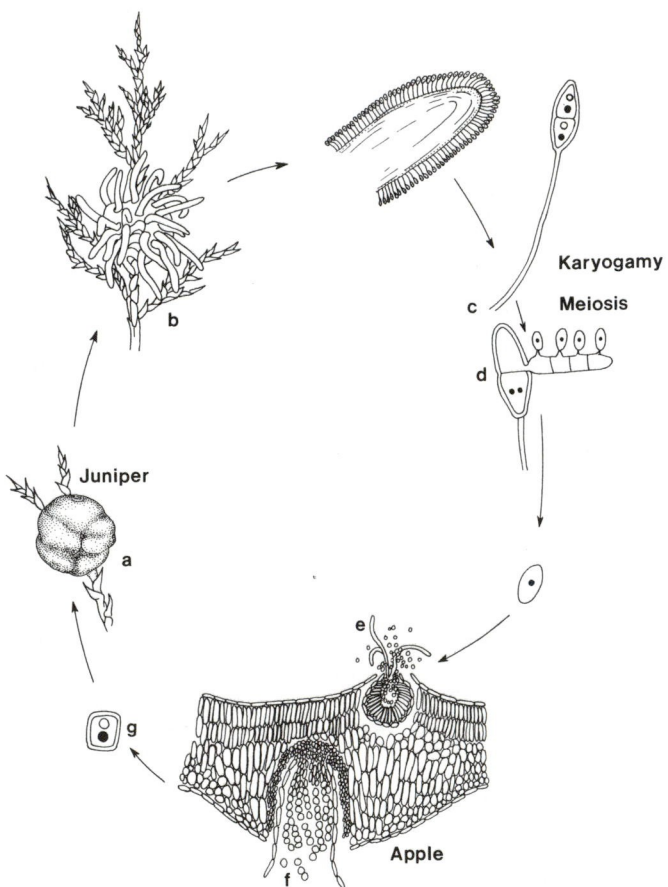

Fig. 10-15. Life cycle of cedar apple rust (*Gymnosporangium juniperi-virginianae*). Gall on juniper (a); in spring, orange gelatinous structures expand from the gall (b) and are covered with teliospores (c); karyogamy and meiosis occur in the teliospore (d), resulting in the production of four haploid basidiospores that infect apple and related plants; basidiospore infections result in the formation of spermogonia (e); if a dikaryotic mycelium is formed, aeciospores are produced from aecia, generally on the lower leaf surface (f); The dikaryotic aeciospores infect junipers (g). There is no uredospore (repeating) stage.

the rust continued to invade the tissues until the tree was girdled and killed, but healthy neighboring pines were not threatened. Individual trees could be pruned and saved, although this was not possible in vast forests. Since 1966, enforcement has lain in the hands of individual states, which decide where *Ribes* may be safely planted without threatening white pines. In addition, rust-immune *Ribes* are now widely available from nurseries for gardeners and jelly-makers.

Ribes eradication has not been so successful in the western United States, particularly in Idaho and western Montana. Replanting of timber stands of the western white pine, *Pinus monticola*, was temporarily halted in 1966 due to white pine blister rust. In 1952, an intensive breeding program was begun to develop rust-resistant western white pines, and resistant cultivars became available in 1972. The decision to replant white pine timber stands is now made using complex plans of integrated management. Models of environmental conditions that simulate uredospores epidemics can be used to predict rust severity on *Ribes*. Combined with measurements of basidiospore concentrations, these models predict the potential rust hazard at a planting site. Rust-resistant cultivars are chosen to reduce disease but maintain genetic diversity. Forest pathologists are concerned that selective

Fig. 10-16. White pine blister rust (*Cronartium ribicola*).

pressures may result in new races of the rust fungus that can infect the new cultivars. New strains capable of overcoming resistance have already been reported in Oregon (1984) and Japan (1983). Once a site has been replanted, cultural practices are integrated to reduce this selective pressure. These may include fertilization for fast growth to get the new trees quickly through the early stages when they are most susceptible to infection. In some cases, infected branches are pruned and infected cankers are excised. Many years of research to develop this integrated approach have allowed the western white pine to become a viable timber species once again.

Other tree rusts are of economic importance. The fusiform rust fungus, *Cronartium quercuum* f.sp. *fusiforme*, is a continuing problem in the southern United States because many southern pines are susceptible and the alternate host is oak, another tree common in the South. Here it is

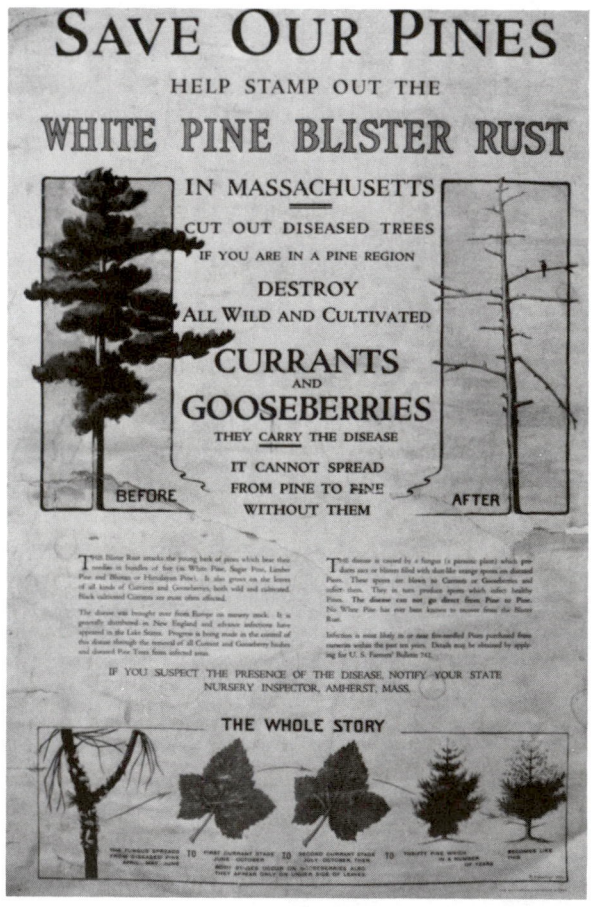

Fig. 10-17. Poster warning that *Ribes* species are the alternate host of the white pine blister rust fungus.

not possible to consider rust control through the removal of the alternate host, although oaks near commercial pine plantations are sometimes removed. Fungicide protection and pruning can help control infection of commercial stands. Genetic resistance is a future goal, but the long generation time of trees makes the breeding of resistant trees a lengthy process. As with the rust-resistant pines, one must also worry about the durability of resistance in the face of the many races produced by rust fungi, especially in long-lived perennials like trees.

A similar problem plagues coffee growers, who wish to find durable resistance to coffee rust within the narrow gene pool that exists in currently available cultivars. With coffee rust (*Hemileia vastatrix*), there is the added problem of having the repeating stage on the host of economic importance. Even without the presence of an alternate host, which has not yet been discovered, the uredospores continually reinfect coffee trees unless they are repeatedly sprayed with protective fungicides.

Biological Controls

For tree rusts and others, a possible biological control is the use of fungi that parasitize rust fungi. Several such fungi, called **hyperparasites**, have been identified. So far no commercial applications have been successful for rust fungi, but research continues. Several rust fungi have been used

Fig. 10-18. Extensive damage in loblolly pine plantations from infection by fusiform rust (*Cronartium quercuum* f.sp. *fusiforme*). The alternate host of this rust fungus is oak.

as biological controls against unwanted plants. Fungi applied in this way are sometimes called **mycoherbicides**. The uredospores of rust fungi are easy to collect, store, and disperse on plants; they then multiply quickly to create an epidemic in the pest plant population. A successful example comes from Chile, where European blackberry plants had become an intrusive weed in pastures. Uredospores of the rust fungus, *Phragmidium violaceum*, were collected from European blackberries and spread by helicopters. Rust infection reduced the competitiveness of the blackberries and slowed their invasion.

Afterword

Tree diseases are some of the most difficult to manage. The loss of a single plant can be aesthetically and emotionally devastating. Replacement of a tree 50 years old or more takes time, so its loss is felt for a long period. The long generation time of most trees makes them, as species, particularly vulnerable to severe damage, as they are unable to rapidly change genetically to respond to the danger of an introduced pathogen. Witness the loss of the chestnut and elm trees in Europe and North America and many native pines in Japan. Introduced insect pests, most recently the gypsy moth and the Japanese beetle, cause continuing damage. Trees are at once particularly vulnerable and amazingly resilient to attack by pests and pathogens in their environment. The next chapter examines some other important tree diseases and how trees, the organisms with the longest life spans, protect themselves from invasion.

Selected Readings

Bardell, D. 1988. The discovery of microorganisms by Robert Hooke. American Society for Microbiology News 54:182-185.

Fulton, R. H., ed. 1984. Coffee Rust in the Americas. American Phytopathological Society, St. Paul, MN.

Hagle, S. K., McDonald, G. I., and Norby, E. A. 1989. White pine blister rust in northern Idaho and western Montana: Alternatives for integrated management. General Technical Report INT-261. U.S. Department of Agriculture, Forest Service. Intermountain Forest and Range Experiment Station, Ogden, UT.

Henderson, D. M. 1976. The living rust fungi. Transactions of the British Mycological Society 67:189-192.

Littlefield, L. J. 1981. Biology of the Plant Rusts: An Introduction. Iowa State University Press, Ames.

Ostofsky, W. D., Rumpf, T., Struble, D., and Bradbury, R. 1988. Incidence of white pine blister rust in Maine after 70 years of a Ribes eradication program. Plant Disease 72:967-970.

Roelfs, A. P. 1982. Effects of barberry eradication on stem rust in the United States. Plant Disease 66:177-181.

Zadoks, J. C. 1982. Cereal rusts, dogs and stars in antiquity. Garcia de Orta, Serie de Estudios Agronomicos, Lisboa 9:13-20.

Dying Trees and Parasitic Plants

Trees live longer than any other form of life, sometimes for 2,000 years or more. In many places, trees serve as memorials to historic events long after the people who participated in the event have died. They remain standing in one place, unable to move, while lightning, insects, fungi, bacteria, and human beings and other animals feed on or invade their tissues. Yet they may be able to withstand this dangerous onslaught and continue to grow for hundreds of years.

The survival of trees in a hostile environment is mainly due to two important features: the protective woody bark and a mechanism for continuous growth despite damage. To understand how trees protect themselves, one must first examine the transition from the soft and relatively vulnerable green stem of a young tree to the woody, bark-enclosed twigs, branches, and trunk of a mature tree.

Transition to Bark and Wood

The young stem of a woody plant is composed of the same kinds of cells as those in the stems of herbaceous plants. The outer epidermal layer encloses a cortex of parenchyma cells. The center of the stem is also composed of parenchyma cells. The vascular tissue is arranged in a cylinder, with xylem (water- and mineral-conducting) cells toward the inside of the stem and phloem (food-conducting) cells toward the outside of the stem. Often, there are also bundles of thick-walled supporting cells called sclerenchyma.

As the stem matures, two important meristems begin to function. Apical meristems, at the growing tips both above and below the ground, were already mentioned in previous chapters. The new meristems, important for the lateral growth of woody plants, are the **vascular cambium**, which provides xylem and phloem cells for the developing woody tissue, and the **cork cambium**, which contributes cells to the developing corky bark. Sometimes the cork cambium develops gradually in patches, initiating a bark layer on some parts of the stem while other parts of the stem remain temporarily green. Eventually, the entire stem is covered with bark. The cork cambium continues to produce cork cells, sometimes for many years, to replace bark sloughed away due to weather or wounds. As the girth of the trunk increases, new cork cambia begin to originate from parenchyma cells deeper in the bark to maintain coverage of the expanding circumference of the tree. Similar changes occur belowground as roots develop bark and secondary vascular tissues.

The vascular cambium produces **secondary xylem** and **secondary phloem**.

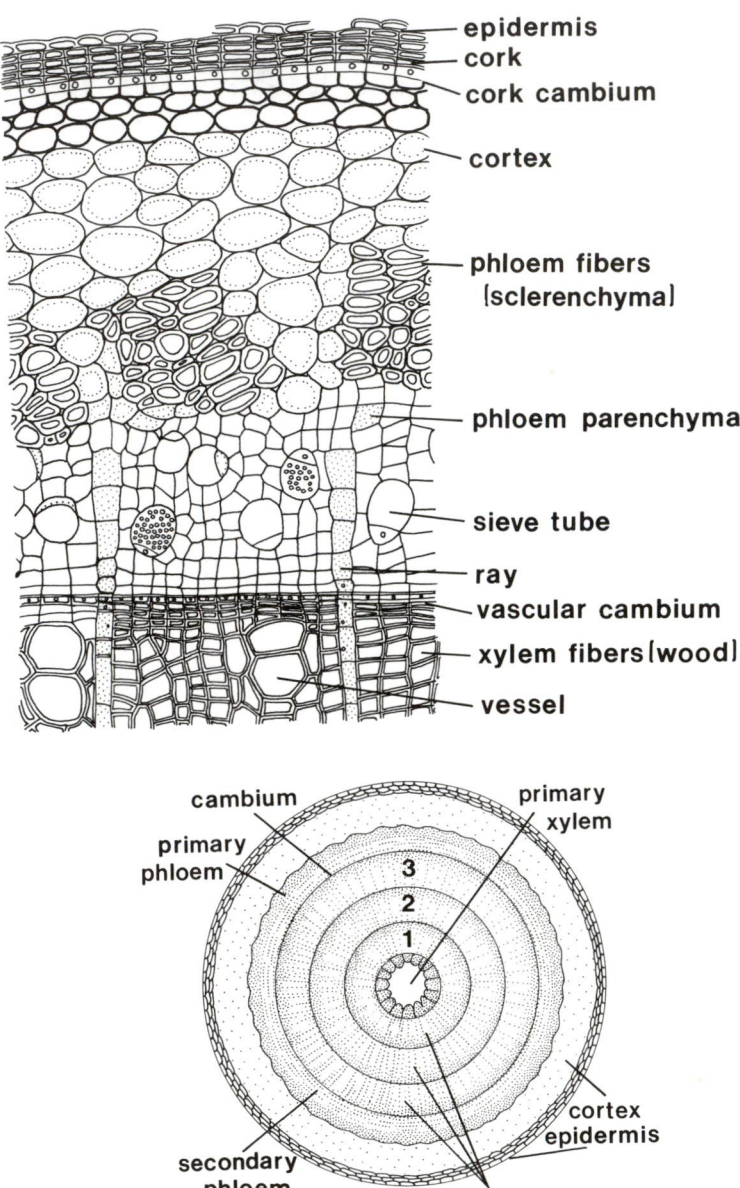

Fig. 11-1. Stem of a young woody plant. **Top,** cross section of a dicotyledon stem in the early stages of secondary growth. The cork cambium has begun to produce cork cells, although parts of the primary epidermis remain. The vascular cambium has also begun to function, producing secondary phloem to the outside and secondary xylem (wood) to the inside. **Bottom,** a young woody stem, showing annual growth rings in the secondary xylem.

The layer of active secondary phloem remains relatively thin as the older phloem becomes compressed by the expanding trunk. Secondary xylem layers are commonly referred to as **wood**. These layers are produced each year, and, unlike the secondary phloem, are preserved for the life of the tree. In many trees, the cells of the secondary xylem produced in the spring are larger than those produced in the summer. The resulting seasonal variation in cell size creates a pattern in which each year's growth can be discerned. These are called **annual rings**. Where rainfall and temperature are uniform and less "seasonal," as in parts of the humid tropics, annual rings may not be as apparent. In many climates, annual rings record the growing conditions in each year of the tree's life. Historical studies of drought periods and other environmental variations have been made by studying the annual rings of ancient trees that have fallen or been cut down or through the study of sample cores taken from living trees.

The chief functions of the secondary xylem are water conduction and support. Secondary xylem is composed of several types of cells. **Tracheids** are elongated, nonliving, conducting cells that consist only of cell walls. Besides having a primary cell wall, tracheids are endowed with a secondary cell wall that adds support. The tracheid wall contains numerous pits where only the primary cell wall, not the secondary cell wall, is present. Where the ends of the tracheids overlap, water and minerals pass from one tracheid to the next through the primary cell wall. Tracheids are the only conducting tissue in conifers, and they also play a conductive role in hardwoods. Hardwood trees, like oaks and maples, also produce larger conducting tubes called **vessels**. The formation of thickened secondary walls and the dissolution of end walls occurs in young living cells, but the protoplast soon dies and most xylem function occurs in nonliving cells. Since connecting vessel elements have perforations at their ends, water flows more easily through the strawlike vessels than through tracheids. Hardwoods also contain more **fibers**, a third kind of nonliving cell that adds support to the wood.

In conifers, tracheids and living parenchyma cells are found in radial layers only one cell thick. In hardwoods, radial layers are composed of tracheids, vessels, and parenchyma cells that vary in thickness. These radial layers are called **rays** and may be visible in the wood pattern. Living parenchyma cells are also present in the vertical plane of the secondary xylem. The role of these living cells is explained later in this chapter. In cross section, the secondary xylem of older trees is often composed of a darker inner cylinder known as **heartwood** and an outer, lighter-colored layer known as **sapwood**. The sapwood remains approximately the same thickness throughout the life of the tree and actively functions in conduction. Each year, new sapwood is produced from the vascular cambium, and the oldest part of the sapwood becomes part of the heartwood. The heartwood is often darker because it has undergone chemical changes that help protect it from decay and increase its strength. Many trees that are entirely hollow, due to decay of the heartwood, continue to function and grow because the sapwood conducts water and minerals for the tree.

Unlike the secondary xylem, which consists mainly of nonliving cells, the secondary phloem is composed of living cells, including **parenchyma**

cells and **sieve elements**, which transport the products of photosynthesis from the leaves to the rest of the plant. This relatively narrow band is vital to the health of the tree because it carries food to the roots. If a tree is "girdled" by having its bark stripped away, the layer of separation is usually the vascular cambium. Technically, bark includes all tissues outside the vascular cambium. When the bark is peeled away, the secondary phloem is often removed as well. The tree may remain alive temporarily using food stored in the roots while the secondary xylem continues to carry water to the leaves, but eventually the tree dies due to a lack of food. The phloem is also very vulnerable to damage when cuts are made in tree bark due to human mischief or accidents.

The protective part of bark is composed of cork cells. Cork is nonliving tissue that helps make the tree trunk impervious to water loss and helps prevent physical damage. The continual production of cork cells is necessary

Fig. 11-2. The American elm. **Top,** healthy tree with characteristic shape; **bottom,** street sign suggesting that elms have been important shade trees, although none exist at this location today.

since bark is constantly subject to weathering and other damage. Commercial cork is peeled from the extra thick bark of the cork oak, *Quercus suber*, which grows in the Mediterranean region. In the thin bark of some trees, such as birch and cherry, **lenticels** may be particularly visible. These are lens-shaped structures composed of loosely packed cells that help in the exchange of air for the living tissues within the woody tissues. Lenticels can be vulnerable to invasion by pathogens and insect pests.

Leaves are the most vulnerable aboveground parts of a tree. The woody trunk and twigs are generally well protected from invasion by micro-organisms and insect pests. Deciduous trees lose their leaves and replace them with new ones each year, shedding not only leaf tissue but many insects and parasites as well. A recommended sanitary practice is to clean up and compost, burn, or dispose of diseased leaves in autumn to help prevent spores from reinfecting the new leaves in spring. Anthracnose fungi sometimes invade woody twigs from infected leaves, especially in stressed trees or during particularly wet years, but generally these diseases do not cause severe damage.

The most devastating tree diseases are those caused by microorganisms imported from foreign lands. Because tree populations are not likely to have natural resistance to an imported pathogen, initial disease outbreaks are often severe. Generally, indigenous pathogens cause little trouble on native trees in the countries of origin because those trees have reached a genetic balance with their attackers over long periods of coexistence.

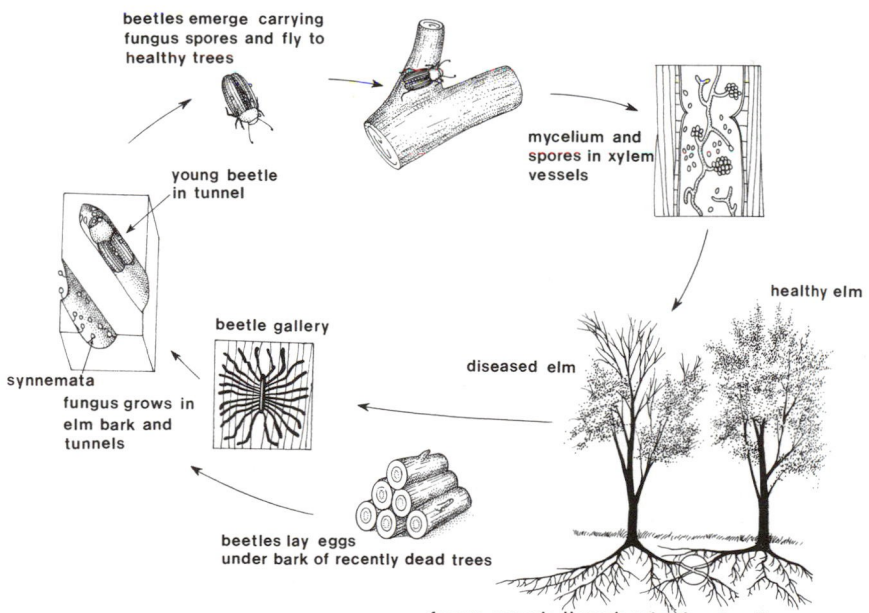

Fig. 11-3. Dutch elm disease, caused by *Ophiostoma ulmi*.

Dutch Elm Disease

A particularly sad example of the effect of an imported pathogen has been the demise of the American elm, *Ulmus americana*, an important shade tree in most parts of the United States. Dutch elm disease, named for the country where many of the important early scientific studies were done, was introduced into the United States on elm logs from Europe on at least three occasions. Its presence was first confirmed in Ohio in 1930. The disease had already caused losses of elm trees in Europe, starting at the time of World War I. Two Dutch plant pathologists, Christine Buisman and Bea Schwarz, played key roles in the discovery of the cause of Dutch elm disease. It is suspected that the fungus originated in Asia, because Japanese and Chinese elms are generally resistant to the disease. In the late 1960s, a renewed epidemic of Dutch elm disease plagued Great Britain and Europe, killing many elms that had survived the initial outbreaks, apparently due to the introduction of a more aggressive strain of the fungus imported from North America. It has been estimated that more than 40 million urban elms have died of Dutch elm disease in the United States alone but that perhaps 136 million landscape elms with a diameter of more than four inches (about 10 centimeters) still exist, primarily in Europe, Asia, and North America.

Fig. 11-4. American elms and the effect of Dutch elm disease. **Left,** American elms were popular because of their overarching growth that provided shade over the streets. This planting pattern increased the demise of the elm due to Dutch elm disease because *Ophiostoma ulmi* moves through xylem connections in root grafts between neighboring trees. Photo taken in 1975 at Longwood Gardens, PA. **Right,** American elm that has died of Dutch elm disease.

Dutch elm disease is a vascular wilt disease in which the fungus, unlike *Verticillium* species and *Fusarium oxysporum*, never lives a life independent of its host. The fungus, *Ophiostoma ulmi* (formerly *Ceratocystis ulmi*), produces several spore stages that help this clever parasite complete its life cycle. The fungus is an Ascomycete, but the sexual stage is relatively rare in the United States because the second mating type necessary for sexual reproduction is not common. The fungus produces two conidial (asexual) stages.

As with other vascular wilts, the primary disease symptom is one of water stress. The invading fungus may produce toxic substances that increase tylose production by parenchyma cells lining the xylem vessels. **Tyloses** are overgrowths of the parenchyma cell protoplasts; these push through the pits in the xylem walls and contribute to vessel blockage. In addition, the fungus produces enzymes that degrade the cellulose, pectin, and other components of the vessel cell walls. The cell walls begin to break down and collapse, which greatly restricts the movement of water in the xylem. In addition, the mycelium and budding conidia of the fungus contribute to vessel blockage. The small, white, oval conidia help spread the fungus throughout the xylem after the fungus is introduced into the tree.

After the tree has died, the fungus becomes saprophytic and grows throughout the trunk of the dying tree. A second kind of conidium is produced in a sticky mass on the tip of a stalk of fused hyphae under the bark of an elm tree that has recently died. These reproductive structures, called **synnemata** or **coremia**, are produced in tiny tunnels carved out beneath the bark by female elm bark beetles. The tunnels are maternal **galleries**

Fig. 11-5. Damage by *Ophiostoma ulmi*, a vascular wilt pathogen. A symptom of the disease is discoloration of the active xylem tissue.

about an inch (two to three centimeters) long, in which the beetles lay their eggs. After the eggs hatch, the larvae feed under the bark, each one making its own side gallery perpendicular to the egg gallery. After a short pupation or resting stage, the new adult beetles emerge through the bark and fly to a healthy elm tree, carrying on their bodies the sticky conidia of *O. ulmi*. The beetles feed on twigs and branches of healthy elm trees and, in the process, deposit the conidia in the xylem of the tree, thus completing the disease cycle.

Two species of elm bark beetles are **vectors** of the Dutch elm disease in North America: *Hylurgopinus rufipes*, a native elm bark beetle, and *Scolytus multistriatus*, a European elm bark beetle that was inadvertently introduced from Europe. The European beetle is more numerous, because it can complete two generations in many parts of the United States, and is an important vector of Dutch elm disease. Beetle galleries are visible on the trunks of dead elm trees when the bark is removed. The European elm bark beetle creates its egg gallery parallel to the wood grain, while the native beetle galleries are more or less perpendicular to the grain. An

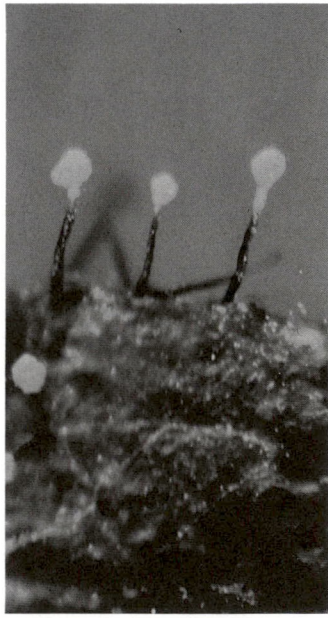

Fig. 11-6. Infection of an elm tree. **Left,** egg gallery of the European elm bark beetle, *Scolytus multistriatus*. The female beetle lays eggs in the maternal tunnel, which runs parallel to the grain of the wood; larval tunnels, produced by feeding larvae, are perpendicular to the maternal tunnel. **Right,** conidia of *Ophiostoma ulmi* are produced on synnemata (coremia) that are produced in egg galleries under the bark of dying or recently dead elm trees. Emerging adult beetles become contaminated with the sticky conidia. (right, ×30)

additional European elm bark beetle exists but, luckily, has not yet been introduced into the United States.

Since fungus transmission by European elm bark beetles is to young twigs in the crown of a healthy tree, the earliest symptom of Dutch elm disease is the flagging or wilting of small branches. Because the xylem fluid moves upward, the initial invasion is fairly slow. Careful inspection of elm trees and pruning out of infected branches at an early stage of disease can sometimes save a tree.

A second means of transmission is more difficult to control. It occurs on the elm-lined streets in many towns and cities in Europe and North America. When trees of the same species are grown close together, roots may graft together underground. Vascular connections may form in the grafts, allowing the passage of the fungus from one tree to the next. The fungus, moving upward in the xylem fluid, may then colonize a tree very quickly. This has led to the death of trees one after another along streets. Monoculture, whether in a wheat field or on a city street, leaves plants vulnerable to diseases. Nearly every town has an Elm Street, but now few elms remain to be seen over much of the United States. Too often, they have been replaced by monocultures of maple, honey locust, or oak that are equally vulnerable to other diseases.

A disease similar to Dutch elm disease and caused by a related fungus is striking oak trees in the Midwest, Michigan, Pennsylvania, Maryland, and West Virginia. Red and black oaks are most susceptible, but all oaks can become infected by *Ceratocystis fagacearum*. This fungus, like *O. ulmi*,

Fig. 11-7. Elm bark beetles transmit *Ophiostoma ulmi* when they feed on healthy elm trees. Their normal length is about 3 millimeters.

causes a vascular wilt disease. Under the bark of dying trees, the oak wilt fungus produces pads of mycelium and conidia that are attractive to insects. Many kinds of insects are attracted by a fermenting odor, and, after leaving the oak, they may move on to another oak and feed in such a way that the fungus is introduced into the xylem, commonly through pruning wounds or bark. Because the vector relationship is not as efficient and specific as in Dutch elm disease, the spread of oak wilt has been slower. However, along tree-lined streets, root graft transmission may increase the rate of infection and death of oak trees.

Control of both Dutch elm disease and oak wilt is difficult. An important sanitary practice is to burn or remove dying and dead trees. In many places, it is illegal to store bark-covered elm logs in wood piles. When Dutch

Fig. 11-8. Oak wilt disease, caused by *Ceratocystis fagacearum*. **Top,** water stress symptoms in leaves of an infected oak tree; **bottom,** fungus pads attractive to insects are produced under the bark of dying and dead oak trees.

elm disease was first discovered in the United States, intensive insecticide spraying was attempted to reduce the spread of the fungus by elm bark beetles. On tree-lined streets, trenching or herbicides were used to break root grafts between trees to try to prevent the fungus from spreading. Today, most remaining elm trees in the United States exist as individuals. These trees can be maintained if carefully watched and pruned when flagging twigs appear. Some trees receive fungicide injections as well. Unfortunately, such injections are relatively expensive and must be repeated every one to three years. Many studies have been made to discover microorganisms that could be injected for biological control of the disease, but so far results have either not been successful or are not yet approved by the U.S. Environmental Protection Agency.

The future of the American elm probably lies in the prolific seed production of the species. Each seed is genetically unique. As young trees grow, many become infected and die, but the introduction of a foreign parasite has never been known to eliminate a species. Over hundreds or perhaps thousands of generations, the species will finally come into balance with its parasite, and mature elms may once again grace the landscape. Tree breeders hope to shorten this time through deliberate crossing and selection for genetic resistance. They have had some success. The earliest resistant trees were selected in Europe and named for the Dutch scientists who participated in the initial studies of the disease. Named Buisman, Schwarz, and Groeneveld, they were, unfortunately, not resistant to all strains of the fungus and also lacked the cold-hardiness of the American elm. Today plant pathologists and breeders continue to search for an elm cultivar that will resist *O. ulmi* and also have the same graceful shape and hardiness that have made the American elm such a popular shade tree. Resistant elms have been released by the U.S. Department of Agriculture and the University of Wisconsin. In North America, breeding programs must take into consideration an additional fatal disease that is threatening elms. It is called elm phloem necrosis or elm yellows and is caused by a bacterium-like pathogen that is described in Chapter 12.

Some beautiful specimens of mature elms still exist, but they require care and protection if they are to survive. The Dutch elm disease story is yet another example of the importance of quarantines for protection of native plants from foreign pests, pathogens, and vectors, but it is not only a historical example. It is a threat today in Australia and New Zealand. Perhaps half a million susceptible elm trees are providing shade and beauty on the Australian continent. An elm bark beetle has found its way to Australia, but, so far, quarantines have successfully barred *O. ulmi*. New Zealand has not been so lucky. Dutch elm disease was discovered in late 1989, and eradication programs have been initiated. It is too soon to know if they will be successful.

Chestnut Blight

The loss of a major shade tree species leaves whole neighborhoods with a bare and blighted look. Earlier in the 20th century, a major forest tree species was nearly destroyed. Chestnut blight was, with citrus canker and white pine blister rust, one of the diseases that triggered the first U.S. quarantine legislation in 1912. Before the blight struck, the American chestnut, *Castanea dentata*, was a major tree species throughout the eastern forests from Maine to Georgia. In the early 1900s, perhaps as many as every fourth tree in the Appalachian forests was a chestnut. The straight, rot-resistant trunks were important for lumber, fence posts, and poles, and the nuts were an important human food crop as well as a major food for a number of wild animals. The dark tannins in the bark were important in the tanning industry.

Chestnut blight is caused by a fungus previously known as *Endothia parasitica* but recently renamed *Cryphonectria parasitica*. Unlike *O. ulmi*, which causes a vascular wilt, chestnut blight is a canker disease. The fungus invades the bark of twigs through small wounds, such as insect feeding sites, and begins to invade the vulnerable vascular cambium of the growing twig. Slowly the mycelium spreads outward until eventually the entire twig is girdled and dies. As the fungus continues growing, it may reach the main trunk, and the mature tree may eventually be killed.

Control of this disease is theoretically possible in a single tree if infected twigs and branches are pruned and destroyed. Fungicide applications are useless, and, of course, pruning trees in forests is not practical. By 1923, the disease had invaded 80% of the chestnut's range, and by the 1950s, 80% of the chestnut trees had died. The fungus, apparently introduced

Fig. 11-9. Chestnut trees. **Left,** forest chestnuts in the Appalachian range before the blight; **right,** chestnut trees in Fairmount Park, Philadelphia, in 1878 and people harvesting the nuts with ladders.

on chestnut seedlings brought to the New York Botanic Garden by a collector, produces sticky conidia and ascospores that are carried to American chestnut trees by birds and insects. The rapid and devastating spread of the disease caused tremendous economic and ecological disruption throughout the Appalachian forests.

The tragedy did not end in North America but spread on to Europe, probably by the export of chestnut wood about 1917. Many European chestnut groves were destroyed following the initial invasion. However,

Fig. 11-10. Nuts of the American chestnut, *Castanea dentata*. **Top,** burr with nuts; **bottom,** harvest from a chestnut plantation in the United States before chestnut blight became widespread.

chestnut trees have a growth habit that keeps the species alive despite the presence of the virulent fungus. From the base of the dead tree trunks, the extensive root system stores enough food reserves to send up new shoots, or "suckers," from the base of the trunk. Year after year, these young shoots allow renewed growth, but eventually they too succumb to the fungus once it has invaded their tissues. Some young trees are able to grow to a sufficient size to allow fruit production, and new chestnuts sprout. Eventually, there is hope that, as with the elm, the chestnut will be able to coexist with *C. parasitica* as a minor parasite. Most people do not expect to live long enough to see this happen.

The continued production of suckers from the base of the dying chestnut trees inspired some to think that perhaps, somehow, the trees could overcome the effects of the deadly parasite. Robert Frost, the famous American poet, even discussed this possibility in his poem "Evil Tendencies Cancel":

> Will the blight end the chestnut?
> The farmers rather guess not.
> It keeps smoldering at the roots
> And sending up new shoots
> Till another parasite
> Shall come to end the blight.

There is now scientific evidence that this scenario may be happening. In Italy, in the 1950s, some orchard chestnuts were found to be surviving despite the presence of extensive cankers. Studies of the *C. parasitica* isolates from these cankers suggested that new strains of the fungus had arisen that were much weaker than the common strains. Such strains have been

Fig. 11-11. An old photograph of dying chestnuts during the initial outbreaks of the blight caused by *Cryphonectria parasitica*.

designated **hypovirulent**. Not only are they weaker but, when they come into contact with virulent strains, the hyphae may fuse (a process called **anastomosis**), and the factor that causes hypovirulence spreads to the virulent strain and weakens it. The phenomenon of hypovirulence seems to be a disease of the fungus. An extra piece of double-stranded RNA in the mycelium of the hypovirulent strains replicates and spreads into the mycelium of virulent strains, weakening them.

While this is not completely understood, the existence of hypovirulent strains has provided a practical biological control to "cure" existing cankers. In Europe, the application of hypovirulent strains of *C. parasitica* to active chestnut blight cankers is now used to stop canker activity. In the United States, there is hope that perhaps this naturally occurring biological control can somehow be used to help reestablish the chestnut in eastern forests.

The practical application of this procedure faces some complications. The hypovirulent strains usually grow more slowly and produce fewer spores, which reduces their ability to spread in nature. The procedure is helpful in stopping the growth of an active canker, but so far no way has been found to prevent the disease in the first place. Also, the parasite has numerous genetic strains. Spread of the hypovirulence factor is possible only if the virulent strains present in a tree are genetically compatible with the hypo-

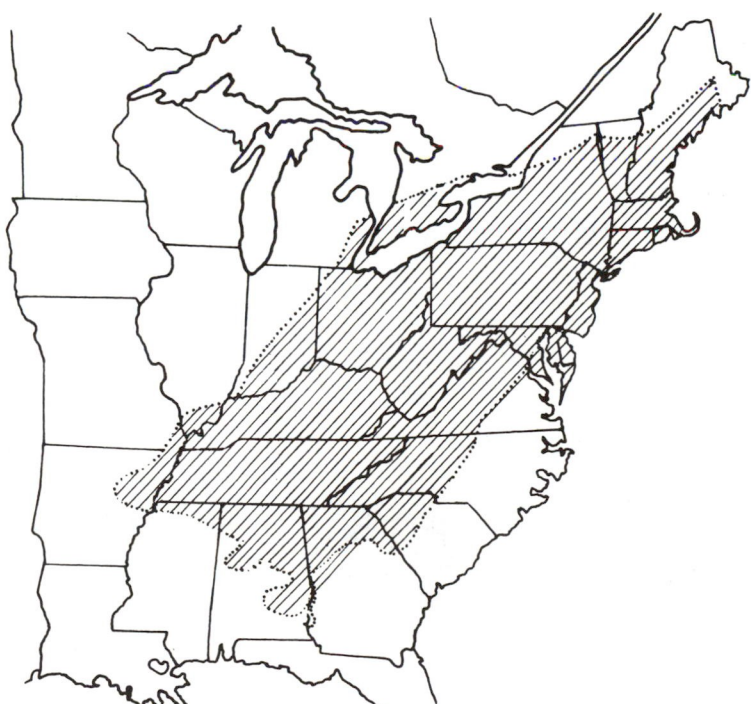

Fig. 11-12. Natural range of the American chestnut.

virulent ones so that anastomosis will occur. In the United States, cankers may contain 10 or more different genetic strains, so biological control may require application of hypovirulent strains that are genetically compatible with all the virulent strains present.

Scientists are optimistic that this method may eventually provide useful results. In the meantime, they are greatly interested in the general phenomenon of hypovirulence. Hypovirulent strains associated with double-stranded RNAs have been found in other fungi, and scientists believe that more knowledge about this interesting phenomenon may lead to a better understanding of how plants and parasites come into genetic balance.

Other Canker Diseases of Trees

Tree species remain the most vulnerable to destruction by introduced pathogens due to the long time needed to replace mature trees. Potential economic and ecological losses make their protection a high priority. The American beech (*Fagus grandifolia*), an important tree in mature eastern forests, is being destroyed by an introduced canker fungus, *Nectria coccinea* var. *faginata* and a native fungus *N. galligena*. By themselves, the fungi cannot penetrate bark, but they can enter through tiny wounds made by the beech scale, *Cryptococcus fagisuga*, a tiny sucking insect. As in chestnut blight, the vascular cambium is killed by the invading fungus. The future of an infected tree depends on the growing conditions and age of the tree. Older trees seem more susceptible to rapid killing, whereas younger trees, although more likely to survive, become defective as timber trees. Following

Fig. 11-13. Living chestnut trees planted outside the natural range in Michigan where they have, so far, escaped infection.

infection, the bright red fruiting bodies of these Ascomycetes erupt through the bark of infected trees. The fungi produce both conidia and ascospores that can be disseminated by air, water, and vectors such as insects, birds, and small mammals.

Control of these canker diseases in forest situations is very difficult and lies mainly in forestry management practices that provide greater species diversity and the culling of infected trees. Beech trees in yards and parks can be protected by pruning infected branches and reduction of scale infestations. Scale insects are usually well protected by a waxy coating, but insecticides can be used to kill them in their crawler stage. Scrubbing of beech bark with a detergent will also help remove the scales. The relatively sedentary life style of the scales reduces the spread of this canker disease. It made very slow progress from the site of its initial introduction in Nova Scotia, Canada, on European beeches in 1890 and did not arrive in the Adirondack Mountains of New York until 1978.

Numerous canker fungi threaten woody plants, including many ornamental trees and shrubs. In some cases, native fungi are an important threat to imported plants. For example, the popularity of the Colorado blue spruce in the eastern United States has increased losses due to Cytospora canker. Colorado blue spruces are adapted to the dry, cold winters of the western

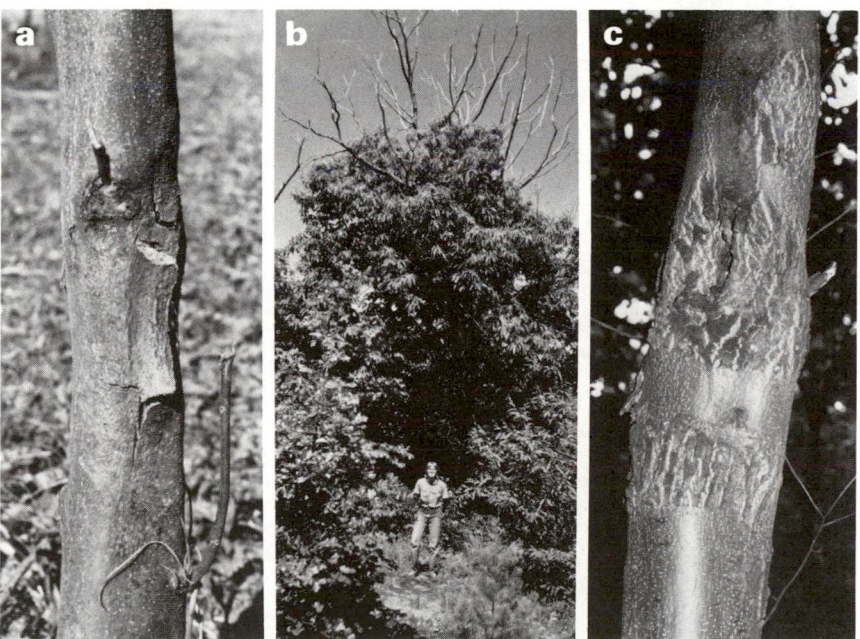

Fig. 11-14. Cankers on chestnut trees. **a,** Lethal canker of *Cryphonectria parasitica.* **b,** Blighted but recovering chestnut tree. Hypovirulent strains of the pathogen have been isolated from the cankers. **c,** A healing canker in which hypovirulent strains of the pathogen have been found.

Rocky Mountain states and become infected by canker fungi in the moist climates in the eastern United States. Many canker diseases are dependent on host plants being concentrated in a small area. For instance, black knot of *Prunus* species is a serious canker disease in uncared-for cherry and

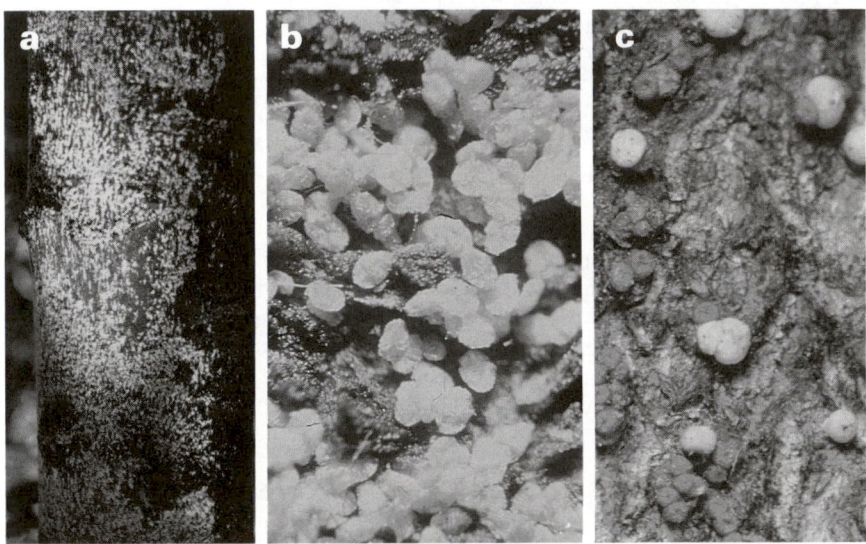

Fig. 11-15. Diseased beech tree. **a,** An American beech, *Fagus grandifolia*, infested with beech scale, *Cryptococcus fagisuga*; **b,** scales with waxy coating removed; **c,** fruiting bodies of the *Nectria* fungi emerging through the bark of an infected tree. Canker fungi are lethal to beech when they gain entry through wounds made by feeding scale insects.

Fig. 11-16. Black knot canker disease of cherry and plum, caused by *Apiosporina morbosa*.

plum orchards but is usually a minor problem on wild cherry trees growing in genetically diverse and species-diverse natural ecosystems.

The damage that a canker fungus causes often depends on the vigor of the tree. Healthy trees may be able to produce callus and new vascular cambium at a rate that will keep the canker fungus from girdling the branch or trunk. In such cases, a canker may show annual rings of callus growth as the tree attempts to contain the growth of the invading fungus. Such a biological defense depends on the vigor of the host, so trees stressed by other environmental factors such as drought or air pollution may not be successful in containing the pathogen. Many canker-forming fungi are "opportunists" and can only attack woody plants stressed by drought, extremes in temperature, other diseases, or a reduced root area.

Woody plants in landscapes and orchards should be pruned to remove canker infections while they are still limited to small branches. Since most canker fungi enter woody plants through wounds, dormant-season pruning is safest because the air is dry and fungi are not active. Pruning wounds are not as likely to be invaded by other canker and decay fungi and bacteria from fall to late winter.

Wood Decay Fungi

A specialized group of fungi is capable of wood decay. Most of these fungi are Basidiomycetes, and many produce relatively large fruiting bodies commonly seen on the trunks and branches of infected trees. To penetrate the normally protected woody tissues, wood decay fungi require wounds that expose the wood. Wounds may be produced by birds, insects, animals, wind or winter, lightning, and many human abuses from lawn mowers and automobiles.

The fruiting bodies of wood decay fungi exist in an amazing array of sizes, shapes, and colors. They may be fleshy or dry, annual or perennial

Fig. 11-17. Fruiting body of a wood decay fungus on an infected tree trunk.

(adding new layers each year). Many have various patterns of gills, pores, and wavy layers that increase the spore-producing area on the surface of the structure. The basidiospores are forcibly discharged in tremendous numbers to be carried on the wind to new victims. Most perish, but a few find their way into a new tree. Saprophytic Basidiomycetes, responsible for much of the decay of dead trees, produce similar fruiting bodies. It is not usually possible to know whether a fruiting body on the trunk of a tree is the pathogen that killed it or just a saprophyte functioning in decay unless the fungus is identified. In either case, removal of the fruiting body will not help the tree because the active vegetative mycelium is what invades the wood. A large reproductive structure simply reflects the massive amount of vegetative mycelium inside the tree, perhaps extending 5-10 feet above or below the fruiting body of the fungus.

Trees that succumb to invasion by wood decay fungi have failed to successfully fight the invasion. One of the secrets of long-lived trees is their ability to prevent invasion by bacteria, fungi, and other microorganisms that might cause decay or other diseases. Rather than fight each possible enemy individually, trees have a mechanism for isolating wounded woody tissue. This mechanism is called **compartmentalization**. It is a biological process that involves the isolation of wounded tissue, the production of antimicrobial compounds by the tree to fight invasion, and a change in the pattern of cell production that allows the tree to continue growth despite the presence of the wound. The type and pattern of cells in woody tissue, described earlier in the chapter, help the tree compartmentalize and "wall off" the invader.

When a wound exposes internal woody tissue to the environment, the tree immediately blocks off the area that has been exposed. Woody tissues consist of numerous small compartments that are bounded by 1) ray parenchyma cells, 2) parenchyma and late season cells of the annual rings, and 3) vascular cambium. Living parenchyma cells allow the tree to respond to wounds by changing biochemical pathways and producing antimicrobial substances, especially phenols, such as tannins. These compounds also help strengthen the cells. The tracheids and vessels become plugged by tyloses, air bubbles, and other substances, which helps prevent microbial invasion.

Cross sections of wounded wood usually consist of a small area of decayed and invaded wood bordered by a darker layer that compartmentalizes the wounded area to prevent further invasion. Vascular cambium produces more parenchyma cells in the area of the wound, which allows a stronger biochemical strengthening of the compartment borders. The final response of a tree to wounding is to keep growing. Unlike animals, which have a healing process that eliminates dead and wounded tissue, trees maintain wounded tissue and continue to grow around it. This natural process, in addition to protecting trees from invasion by microorganisms, allows trees to self-prune lower or damaged branches. The branch stub in the main trunk becomes darker and stronger after the loss of the branch, but the dead stub tissue is compartmentalized to prevent invasion of the main trunk by microorganisms. This phenomenon is easily seen in knotty pine lumber.

An understanding of compartmentalization in woody plants has changed some tree maintenance practices. For instance, some tree surgeons no longer "clean out" decayed areas of hard, darkened wood but leave it as a protective layer. Tree branches should be cut without damaging the branch collar rather than flush with the trunk. This allows the tree to compartmentalize the branch stub more successfully. Also, many tree specialists no longer "paint" a tree wound after pruning because compartmentalization processes should normally protect the tree from invasion.

It is now clear that trees that react quickly to a wound may be able to isolate the tissue before invading microorganisms can cause further damage. The success with which the tree can compartmentalize and prevent infection depends on several factors. The size, depth, and location of wounds are physical factors that affect successful compartmentalization. In addition, because the process is biological, the health and vigor of the tree affects the rate and strength of the chemical and biological reactions of the healing process. Trees subjected to other stresses are less likely to be successful.

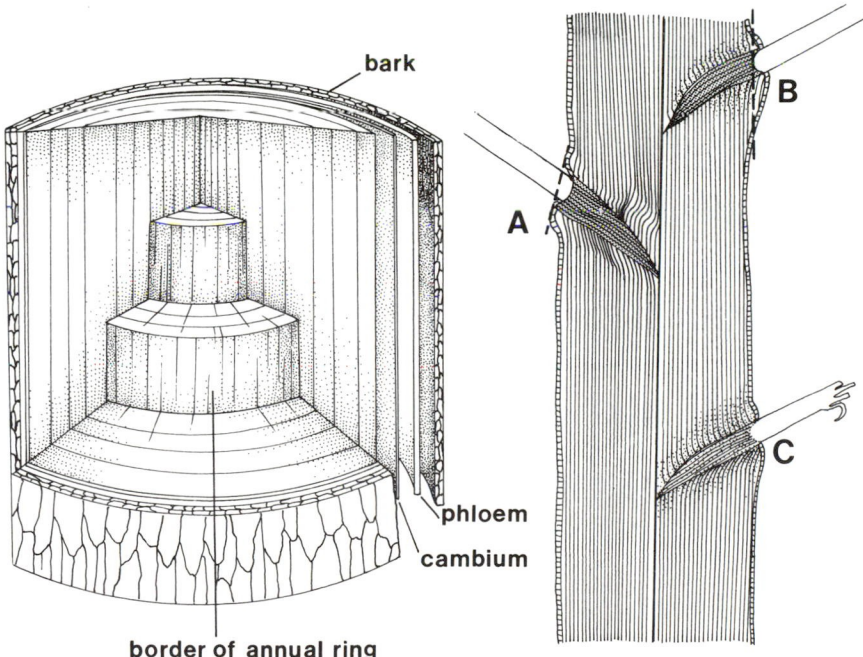

Fig. 11-18. Natural compartments in trees help reduce the spread of decay. **Left,** compartments in a mature tree. **Right,** branch stubs may or may not cause further decay in a tree trunk. A properly pruned branch (A) does not disturb the branch collar, allowing the branch stub to compartmentalize. Improper pruning (B) cuts open larger areas of the trunk to potential decay. Broken branches (C) may encourage the growth of decay fungi that can overcome compartmentalization of the branch stub.

Besides avoiding wounds and using proper pruning techniques, the best way to assure the long life of a tree is to keep it healthy and vigorous through fertilization, watering, spacing, and site selection, so it will be able to compartmentalize successfully.

Mushrooms and Trees

It is relatively common to find mushrooms beneath many kinds of trees. An important parasite that causes root rot of many tree species is *Armillaria mellea*. It is an aggressive parasite of conifers in the western United States and a parasite of stressed hardwoods in the eastern United States. In autumn, clumps of the spore-producing mushrooms can often be found at the base of infected trees. *A. mellea* commonly infects tree stumps and then spreads to neighboring healthy trees with stringlike masses of mycelium called **rhizomorphs**, giving it the common name "shoe string" fungus.

Other types of mushrooms are commonly found in forests. While some are saprophytes important for the decay of dead and dying trees, many have an intimate **mutualistic** relationship with a nearby tree. Such fungi are called **mycorrhizae**. The word *mutualistic* means that both organisms benefit from the relationship. The fungus obtains nutrients from the tree, and the tree receives protection from root pathogens as well as increased surface area for absorption of water and certain minerals, especially phosphorus. Nearly all plants, woody and herbaceous, have mycorrhizae associated with their roots. Some, like orchids and maples, host **endomycorrhizae** that grow into the root cortical cells. Many trees, such as pines and oaks, commonly host **ectomycorrhizae** that form a thick outer layer of mycelium and grow around the cortical cells, forming a layer called

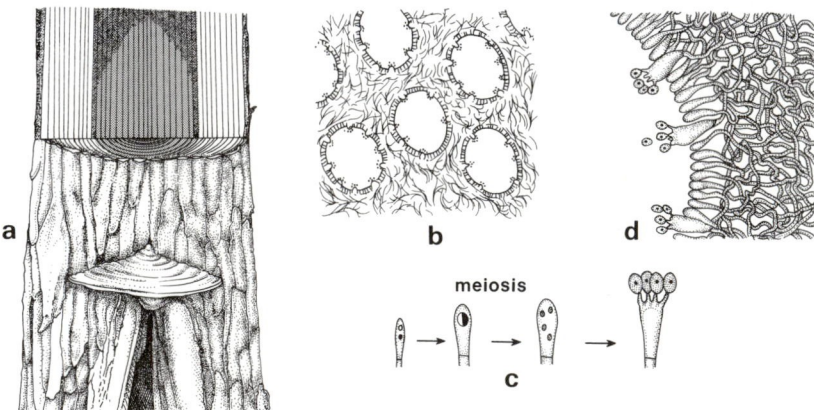

Fig. 11-19. Reproduction of a wood decay fungus. **a,** A fruiting body is produced on an infected tree trunk. **b,** The lower surface of the fruiting body is composed of numerous tiny pores, each lined with spore-producing cells (basidia). **c,** Karyogamy and meiosis take place in each basidium, resulting in the production of four haploid basidiospores. **d,** Basidiospores are forcibly discharged to be wind-dispersed.

the **Hartig net**. Early attempts to plant trees in treeless regions, such as the U.S. Great Plains, Southern Africa, and parts of Australia, were not very successful because the proper mycorrhizal fungi were missing. The seedling trees were grown in sterile soil, and many died when transplanted outdoors without their mycorrhizae. Modern forest plantings include inoculation of sterilized soil and tree seedlings with mycorrhizae.

Mycophiles (mushroom lovers) and mycophagists (mushroom eaters) quickly learn to search for their favorite fungi under the appropriate tree hosts. Flushes of mushrooms are especially profuse under dead and dying trees as the mycorrhizal fungi move on to new host trees via airborne spores. As in other Basidiomycetes, each basidiospore of a mushroom is usually haploid and of a single mating type. The primary mycelium produced from the basidiospore must then fuse with the primary mycelium of a

Fig. 11-20. Root rot caused by *Armillaria mellea*. **Top,** mycelial fans beneath the bark of an infected tree; **bottom,** mushrooms produced at the base of an infected tree.

basidiospore of the other mating type to form a dikaryotic mycelium capable of continued growth and reproduction. Karyogamy and meiosis take place in specialized cells, the **basidia**, from which the **basidiospores** are produced. Each basidiospore forms on a tiny projection, from which it is forcibly discharged. Some related Basidiomycetes use a more passive dispersal mechanism such as puffballs, which produce millions of basidiospores in a powdery mass that is dispersed by raindrops or other bumps.

Parasitic Higher Plants

Another kind of parasite may infect many kinds of trees. This type of pathogen is much larger than the others discussed so far. More than 2,500 species of flowering plants that parasitize other plants are found among many botanical families. Most are not of economic importance but are interesting curiosities. The largest known flower is that of the parasitic plant *Rafflesia arnoldii*, which grows only in a few areas of Indonesia and Malaysia. The dark red, mottled flowers can be more than three feet (more than one meter) across, can weigh 15 pounds (six kilograms), and smell like rotten flesh to attract flies that assist in pollination. The species is **dioecious**, which means that individual plants produce either male or female flowers. The plants produce no photosynthetic tissue but rely totally on a tropical vine for all water and food. In North America, beech-drops (*Epiphagus virginiana*) and squaw-root (*Conopholis americana*) are common parasites of beech and oaks, respectively.

Many trees host various species of parasitic mistletoes. True mistletoes

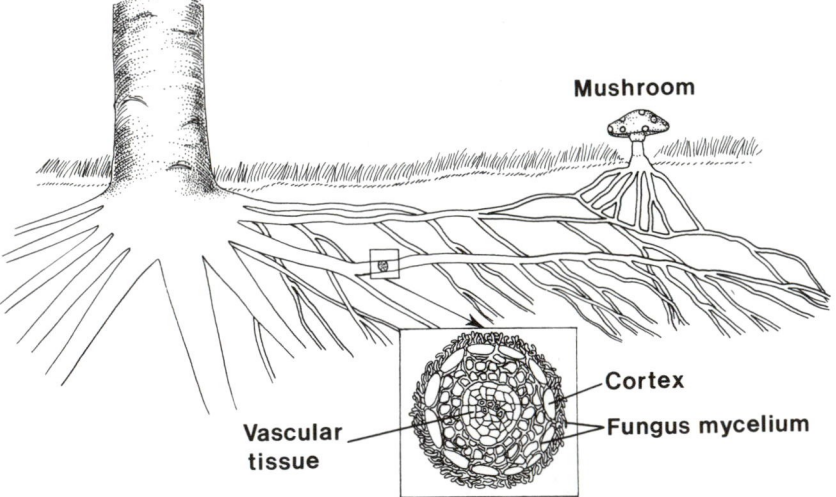

Fig. 11-21. An ectomycorrhizal fungus in association with the roots of a forest tree. Mushrooms of the fungus may be found beneath the host tree. A cross section of an infected root reveals the netlike layer of mycelium on the root surface and hyphae between the cortical cells.

are large evergreen plants that invade the xylem of hardwoods trees, absorbing water and minerals through haustoria. In North America, true mistletoes, *Phoradendron* species, exist mostly in the South. The name comes from the Greek for "tree thief." In Europe, true mistletoes are classified in the genus *Viscum*. Like *Rafflesia*, all mistletoes are dioecious. Female mistletoe plants produce white berries that attract birds, which aid in seed dispersal.

Mistletoes have fascinated people for many centuries. Their ability to stay green in winter and their seemingly magical ability to live without roots on trees, never touching the ground, led ancient people to believe they were divine gifts, perhaps created by thunderbolts. Mistletoes had an important place in many ancient religious rituals among tree worshippers of western Europe, including the Celtic Druids and early Scandinavians and Germans. Ceremonies of both the summer and winter solstice holidays frequently involved the harvest of mistletoes. Druid priests cut the mistletoe with a golden sickle onto a white cloth to prevent the plant from touching the ground. Mistletoe harvest was also part of Norse ceremonies involving Balder, the son of Odin. Because of its role as a parasite of the all-powerful oak, mistletoe became a symbolic source of protective and medicinal powers. The burning of oak logs, human sacrifice, and the harvest of mistletoe were intertwining solstice ceremonies. Mistletoe was brought into houses for protection, and enemies would hug beneath the mistletoe to make a truce. The Christmas custom of hanging mistletoe and kissing under it probably originated in these ancient practices.

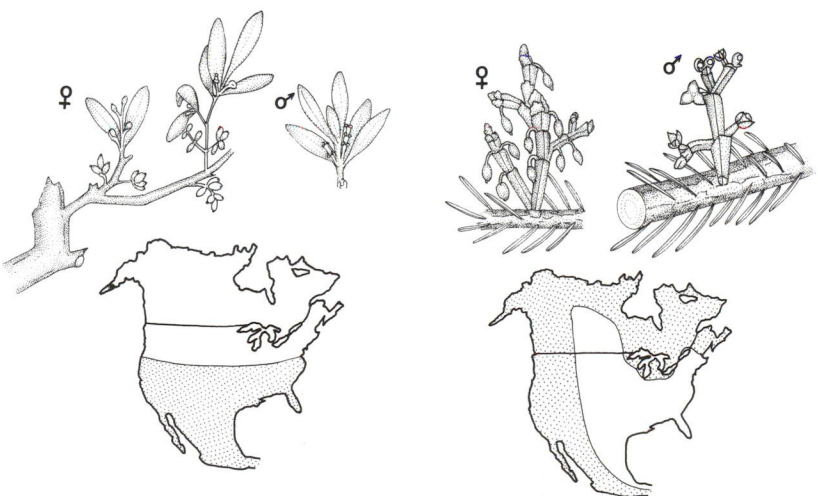

Fig. 11-22. Mistletoes, dioecious plant parasites. **Left,** true mistletoes, *Phoradendron* species, infect hardwood trees and are found in the southern half of the United States. **Right,** dwarf mistletoes, *Arceuthobium* species, infect conifers and are found in the range indicated on the map, primarily the western states, the Great Lakes region, and northern New England.

Another type of mistletoe is an important parasite of conifers. The dwarf mistletoes, *Arceuthobium* species, grow up to several inches tall and invade both the xylem and phloem of host trees. They produce tiny sticky seeds that are shot from mature fruits, landing on the needles and twigs of host trees. They germinate there and penetrate, then grow deep into the woody tissues. Conifers heavily infected by dwarf mistletoes become distorted and stunted, resulting in poor-quality timber. Infected trees often develop clusters of branches known as **witches' brooms**. Many modern forestry practices, in particular improved fire control and the reduction of "clear cutting" of timber lands to help prevent soil erosion, have, unfortunately, increased problems due to mistletoes. Fires reduce many parasites and insect pests. Selective cutting and replanting over a period of time provide trees of many ages in the same areas and allow the parasites to flourish. Control of these parasites is very difficult.

Some parasitic plants cause problems in agronomic crops as well. **Dodders**, *Cuscuta* species, parasitize many crops, including clovers, alfalfa, and

Fig. 11-23. Spruce tree with distorted growth resulting from infection by dwarf mistletoe in northern Michigan.

cranberries. Their common names include strangleweed, pull-down, and devil's hair. Dodder seeds germinate to form rootless shoots that grow upward in search of a host. Once a host plant is contacted, dodder plants sink haustoria into the plant tissue to form both xylem and phloem connections. The tangled yellow strands of dodder reduce the growth of host plants and can make harvest extremely difficult.

A major agricultural pest in many parts of Africa and Asia is witchweed (*Striga* species). These plants get their name because they are root parasites. The aboveground green shoots produce beautiful yellow and red flowers that deceptively mask the deadly parasitism below the ground. Witchweed parasitizes important food crops, including corn, sugarcane, rice, and sorghum. It is a particularly difficult parasite to control because a single plant can produce over 50,000 tiny seeds that can remain viable in soil for up to 20 years. The seeds will not germinate until they detect a chemical

Fig. 11-24. Dodder (*Cuscuta* species) parasitizing cranberries.

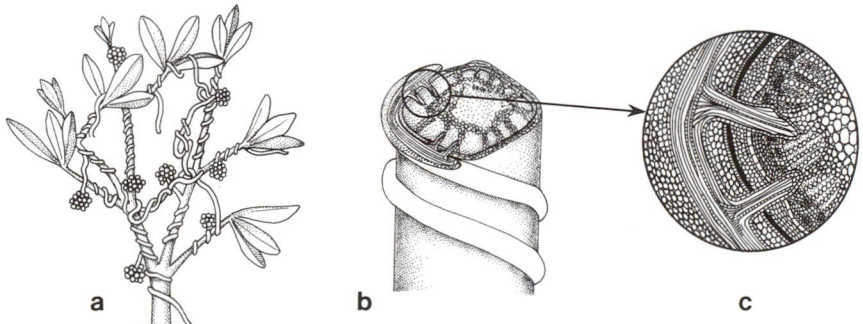

Fig. 11-25. Diagrammatic view of dodder on a host plant (**a**). Dodder winds around the host stem as it grows (**b**), invading with haustoria to absorb water and nutrients from the vascular tissue of the host (**c**).

stimulus from root exudates of host plants. In recent years, the return of rain to many drought-ridden areas of Africa resulted in the germination of many *Striga* seeds along with those of the food crops. Badly infested fields can lose 90% of the crop to witchweed, which has led to the abandonment of farmlands in areas of the world that can least afford the loss.

Witchweed was accidentally introduced into the United States in 1956. It is under strict quarantine in North and South Carolina, where eradication programs are in progress. The potential ecological range of this important weed parasite includes most of the eastern half of the United States. The most effective means of control involves stimulation of seed germination without allowing the plants to mature and produce new seed. In some cases, sacrificial "trap crops" have been planted to stimulate witchweed

Fig. 11-26. Witchweed, *Striga* species, a parasite of host roots. **Top,** at the base of a corn plant; **bottom,** corn growing poorly due to witchweed infection. Witchweed plants can be seen between the corn rows.

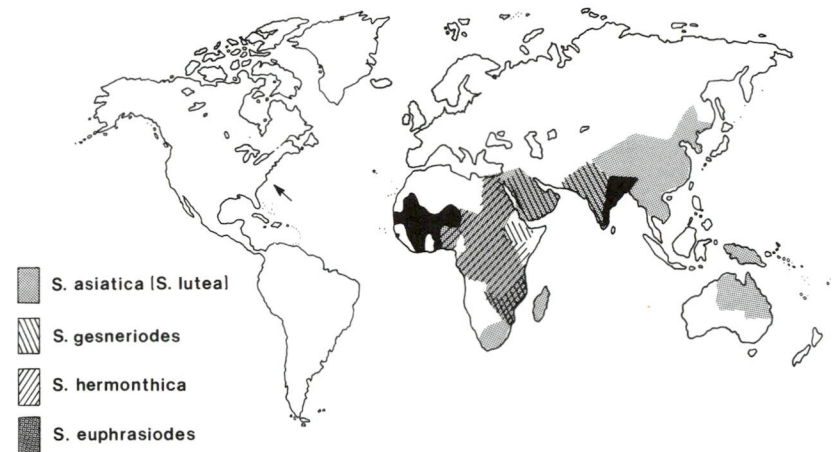

S. asiatica (S. lutea)

S. gesneriodes

S. hermonthica

S. euphrasiodes

Fig. 11-27. Areas of witchweed infection. Witchweed is under quarantine in North and South Carolina, the only area of infestation in the Western Hemisphere. Several *Striga* species are important agricultural pests in other parts of the world, particularly in Africa and Asia.

seed germination and then plowed under to destroy the parasites before they set seed. Ethylene, the growth hormone that aids in fruit ripening, can be applied to soils to stimulate witchweed seed germination before crops are planted. In the United States, the hope remains that witchweed can be eradicated. In Africa and Asia, it remains a serious agricultural pest for the world's poorest farmers.

Selected Readings

Anagnostakis, S. L. 1987. Chestnut blight: The classical problem of an introduced pathogen. Mycologica 79:23-37.

Burnham, C. R. 1988. The restoration of the American chestnut. American Scientist 76:478-487.

Cochran, M. F. 1990. Chestnuts—Making a comeback? National Geographic 177(2):128-140.

Eplee, R. E. 1981. *Striga*'s status as a plant parasite in the United States. Plant Disease 65:951-954.

Harley, J. L. 1989. The significance of mycorrhizae. Mycological Research 92:129-139.

Holmes, F. W. 1976. The American elm fights back. Horticulture 54(1):72-78.

Holmes, F. W., and Heybroek, H. M. 1990. Dutch Elm Disease—The Early Papers. The American Phytopathological Society, St. Paul, MN.

Houston, D. R. 1984. What is happening to the American beech? The Conservationist 38(6):22-25.

Lanier, G. N., Schubert, D. C., and Manion, P. D. 1988. Dutch elm disease and elm yellows in central New York. Plant Disease 72:189-194.

Newhouse, J. R. 1990. Chestnut blight. Scientific American 263:106-111.

Nour, J., Press, M., Stewart, G., and Tuohy, J. 1986. Africa in the grip of witchweed. New Scientist 109(1490):44-48.

Shigo, A.L. 1985. Compartmentalization of decay in trees. Scientific American 252(4):96-103.

Sinclair, W. A., Lyon, H. H., and Johnson, W. T. 1987. Diseases of Trees and Shrubs. Cornell University Press, Ithaca, NY.

Viruses and More Recently Discovered Pathogens

All living organisms support parasites, and parasites are found in every taxonomic group. Some of those discussed in previous chapters include fungi, bacteria, nematodes, and even flowering plants (such as mistletoe). These parasites are all relatively easy to see, although many require magnification, and the symptoms they cause can be related to the interaction between the parasite and the host plant. There is a group of parasites, however, that causes plant diseases and symptoms (including distortions, off-color, stunting, and reduced crop quality and yield) that can also be caused by such diverse factors as poor soil fertility, soilborne pathogens, herbicide residues in soil, and even air pollution. They are not easily visible in the vascular tissue where they reside, and some are even submicroscopic. The most important pathogens in this group are the viruses. Viruses also have the longest scientific history. The other pathogens discussed in this chapter were discovered very recently.

Viruses as Plant Pathogens

A virus contains the minimum characteristics necessary for parasitism. They are mainly nucleic acid and protective coat protein. The nucleic acid may be single-stranded (ss) or double-stranded (ds) RNA or DNA, although the vast majority of plant viruses are composed of ssRNA.

A. Mayer, a German scientist, first identified mosaic (a random pattern of light and dark green color in leaf tissue) in tobacco as a disease in 1886, but he considered the causal agent to be a bacterium. The first indication that a new kind of pathogen was involved came from experiments by the Russian scientist D. Ivanovski in 1892. He passed sap from an infected tobacco plant through a filter that would remove bacterial cells, and the sap remained infective. This demonstrated that whatever caused tobacco mosaic disease was smaller than bacteria. In 1898, M. W. Beijerinck, a Dutch scientist, was able to prove that this infectious sap contained a pathogen that multiplied. After healthy plants were inoculated, a period of time was required, during which the pathogen increased, before the sap in newly infected plants became infective.

Further understanding of the structure and function of viruses had to await new technologies of the 20th century, which allowed chemical analysis and visualization of the viruses. In fact, many of these techniques were invented as new tools to unravel the mysteries of these pathogens. Diseases caused by viruses were particularly frustrating to scientists and physicians. Unlike fungal and bacterial diseases, virus diseases had no visible pathogen, and common culturing techniques failed to isolate a causal agent. Virus diseases of animals and humans were not cured by the "miracle drug" antibiotics that were so successful against bacterial pathogens. Humans and animals might recover from virus diseases through the production of anti-

Fig. 12-1. Detail of a painting by Dutch artist Jacob Marrel (1614–1681) includes "bizarre" tulips with color breaking due to virus infection.

bodies by their immune systems; virus-infected plants had to be destroyed.

Virus diseases of plants had been recognized for many centuries even though no one knew anything about their causal agents. Virus-infected flower bulbs caused economic mayhem in Western Europe, reaching a peak in 1634–1637. About 80 years earlier, a Flemish diplomat, Ogier Ghiselin de Busbecq, had introduced tulips to Europe after he discovered them as wildflowers in Turkey. Tulips became a craze, and an unstable speculative market grew as people tried to make their fortune on the new flowers. The most valuable bulbs were those that produced flowers in which streaks, flames, and feathers of contrasting colors were produced. Known as "bizarres," these bulbs were particularly valuable because the seed produced by the flowers usually grew only into plants with normal, solid-color flowers.

It is now known that such color "breaking" is a common symptom in virus-infected plants. These plants may transmit a virus to their seed, but usually at a low rate, perhaps 1–10%. This contributed to the value of bizarre tulip bulbs, one of which was traded for the exorbitant price of $8,395. The market finally crashed, leaving many dealers bankrupt, but The Netherlands remains the producer of 95% of the world's tulip bulbs.

During the so-called "tulipomania" period, these flower symptoms were shown in paintings by many Dutch and Flemish artists, and even today one popular tulip cultivar is known as the Rembrandt. Modern tulips are available with genetic mutations that consistently give beautiful color patterns. Mutations were needed to accomplish this because virus-infected plants eventually become stunted and produce small and weakened bulbs. As described in Chapter 1, potato growers saw the same "running out" in virus-infected potato tubers. By periodically harvesting the seeds of sexual reproduction, they could obtain virus-free plants and restore the yield that is greatly reduced by virus infection.

Virus Transmission in Plants

Because viruses are very small and unable to grow or move, they must enter plant cells through a wound. The only exceptions to this are viruses that enter an ovule during fertilization by virus-infected pollen and viruses that pass through root grafts between neighboring plants. Many agricultural practices result in the **sap-transmission** of viruses when wounds are produced during cultivation, touching, grafting, or pruning. A few viruses may also be transmitted when infected plants are rubbed against neighboring plants during windy weather and storms.

In nature, various **vectors** are responsible for virus transmission, the most important of which are arthropods, especially insects with sucking mouthparts. They may carry viruses on their mouthparts, in the sap drawn up and carried in or on the **stylet** (a hollow feeding tube), or in their bodies as they move on to a healthy plant. Aphids and leafhoppers, in particular, transmit many viruses.

The relationship between vectors, viruses, and host plants is complex. One kind of transmission is called **nonpersistent** (or styletborne); in this, the insect becomes capable of transmitting a virus after only a few seconds of feeding on an infected plant but remains viruliferous (capable of transmitting the virus) for only a short time. Transmission that is called **persistent** (or circulative) involves a longer feeding period, in which the insect acquires more virus, and usually a **latent period** before the insect becomes able to transmit the virus. During this time, the virus moves through the digestive system of the insect, circulates in the blood, and eventually passes into the salivary gland. The virus can then be excreted along with salivary secretions into plants when the aphid feeds. After the latent period, which is at least 12 hours for aphids and 1–2 weeks for leafhoppers, the insect remains viruliferous for some time. In a few cases, the virus even infects and replicates in the insect, which can greatly increase the transmission

Fig. 12-2. Symptoms of cucumber mosaic virus infection. **Top,** mosaic in pumpkin leaves; **bottom,** irregular green pattern on yellow squash.

potential. Other categories of transmission fall between persistent and nonpersistent.

Sometimes the virus-vector relationship is specific to one insect species, and other times it is quite nonspecific. A single aphid can feed on a plant

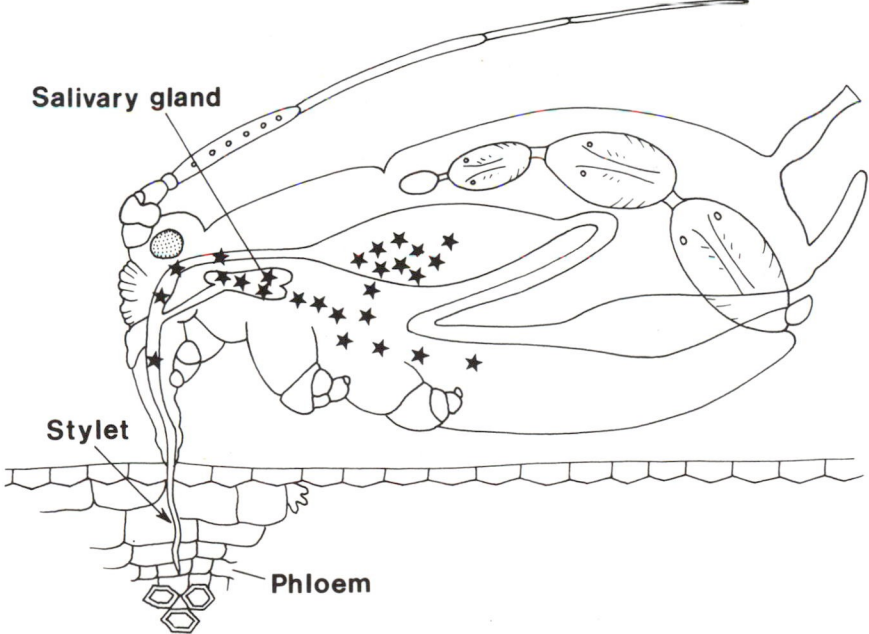

Fig. 12-3. Virus transmission by an insect vector. **Top,** aphid; **bottom,** diagrammatic view of virus transmission by an aphid. In persistent transmission, the virus enters the digestive tract of the aphid and eventually arrives in the salivary glands, as indicated by the stars. Aphids and other sucking insects frequently probe deep into plant tissue and feed in phloem.

infected with three different viruses and transmit one virus in a persistent manner, transmit a second virus in a nonpersistent manner, and be unable to transmit the third virus. Other viruses, known as **soilborne viruses**, must be transmitted by vectors in the soil since they are incapable of entering the plant on their own. Soilborne primitive fungi and ectoparasitic, migratory nematodes are known to vector these viruses.

Structure and Replication of Viruses

Not until the 20th century were the structure and chemistry of viruses understood. The electron microscope and the techniques that allow detailed visualization of viruses were not available until about 1945. Plant viruses may be short to long rigid or flexuous rods, spherical (or more commonly polyhedral), or shaped somewhat like rod-shaped bacteria only smaller. They are commonly approximately 10 times shorter in length than bacteria, however, and are measured in nanometers (1 nm = 10^{-9} meters). A typical rod-shaped virus might be about 15×300 nm.

W. M. Stanley, an American scientist, received the Nobel Prize in 1946 for the first crystallization of a virus, which was tobacco mosaic virus. For the first time, a relatively large amount of purified virus was available for analysis. Shortly thereafter, F. C. Bawden and N. W. Pirie, in England, were able to demonstrate that the virus was a nucleoprotein. Bawden was responsible for much of the fundamental work with potato viruses. Later work showed that the nucleic acid was necessary for virus infection, and the protein served only in a protective role. Of course, further understanding of viruses was dependent on the publication of the structure of DNA by J. D. Watson and F. Crick in 1953, as discussed in Chapter 5. These scientists also discovered some structural aspects of virus coat proteins.

Plant viruses have not been given Latin binomials. Instead, they are placed in groups based on several characteristics including their size, shape, and nucleic acid components. Viruses are often named according to one of their host plants and the symptoms they cause and are abbreviated with initials. For example, tobacco mosaic virus is commonly written as TMV.

All viruses are composed of either DNA or RNA. Many viruses consist of one type of **virion** (virus particle), which contains the complete genome, or genetic information, of the virus; however, the genome in a number of plant viruses consist of several different parts. In such viruses, all parts of the genome must be present in a plant cell before the virus can function. The need to infect the cell with several different types of particles does not necessarily reduce transmission efficiency. Many easily transmitted plant viruses are of this type. One is cucumber mosaic virus (CMV), a common cause of mosaic symptoms in cucumbers and squash.

Chapter 5 explained that genetic information in the DNA is transcribed to messenger RNA (mRNA), which moves from the nucleus into the cytoplasm. The genetic code of the mRNA is then translated through **triplet codons**. Transfer RNAs (tRNAs) carry amino acids to the mRNA, and, with the help of the ribosomes, a protein is eventually produced. These same mechanisms are used in the replication of plant viruses.

When plant viruses enter the plant cell, the protective protein coat is removed. In most plant viruses, the nucleic acid is a ssRNA that serves as a mRNA, using the same genetic code that functions in all living organisms. One section of the virus genome codes for RNA polymerase, an enzyme necessary for replication, or multiplication, of the virus RNA. A "mirror-image" copy of RNA is made from the original virus RNA, which then serves as a template to produce many copies of the virus RNA.

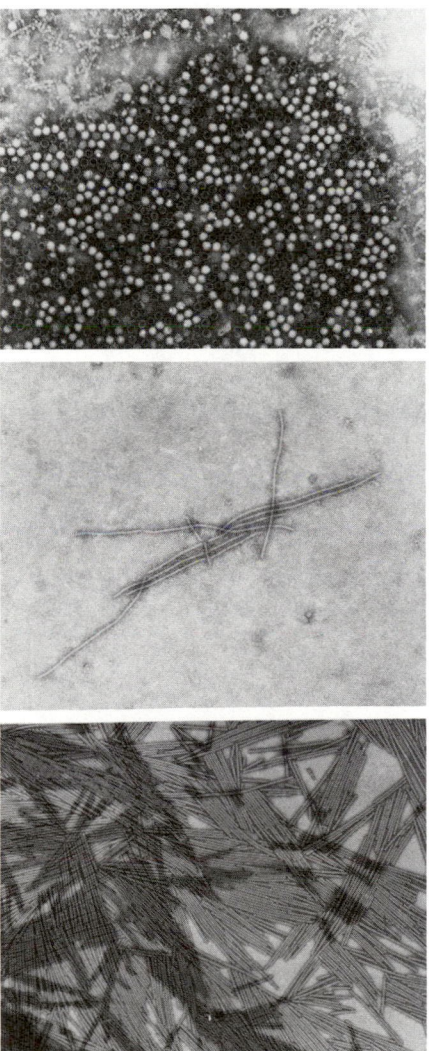

Fig. 12-4. Common shapes of plant-pathogenic viruses: **Top,** isometric (cowpea mosaic virus, ×81,500); **center,** flexuous rods (maize dwarf mosaic virus, ×42,500); **bottom,** rigid rods (tobacco mosaic virus, ×35,000).

After the original ssRNA is replicated, another section of the RNA codes for coat protein. The coat protein then coats or encapsidates the RNA, and new virions are formed.

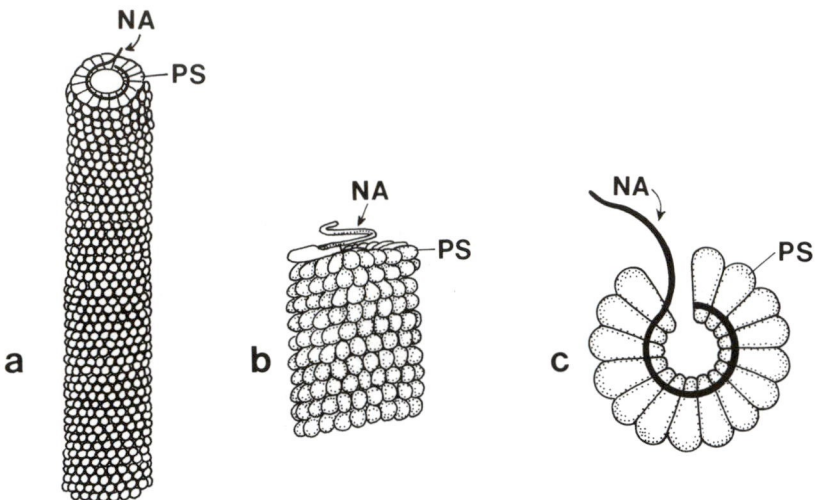

Fig. 12-5. Rigid rod-shaped virus. **a,** Structure, showing protein subunits (**PS**) and nucleic acid (**NA**); **b,** side arrangement; **c,** cross section of the same virus.

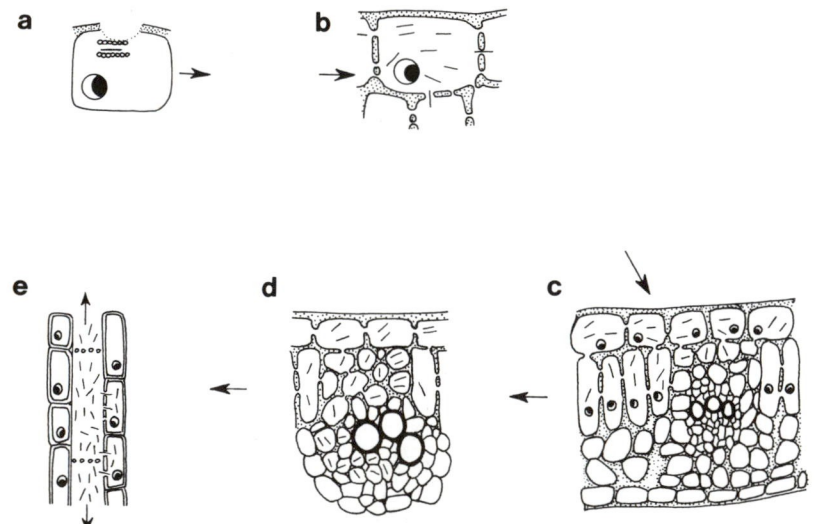

Fig. 12-6. Mechanical inoculation and early stages in the systemic distribution of viruses in plants. Note that viruses must enter a cell through a wound (**a**). They move between cells through plasmodesmata (**b–d**). Once they reach the phloem (**e**), distribution throughout the plant may be quite rapid.

The entire genome of some viruses is known. For instance, TMV consists of 6,400 nucleotides, which include 2,130 triplet codons. Only 158 of the codons are used for coat protein. Some codons are used for production of RNA polymerase, while other codons presumably code for other proteins. According to recent research, one protein coded for in TMV, called the "movement protein," appears to facilitate the movement of virus within a plant. It is not yet clear exactly how viruses cause disease in plants. No obvious toxins have been detected, but viruses do disrupt normal metabolic processes since virus replication itself requires substantial contributions from the host cell, including proteins and nucleic acid as well as the use of host ribosomes and tRNAs.

Virus Movement in Plants

The production of the first new virions is usually accomplished about 10 hours after a cell is infected. As the virions accumulate, they begin to move into adjacent cells via cytoplasmic channels known as **plasmodesmata**. Movement of virions into new tissues is relatively slow until the vascular tissue is reached. Because viruses require living cells for replication, movement in the vascular tissue takes place in the living cells of the phloem rather than in the nonliving xylem vessels. Viruses follow the same pattern of movement in the translocation stream as do the products of photosynthesis.

Viruses are found in various tissues of plants and in different patterns. They can be found in almost all cells of a small herbaceous plant only 30 days after infection, whereas systemic infection of larger woody plants may take years to accomplish. Such long time lags can have important economic implications. Many woody ornamentals and fruit trees are composed of grafted root stocks and scions (shoots). If one part of the graft is virus-infected, some time may pass before the virus reaches the other part. If the healthy part reacts strongly to the invasion by the virus, the reaction may be a "hypersensitive" reaction that results in a layer of dead cells at the graft union. Virus infection is the most common cause of graft failure, which results in dead trees.

Control of Virus Diseases

Virus diseases cause tremendous losses each year, and their control is difficult. Because there is no cure for virus infection, infected plants must be destroyed. Prevention is thus the most effective means of virus control. In some cases, control of the vector can prevent virus transmission. If an important virus is transmitted by an insect vector in a persistent manner, insecticides are sometimes applied to high-value crops to prevent infection. Unfortunately, most viruses are transmitted by insects in a nonpersistent manner. Insecticides are then ineffective because they do not kill fast enough to prevent transmission. Oil sprays that interfere with nonpersistent transmission have been investigated but are not widely used. Soil may be fumigated before planting of certain high-value crops, such as raspberries, to reduce virus transmission by nematodes and some primitive soil fungi. Weed control is also important, especially in annual flower and vegetable production, because many viruses overwinter in common weeds.

Exclusion of viruses by the production of virus-free seed and propagative parts is a common and economically sound means of virus control. Most flower and vegetable seed is "certified" as meeting acceptable standards so that very few, if any, of the resulting plants will be virus-infected. The seed is carefully tested to determine the percent of virus infection. In some cases, even 1% virus infection in seed is unacceptable for commercial use. Commercial growers commonly walk through their fields early in the growing season looking for potentially virus-infected plants and "rogue" them out before the virus can be transmitted to healthy plants.

Seed- and pollenborne viruses are usually found at relatively low levels in seed, but propagative parts such as cuttings, tubers, bulbs, and corms will all be systemically virus-infected if produced by a virus-infected plant. Thus, commercial producers of such propagative parts have elaborate schemes of virus testing and clean propagation so that virus-free stock

Fig. 12-7. "Seed" potato facility in New York State. The facility, isolated from potato farms, is situated in the Adirondak Mountains, where the higher elevation reduces aphid populations. The farm is surrounded by a high security fence, and visitors must have their feet cleaned with a bacteriocide to prevent transmission of pathogens to the site.

will be available to growers. This is especially important for high-value perennial plants that require a major investment in time and money before the plants mature. Examples include many woody ornamentals, fruit trees, strawberries, and bramble fruits. Virus-free propagation of many flower crops, including carnations, chrysanthemums, and geraniums, is also necessary.

Potato "seed" production begins in laboratories, where the first plants are grown in a completely protected environment. The resulting tubers from these plants are planted by "foundation" growers, who repeatedly inspect their fields and remove virus-infected plants. Government inspectors certify the resulting tubers as meeting acceptable levels of virus infection based on several visual inspections and laboratory tests. A sneak preview of the next season's results is provided by the "Florida test." Samples of the harvested tubers are planted during the winter months in south Florida to allow inspectors to determine how many plants are virus-infected. This test was developed because, although the mother plants may appear healthy, aphids may have transmitted a virus to the plants late in the season, and the virus may have been transported through the phloem to the developing tubers. Commercial growers, and even home gardeners, find the cost of "certified seed" well justified. Potatoes grown from supermarket tubers will commonly all be virus-infected and produce much lower yields than those grown from certified seed.

Detection of Viruses

In some cases, viruses cause distinct and obvious symptoms in plants, including distortions, mosaics, ringspots, stunting, and flower color changes. Many viruses produce no obvious symptoms even though they can cause significant yield losses. It is important to detect viruses in plants before the plants are used for plant breeding, seed production, propagation, or agricultural production. Because not all plants produce obvious symptoms of infection, **indicator plants** are used to detect the presence of certain viruses. Scientists discovered through trial and error that when sap from a test plant is rubbed on the leaves of the indicator plant, distinct symptoms, which usually consist of small necrotic spots called **local lesions**, appear within about 1–2 weeks if a virus is present.

Sap transmission of many viruses is accomplished by gently rubbing sap from an infected plant onto the leaves of a healthy plant. Usually the leaves are lightly dusted with an abrasive powder so that the rubbing creates tiny wounds for virus entry. Some viruses can only be transmitted by a vector or through a living plant connection. Dodder, a parasitic plant with haustoria, is sometimes used as a living bridge between plants for experimental virus transmission. Some viruses are transmitted using insect vectors or by grafting a piece of the test plant to the indicator plant. Indicator plants are often used commercially and in many research laboratories. The disadvantages of such a test include many of the factors associated with any biological system. These include symptom variation due to environmental conditions, low virus titre (or concentration), misses, cross-contamination during inoculation of large numbers of test plants, and, of course, the time spent waiting for the expression of symptoms.

Serology

Scientists have long searched for methods of virus detection that are more specific, faster, and better suited for the multiple tests necessary in commercial plant production. Some of the newer methods make use of **serology**; these rely on the immune systems of animals. When a foreign protein (an **antigen**), such as those associated with viruses, invades an animal, the animal produces **antibodies** against that antigen. Antibodies are themselves proteins that physically fit around the foreign protein to help prevent an invading virus from functioning and expedite its removal from the body. If the antibodies produced by the immune system successfully inactivate the viruses and stop their replication, the individual may recover from the disease. Antibodies are very specific and unique for each type

Fig. 12-8. Bioassay for virus infection of a plant. **Top,** sap from a test plant is rubbed on a leaf of an "indicator" plant, which shows distinct symptoms if the virus is present. Leaves are usually dusted with a fine abrasive powder so that small wounds are produced when the sap is applied. **Bottom,** about one week later, small necrotic "local" lesions indicate a positive test for the presence of the virus.

of virus. Each time infection by a new virus occurs, new antibodies must be produced. Because the viruses that cause the common cold change frequently, many people get colds each year and must wait for new antibodies to be produced before they recover.

To obtain antibodies for serological tests, scientists first inject purified plant viruses into animals such as rabbits and mice. Their immune systems recognize the plant virus proteins as foreign and produce antibodies against them, even though plant viruses cannot infect the animals. Several weeks later, some blood is drawn from the animals, and the antibodies are separated from the blood cells and the **serum** (blood fluid). Concentrated quantities of these antibodies can then be mixed with plant sap from a test plant. If the virus is present, a cloudy **agglutination** is visible in the liquid where the antibody and the matching virus meet and adhere together.

Commercial antibody preparations for many common plant viruses are available for laboratory testing of plants for virus infection. Such a system has some limitations. When individual animals must be bled to "harvest" the antibodies, considerable expense and animal care is involved. In addition, a mixture of antibodies is collected because the original "purified" virus extract usually contains certain plant protein contaminants. Also, each virus has several antibody-binding sites, so individual viruses trigger the production of several different kinds of antibodies (such antibodies are called **polyclonal**). These serological tests also rely on sufficient virus titre for visual detection of the agglutination. Two important new techniques solve the problem of mass production of specific antibodies for commercial use and also the problem of detecting low levels of virus.

One effective solution to the problem of sensitive virus detection in small sap samples is a method known as an **enzyme-linked immunosorbent assay**, or **ELISA**. Several different types of ELISA tests have been created. Figure 12-9 is a diagram of the double-antibody sandwich ELISA, so called because it involves two different kinds of antibodies that have been produced against the same virus, generally from two different animals.

The first kind of antibody is adsorbed to the plastic ELISA plate in a small cylindrical well. Excess antibody and other materials are washed away. In the second step, sap from the test plant is added. If the virus is present, it will be "caught" by the antibody and held in the well while other material is washed away. In the third step, a second antibody to which an enzyme has been attached is added. If the expected virus is present, the second antibody also fits onto the virus, completing the "sandwich," and remains in the well while all excess material is washed away. In the final step, a substrate is added that changes color if the enzyme is present. If the virus were not present in the plant sap, the enzyme-attached antibody would have been washed away and lost. If color begins to appear after the substrate is added, then the enzyme and virus must be present.

Agglutination tests require that enough antibody and virus be present to make a visible agglutination, but in the ELISA test, the substrate begins to turn color due to the presence of the enzyme, and even very low virus titres are detectable. Virus concentration also can be quantified by measuring the color intensity. ELISA allows rapid testing of numerous plants using only small sap samples once antibodies are available.

The problem of specificity in antibody production is solved by a new technique using **monoclonal antibodies**. The inventors of this technique received the 1984 Nobel Prize in Medicine. As seen in Figure 12-11, the first step involves injection of the purified virus into an animal. Unlike plant cells, animal cells cannot be grown in culture indefinitely, so antibody-producing cells cannot be cultured for long periods. The monoclonal technique, however, circumvents this problem by fusing the antibody-producing cells from the animal's spleen with lymphoma cells (cancer cells) that divide continuously. The resulting **hybridoma** can be cultured indefinitely and produces a continuous supply of antibodies. Pure cultures of cells from a single hybridoma can be grown so that each culture produces a pure supply of a single kind of antibody, thus the name "monoclonal" antibodies. This results in an inexpensive source of very specific antibodies that eliminates the maintenance and repeated bleeding of living animals. Yet another technique for antibody production using bacteria, first described in 1990, may circumvent the use of animals altogether.

Because a continuous supply of monoclonal antibodies can be produced, it is now possible to use **immunoassays** to detect more complex pathogens

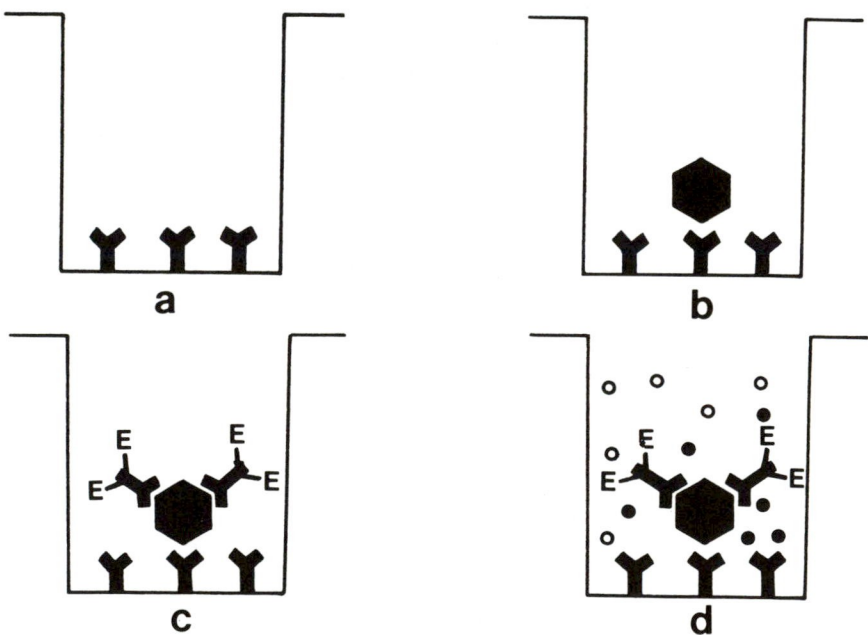

Fig. 12-9. Procedures used for enzyme-linked immunosorbent assay (ELISA). **a,** Specific antibody is adsorbed to the plate. After a short waiting period, the remaining liquid is washed away. **b,** The test sample containing the virus is added, and remaining liquid is washed away. **c,** The enzyme-labeled (**E**) specific antibody is added. The remaining liquid is washed away. **d,** The enzyme substrate (circles) is added and reacts to give the color (dark circles).

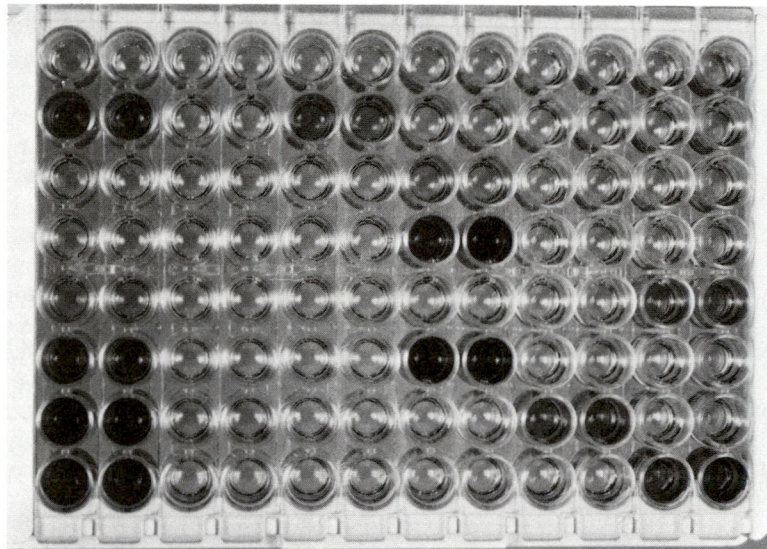

Fig. 12-10. Multiwell plate for enzyme-linked immunosorbent assay. Each well contains a test sample. Wells showing dark color are positive; antibodies have matched with antigens. The enzyme linked to an antibody has caused a color change when the substrated was added. Relative amounts of antigen can be detected by measuring color intensity.

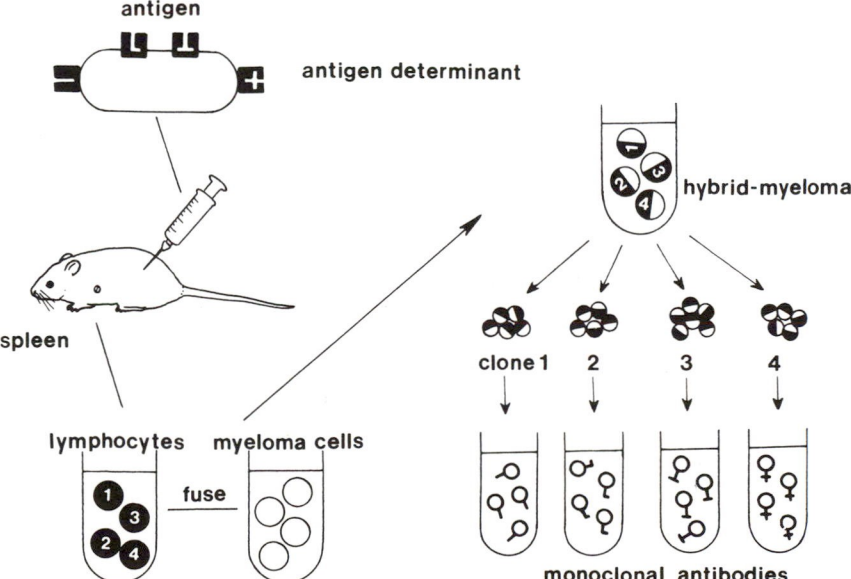

Fig. 12-11. Steps in the process of generating monoclonal antibodies.

and nearly any protein. If an identifying protein can be found that is specific to a fungus, nematode, or bacterium, monoclonal antibodies can be produced to rapidly detect and identify its presence in plant tissue without the complex and time-consuming isolation techniques in present use. Commercial immunoassay kits are available for a number of diseases of turfgrass, ornamental and woody plants, flowers, and vegetables. Many biotechnology companies are actively researching new products using monoclonal antibodies to detect not only pathogens but other important compounds such as aflatoxin and various pesticides.

For rapid and sensitive assays, both polyclonal antibodies (with mixed sources) and monoclonal antibodies (with a single source) can be combined with enzymes for use in ELISA tests. Although the subject of this book is primarily plants and their diseases, it is interesting to note that the discovery of immunoassays, in particular the ELISA technique, has revolutionized many aspects of biology and medicine. Immunoassays are now produced for many human pathogens so that medical laboratories can quickly and accurately provide diagnostic information. Previously, diagnosis of bacterial diseases required several days for cultures to grow on selective media, but ELISA tests can be completed in minutes or hours. Rapid pregnancy tests are based on monoclonal antibody immunoassays that detect a hormone in the urine of pregnant women. In the case of acquired immune deficiency syndrome (AIDS), the virus itself is very difficult to detect, so AIDS tests detect the antibodies produced in people who have been exposed to the virus. Thus, this test is done using antibodies produced against antibodies.

Small amounts of the tumors of some cancer patients are used to create antibodies to which powerful chemicals are attached. When reinjected into the patient's body, the antibodies travel through the bloodstream until they attach specifically at the tumor site. This specific delivery system reduces the amount of toxic chemicals that must be used in chemotherapy. Because all of this progress has occurred so rapidly, it is difficult to predict accomplishments that may occur in the next few years. Certainly, significant improvements in disease diagnosis and detection in both plants and animals will result.

Pathogen-Free Propagation

If a plant is virus-infected, the only recourse in many cases is to destroy it. However, in breeding programs only a single plant of a particular genotype may exist. Also, even if all virus tests have come out negative, one can rarely be certain that a plant is truly virus-free before propagation begins. There is always the chance that sometime during the many generations of propagation, even in insect-free greenhouses, viruses may be introduced, contaminating many thousands of plants. The solution to this problem lies in some scientific research that paralleled the studies of plant virology in the 1950s.

The solution, briefly mentioned in Chapter 5, was F. C. Steward's successful application of theoretical "totipotency" through the regeneration of carrots from mature cells grown in coconut milk. In theory, an entire plant can be regenerated from a single cell. In practice, success is most likely when a tiny sample of cells from the meristem of a plant is removed. In 1952, G. M. Morel and C. Martin first regenerated virus-free plants from meristem tips taken from virus-infected plants. Even though studies have shown that, in general, meristems are relatively virus-free, due perhaps to the lack of well-developed plasmodesmata in cells, testing regenerated plants for virus infection is always prudent. Whether the meristem is virus-

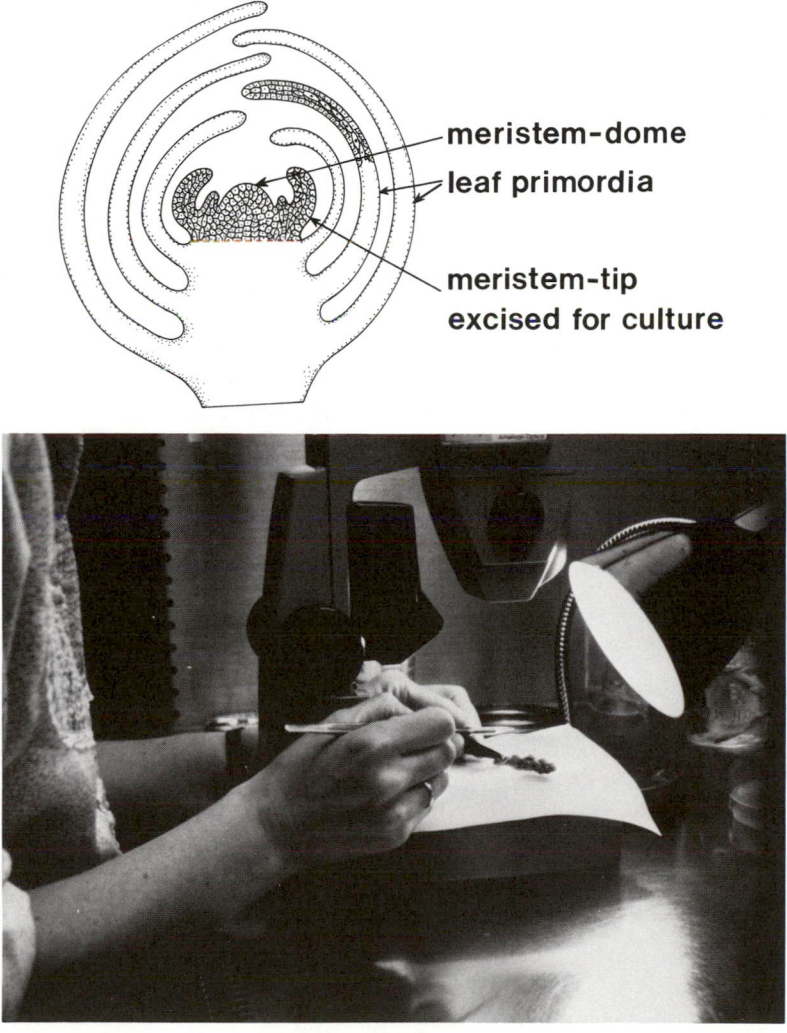

Fig. 12-12. Meristems for tissue culture. **Top,** diagram of a bud showing the meristem-tip region. The dotted line indicates the section that is usually removed for tissue culture. **Bottom,** removal of the meristem tip requires sterile conditions, good light, and a steady hand.

free or not is still somewhat a matter of chance. Plants grown at high temperatures (30–40°C) are more likely to have virus-free meristems, probably because virus replication is inhibited.

The technology of tissue culture had to make certain advances before this means of virus-free propagation could become economically feasible. Tiny meristem tips taken from mature plants cannot photosynthesize and must be grown on a sterile nutrient medium just as microorganisms are. The early use of coconut milk was successful because it contained both sugars and plant hormones. As an understanding of specific hormones and their functions developed, more precise control of plant development became possible.

The word **hormone** is derived from the Greek word for "to excite." Hormones regulate plant growth and are now commonly called **growth regulators**. The role of various growth regulators depends on the type of plant tissue and their relative concentrations. Without the addition of appropriate growth regulators, simply providing nutrients to excised meristems does not allow regeneration of a normal plant.

Three main types of growth regulators function in the regulation of elongation and growth of plants: these are **auxins, gibberellins**, and **cytokinins**. A natural plant auxin, indoleacetic acid (IAA), controls cell elongation, which allows plants to curve toward light sources. It also functions in apical dominance, in which the growth of lateral buds is inhibited as long as the IAA-producing apical bud is present. Auxins also play a role in the development of adventitious roots. Plant cuttings are often dipped in auxin solutions or powders to make them develop roots more quickly.

An important characteristic of plant growth regulators is that they are active at extremely low concentrations. Many plant pathogens produce

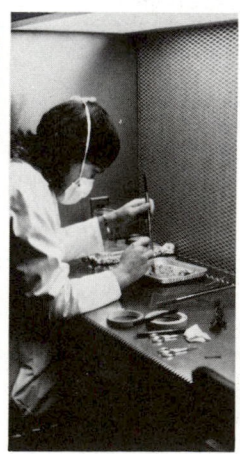

Fig. 12-13. Regeneration of plants from a meristem. **Left,** growth of the meristem tissue on nutrient medium containing appropriate growth regulators results in numerous shoots. **Center** and **right,** under sterile conditions, the tiny shoots can be separated and transferred to another medium for root development.

auxinlike compounds and/or stimulate host plants to increase IAA production, causing growth effects, especially gall formation, in plants. These include root-knot nematodes, crown gall bacteria, and the fungal pathogens responsible for corn smut and cedar apple rust.

A second group of growth regulators is the gibberellins. They were first discovered because of their production by a plant pathogen. In the 1930s, the fungus *Gibberella fujikuroi* was determined to be the cause of the "foolish seedling disease" of rice, which caused excessive elongation of the young plants. The active compound isolated from the fungus was named gibberellin and, in 1956, plant-produced gibberellins that play a role in cell division and elongation were also discovered. Application of gibberellins to seedless fruits, particularly grapes, results in larger fruits. Auxins and gibberellins both function in the control of cambium development for the production of secondary plant tissues.

A third important group of growth regulators has been named cytokinins because of their influence on cell division (also called cytokinesis). In many plant tissues, normal growth appears to result from a balance between auxin and cytokinins. In tissue culture, meristems or protoplasts are typically first regenerated to a **callus** stage, which consists of a mass of undifferentiated cells. This occurs when the cells are grown on a nutrient medium containing sugars, nutrient salts, and a balance between auxin and cytokinins. After callus growth on this **initiation medium**, the tissue can be transferred to a medium in which cytokinins dominate to stimulate shoot formation. Many shoots may form from a small amount of callus. The shoots are then transferred to a third medium in which auxin dominates to induce root

Fig. 12-14. Tissue culture allows thousands of plants to be rapidly propagated under pathogen-free conditions in a relatively small area.

formation. At this point, the tiny plant may be transferred to a soil medium and normal environmental conditions.

The previous description is a highly simplified account of the regeneration of whole plants from tiny bits of tissue. Certain steps are combined or modified depending on the requirements of specific plant species. Monocots, in particular, have been very difficult to culture, although rice has been successfully regenerated from protoplasts. Even within species, modifications of nutrient and hormone levels must be experimentally determined for various cultivars before regeneration can be accomplished.

Despite these complications, regeneration of plants is an area of intense research activity for two main reasons. First, pathogen-free plants can be multiplied rapidly in a completely protected environment at a cost that approaches traditional propagative costs. The value of such plants lies in their increased yield and vigor and the longer production time before replacement is necessary. In the past, propagative parts had to be grown in natural soils for several years, during which they were exposed to various virus vectors and soilborne pathogens before they were sold for commercial production. Plants may now be sold directly from pathogen-free laboratories and greenhouses to commercial growers. Second, regeneration of plants from protoplasts or single cells is necessary for successful genetic engineering.

Fig. 12-15. Apical cuttings from pathogen-free, tissue-culture potato plants are rooted in rolled banana leaves and later sold to other Vietnamese farmers for planting directly into the field.

Once gene transfer has occurred, the transformed cell must be regenerated into a plant in which each cell contains the new genetic information. There is great interest in the use of viruses as gene vectors. Scientists hope to be able to eliminate the codons responsible for disease without destroying the infectivity of a virus. The "disarmed" viruses may then be used to carry new genes into plant cells, especially in plants not susceptible to infection by the bacterium now used in genetic engineering (*Agrobacterium tumefaciens*).

Resistance to Virus Diseases

Genetic resistance to viruses, selected through traditional breeding programs consisting of repeated crosses between plants, has been quite successful for many important viruses. Scientists and breeders are always concerned, however, that a new virus strain may overcome the resistance. It is also difficult to provide resistance to all the viruses that commonly infect important crop plants. Some plants are deliberately inoculated with mild strains of viruses to protect against infection by severe virus strains, making use of a phenomenon called **cross-resistance**. Several potential problems with this technique, including the fear that the mild strain might mutate into a severe one, make virologists wary.

An amazing virus-resistance success story, mentioned in Chapter 5, may be more easily understood now. The Ti plasmid of *A. tumefaciens* was used to transfer the coat protein gene of TMV to the nucleus of a tomato cell. Tomato plants regenerated from the transformed cells are resistant to TMV. The theory currently used to explain the resistance is that the coat protein produced by the plant cells prevents TMV from replicating if it is introduced into the plant. Similar experiments are now being done to transform the cells of other crop plants so they will produce coat protein of important viruses. Once the coat protein gene of a virus has been identified, this may present a rapid means of producing virus-resistant plants.

Viroids

Other types of pathogenic, replicating molecules have been discovered fairly recently. Viroids, first named in 1973, are small, circular RNAs that lack coat protein. They cause a number of economically important diseases, including potato spindle tuber disease, two different viroid diseases in chrysanthemum, and cadang-cadang (dying-dying) disease of coconut palm. More than 30 million coconut palms have died of this viroid disease in the Philippines.

Viroids are so small that they do not contain enough genetic information to code for a protein, so their function as pathogens is not well understood. Their origin may be RNA from host plants, a theory supported by the fact that viroids have mostly been found in plants that are intensively vegetatively propagated. No vectors are yet known, and control of these diseases is very difficult. Viroids are the only pathogen that cannot be detected by immunoassays because they have no protein coat. Other pathogenic RNAs and viruslike particles that require accompanying viruses for transmission have also recently been discovered. Some modify the disease symptoms of virus infection in plants, either increasing or decreasing their severity.

Mycoplasmalike Organisms and Other Prokaryotic Pathogens

It may seem strange to include plant-pathogenic prokaryotes in a chapter that concentrates on virus diseases. However, most of these diseases they cause were considered to be caused by viruses until 1967. The symptoms usually include stunting, yellowing, and various growth distortions. As with virus diseases, no pathogen can easily be cultured from diseased tissue, and transmission usually requires insect vectors or a living bridge such as dodder or grafting.

The most important pathogens in this group are the **mycoplasmalike organisms** (MLOs), which cause over 200 plant diseases. They received this somewhat clumsy designation because mycoplasmas were first discovered as animal pathogens. Plant mycoplasmas appear to be very similar but, because so little is yet known about them, they are called MLOs. MLOs are obligate pathogens that have not, so far, been grown in culture. They are single-celled prokaryotes that lack a cell wall. Such wall-less

Fig. 12-16. Cadang-cadang disease of coconut palm. This viroid disease has been estimated to cause a yearly loss of nearly one million palms in the Philippines. This photograph illustrates the destruction on San Miguel Island off Luzon. These trees were not the ones originally killed but were palms replanted after World War II. In 1960, not one healthy palm could be found.

prokaryotes are known as **mollicutes**. In plants, they are restricted to the phloem. Many infected plants are induced to form a proliferation of branches in certain areas, a symptom described as **witches' broom**. Other pathogens and insect pests, including mistletoes, can also stimulate production of witches' brooms.

When MLOs were discovered in 1967, an entire group of "virus" diseases with "yellows" as a characteristic symptom had to be redefined. Another common symptom of MLO infection is an off-taste in fruits and vegetables produced from infected plants. Two economically important MLO diseases are aster yellows, which causes problems in many annual flowers and vegetables, and X disease of peach and cherry trees. Control of MLO diseases depends on the specific host plant that is threatened. For annual plants, vector control with insecticides may sufficiently reduce leafhopper

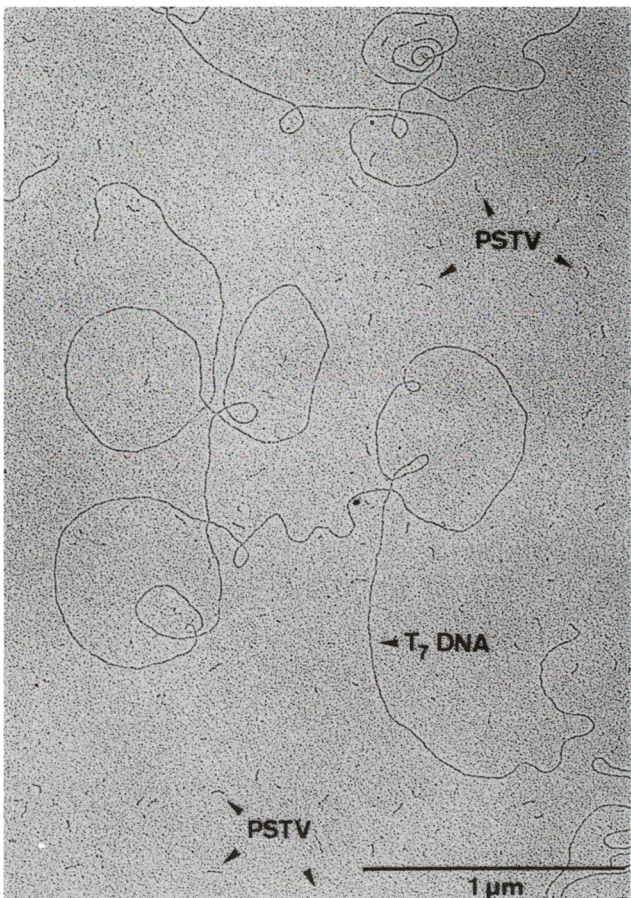

Fig. 12-17. Electron micrograph of potato spindle tuber viroid (PSTV) in relation to the DNA of a bacteriophage T7. Arrows indicate the tiny viroid RNAs.

populations to prevent economic damage. MLOs are transmitted in a persistent manner and reproduce in the vector. In northern areas, MLOs are carried by leafhoppers carried in air masses from warmer southern areas.

MLOs are sensitive to tetracycline antibiotics. Trees may be injected with tetracyclines to prolong their life, although treatments must be repeated because they cause only a remission in symptoms. This may be a short-term, cost-effective practice in high-value cherry or peach orchards. Another effective practice involves the removal of wild cherry and plum trees near cherry and plum orchards because the wild trees may serve as a source of MLO that can be transmitted by leafhopper vectors.

In the Western Hemisphere, lethal yellowing of coconuts is a severe MLO disease for which few controls exist. High-value trees may receive tetracycline

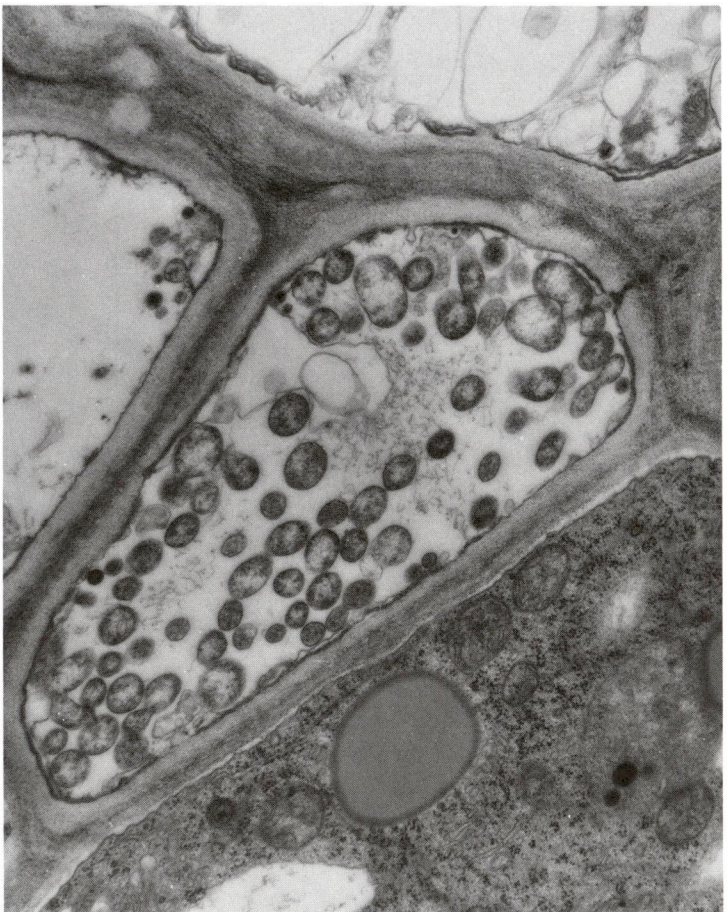

Fig. 12-18. Phloem cells of a plant infected with a mycoplasmalike organism (×40,500).

injections, but such expensive treatment is not practical for the numerous trees in the poorer areas of some Caribbean islands and coastal Mexico. In the Miami, Florida, area, over 90% of the coconut palms have been eradicated. For many tourists, these palms are part of the atmosphere of a semitropical vacation, but for many poor people, coconut palms are called the "trees of life." They provide shade, food, and shelter; they are also a source of income from copra, oil, and charcoal used in air purification systems, which is produced from the shells. Many trees have already been lost and must be replaced by resistant types such as the Malayan Dwarf, but these take several years to mature. This disease will continue to cause hardships as the transition to resistant cultivars is made.

Several other prokaryotic pathogens have been discovered since the MLOs. These include another kind of mollicute, the **spiroplasma**, which, as the name suggests, exists at times in a helical form. Spiroplasmas have been found to be the causal agents of several plant diseases, but their relation to MLOs is not yet clear. Some spiroplasmas have been cultured on nutrient media. Corn stunt, caused by *Spiroplasma kunkelii*, is a relatively minor spiroplasma disease in the United States, but the disease has modern and, perhaps historical, significance in Central America.

More than 50 large cities in Quintana Roo had already been abandoned by the time Cortez and the Spanish Conquistadors arrived at the end of the 15th and the beginning of the 16th century. The disappearance of the inhabitants in Quintana Roo and in the Yucatan region has been the subject of much speculation. In 1973, J. L. Brewbaker, of the University of Hawaii, suggested that maize mosaic virus led to starvation and the collapse of the Mayan empire before the Spanish invasion. L. R. Nault of Ohio State University has provided evidence that the virus and its vector adapted to maize in Central America in post-Columbian times. He has suggested that corn stunt is a much more severe disease in the Yucatan Peninsula of Mexico,

Fig. 12-19. Diseased coconut palms. **Left,** trees dying of lethal yellowing, a disease caused by mycoplasmalike organisms in the Western Hemisphere. **Right,** coconut tree with an injection port used to treat the tree with tetracycline antibiotics. Injections cause a remission in symptoms but do not cure a tree and must be repeated.

where a leafhopper, *Dalbulus maidis*, thrives and vectors the spiroplasma so efficiently that corn can still not be grown in that region. This vector does not survive well in the colder winters of the United States.

Fig. 12-20. Corn stunt, a spiroplasma disease (*Spiroplasma kunkelii*). **a,** Yellowing and stunting, field symptoms of the infection; **b,** spiroplasmas in the phloem of an infected corn plant.

K. Maramorosch of Rutgers University concurs with Nault's opinion and has suggested that corn stunt disease destroyed the staple food of the Mayans and thereby opened up Mexico to Spanish conquest.

Other prokaryotes exist in addition to spiroplasmas. Some have cell walls like common bacteria do, although their structure is somewhat different. Because they are difficult to isolate and grow in culture, they are referred to as **fastidious bacteria**. Some appear to be restricted to xylem tissue and others to phloem, but relatively little is known about their role as pathogens.

Even eukaryotic **protozoa** have been found to cause a few plant diseases. Thus, the infectious pathogens of plants range from viroids (replicating molecules) to single-celled prokaryotes of various types, single-celled eukaryotic protozoa, microscopic and larger fungi, and even animals (nematodes) and parasitic plants. Some diseases are not caused by infectious pathogens, and these are the subject of the next chapter.

Selected Readings

Diener, T. O. 1983. The Viroid—A Subviral Pathogen. American Scientist 71:481-489.

Maramorosch, K. 1987. The Curse of Cadang-Cadang. Natural History 96(7):20-22.

McCoy, R. E. 1988. What's killing the palm tress? National Geographic 174(1):120-130.

Nault, L. R. 1983. Origins of leafhopper vectors of maize pathogens in Mesoamerica. Pages 75-82 in: Proceedings of the International Maize Virus Disease Colloquium and Workshop, 2–6 August 1982. D. T. Gordon, J. K. Knoke, L. R. Nault, and R. M. Ritter, eds. The Ohio State Agricultural Research and Development Center, Wooster, Ohio. 266 pp.

Walkey, D. G. A. 1985. Applied Plant Virology. John Wiley and Sons, New York.

Yeh, S.-D., Gonsalves, D., Wang, H.-L., Namba, R., and Chiu, R.-J. 1988. Control of Papaya Ringspot Virus by Cross Protection. Plant Disease 72:375-380.

Environmental Diseases and Problems

CHAPTER 13

Environmental factors that can cause disease in plants are the subject of this chapter. Many environmental factors affect all living organisms, including humans and plants, each day. Although necessary for our existence, these same factors, when deficient or excessive, may become life-threatening. For example, sunstroke and hypothermia are commonly reported diseases of very hot and very cold weather, respectively. Certain vitamin deficiencies are associated with well-known diseases: vitamin C deficiency with scurvy, vitamin D deficiency with rickets, and thiamin (B_1) deficiency with beriberi.

Vitamins play a very specific metabolic role and are necessary in only small amounts. Many vitamins are produced synthetically in chemical reactions. Others are produced by microorganisms and harvested for use as dietary supplements much as penicillin and other antibiotics are. For example, riboflavin (B_2) is commercially produced by *Ashbya gossypii*, a fungus found on cotton. The actinomycete *Streptomyces* is used in the production of vitamin B_{12} as well as the antibiotic streptomycin. Synthetic sources are an inexpensive means of ensuring adequate doses of these necessary vitamins, particularly for those who do not consume a varied diet or who, like pregnant women, temporarily require extra-normal doses of certain vitamins. On the other hand, excessive doses of fat-soluble vitamins (A, D, E, and K), which accumulate in the body, can cause toxic symptoms in humans. Similarly, environmental diseases of plants often involve deficiencies or excesses of factors necessary for growth, including temperature, water, light, and nutrients.

Winter Injury

Because of their immobility, outdoor plants must often withstand tremendous environmental variation. In temperate climates, an important source of environmental stress is winter weather. Plants are subjected not only to low temperatures, which slow metabolic processes and inhibit growth, but to temperatures near the freezing point, which can lead to the formation of ice crystals that puncture cells. Even lower temperatures may lead to the actual freezing of tissues and to cellular death.

An additional risk to plant tissue in winter is known by several common names including **winter burn**, "winter scorch," or "winter desiccation." The damage has nothing to do with the actual burning of tissues, although damaged plants have the same appearance as plants that receive insufficient water in hot weather. During sunny, relatively warm winter days, the upper parts of plants transpire (evaporate water), but frozen soil prevents the uptake of replacement water through the roots. As a result, exposed plant tissue dies of water stress. Damage is especially pronounced in areas of excessive air movement, where the dry winter winds increase the rate of transpiration.

Winter injury to plants takes several forms. Some plants appear wilted, whereas other plants show marginal necrosis on the leaves, which is known as "scorch." Particularly vulnerable areas are the apical tips of plants, where the buds, which are usually protected from water loss by several layers of bud scales, may be killed by water deprivation. Tissues that have been killed by freezing are likely to appear watery and mushy due to the destruction of the cells by ice formation.

Perennial plants in temperate climates have evolved many protective adaptations. For example, transpiration in trees and shrubs is predominantly from leaf surfaces, and many woody plants are deciduous, shedding their leaves in the fall. Without leaves, woody tissues remain well protected from excessive water loss by the impervious cork cells of the bark. Buds are

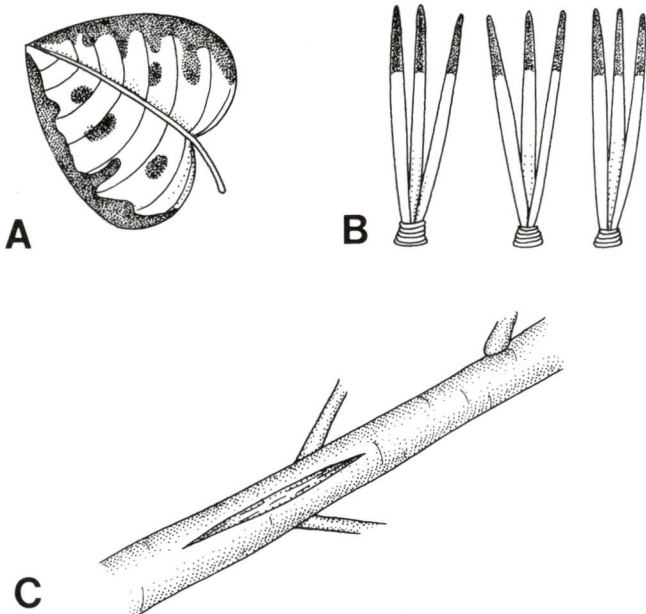

Fig. 13-1. Winter injury. **A,** "Scorch" symptoms on leaves due to winter "burn"; **B,** conifer needles may have tip bleaching in areas where prevailing winds of winter cause greater transpiration and water is not replaced because soil is frozen; **C,** in severe cases of freezing, twigs and stems may crack.

protected from desiccation by bud scales. "Evergreens" produce leaves that remain on the plant throughout the winter, but they also have ways to reduce transpiration during winter. In general, their leaves are smaller, often needlelike, and have thicker waxy layers that reduce transpiration. In addition, the stomata are recessed into the epidermis, unlike those in deciduous plants, so that air movement over the leaf surfaces results in less water loss. In very cold weather, many evergreen plants roll their leaves, an additional means of temporarily reducing surface area during severe conditions. This is a common phenomenon in rhododendrons and other broad-leafed evergreen plants that do not have the needlelike leaves of conifers.

To help prevent winter injury, commercial sprays are available that increase the waxy protective coating on leaf surfaces. Planting sites for evergreens should be chosen such that the trees are protected from the prevailing wind, which will also reduce winter burn. It is also important not to place plants too near a building, where the warmth from the foundation will increase transpiration while the ground is still frozen. Thin-barked and young trees, are sometimes subject to sunscald and should also be protected from southern exposures.

Water and Light

Excessive water is as harmful to plants as water deficiency. This problem may occur following heavy rains that leave water temporarily standing in

Fig. 13-2. Severe winter burn on an arborvitae bush planted in an exposed area.

fields, or chronically, in heavy soils that do not drain properly. Plants potted in containers without drainage holes often suffer from excessive soil water. Excessive water fills the pore spaces between soil particles, reducing the oxygen supply to plant roots and inhibiting growth. If oxygen is greatly reduced and anaerobic conditions occur, the roots may die and decay. Because continual root growth is necessary for adequate water and mineral uptake, any inhibition of root growth may be expressed by yellowing, stunting, and wilting aboveground. Because many factors, including various root-infecting pathogens, can cause these symptoms, further investigation is always necessary before an accurate diagnosis of the problem can be made. In most cases, belowground problems in plants are due to a complex of several factors.

Occasionally, light exposure causes damage to plants. Sunscald is not likely in plants used to full sun but may occur when plant tissue is suddenly exposed to the full intensity of natural sunlight. Examples include seedlings transplanted into vegetable gardens or house plants moved outdoors for the summer from areas of lower light intensity such as greenhouses or window sills. Gradual exposure to high light intensity over a period of time will prevent damage. Beans, bell peppers, other vegetables, and fruits may also suffer sunscald when wind storms rearrange the protective leaf canopy of the plant and suddenly expose fruits to full sunlight.

Plants vary in their light requirements. Some prefer full sun and others are moderately or very shade-tolerant. Plants that are choked by heavy weed growth or planted at too high a density not only compete for water and nutrients but for light as well. Low light intensity often inhibits growth of plants in greenhouses, particularly in winter, unless artificial supplemental light is provided. In insufficient light, plants become spindly and yellowish, a condition described as **etiolated**. When plants are suddenly moved from full sunlight to a lower light intensity, such as when hanging porch plants are brought back into the house at the end of summer, it is common to see a loss of lower leaves as the plant reduces its leaf area in response to the reduced light supply.

Storms and unusual weather can damage plants. Each year lightning, wind, hail, ice, and snow loads cause plant injuries. Besides striking trees, lightning may strike agricultural fields in flat areas, leaving a somewhat circular area of damage that may be difficult to diagnose unless the timing of its appearance can be correlated with that of a storm. Hail is particularly damaging to young fruits on trees, although the full extent of the injury may become apparent only later in the season.

Pesticide Injury

A common injury to plants can result from the misuse, inadvertent drift, or excessive persistence of pesticides. Herbicide injury is probably the most common pesticide problem because many herbicides cause **phytotoxicity** or **growth regulator** (hormonal) **effects** in desirable plants as well as in weeds. Drift of even low concentrations of herbicides to nontarget areas can damage plants. When herbicides fail to break down in the soil at the expected rate, new crops may show effects of the residues. Following the 1988 midwestern drought, herbicide residues used on soybean crops the previous year threatened corn as a rotation crop because the normal microbial breakdown of the herbicide residues did not occur in the excessively dry soils. Farmers were warned to test soils for herbicide residues by planting seeds in soil samples in pots on window sills in advance of the planting season to determine whether injurious residues were present. New immunoassays are being developed to allow rapid, quantifiable measurement of such residues in agricultural soils. These assays will be useful for practical matters such as to accurately measure potential herbicide damage to new crops and to study the fate of pesticides in soil and water environments.

Many pesticides are insoluble in water, so they are produced as "emulsifiable concentrates," which means that the pesticides are dissolved in carrier compounds that can be mixed with water before application. However, many carriers are organic compounds that can be phytotoxic if applied at high concentrations or in hot weather. In addition, dormant oils are sometimes used to control certain insects and mites. These compounds are not toxic pesticides but effectively suffocate the pests by coating them with oil. Because the oil is phytotoxic, it is usually applied to plants during the dormant season just before the buds open, while most vulnerable plant tissue is still protected by bud scales or bark. If the growing buds develop more quickly than expected in suddenly warm spring weather or if oils are sprayed at high temperatures, plant damage is likely.

Essential Plant Nutrients

Just as certain vitamins are necessary for good health in humans, certain elements are necessary for plant growth. The nutrients are not food for

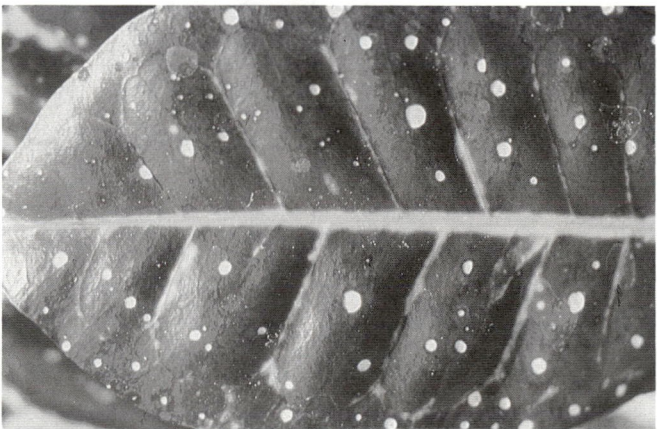

Fig. 13-3. Contact phytotoxicity caused by an insecticide applied to the foliage of *Codiaeum variegatum.*

the plant, although they are often sold as "plant food." Plants are autotrophs, producing their own food by trapping the energy of the sun through photosynthesis, but they still require certain elements to build proteins, carbohydrates, and all the other compounds necessary for their growth and development. These necessary elements are divided into two categories, depending primarily on the amount necessary for adequate growth. Those needed in larger quantities are called **macronutrients**, and those necessary only in small amounts are called **micronutrients**. Macronutrients include carbon (C), hydrogen (H), and oxygen (O), which are supplied by CO_2 and H_2O. Other macronutrients are dissolved in the soil water in various chemical forms and are absorbed primarily through the roots. These are nitrogen (N), phosphorous (P), potassium (K), calcium (Ca), magnesium (Mg), and sulfur (S).

Important micronutrients, which are also called "minor" or "trace" elements, include iron (Fe), boron (B), manganese (Mn), zinc (Zn), copper (Cu), molybdenum (Mo), and chlorine (Cl). Although micronutrients are necessary elements, they can cause toxicity symptoms in plants when present in excessive quantities. Generally, a relatively narrow range separates deficiency and toxicity levels of micronutrients. While toxicity due to an excessive supply of a macronutrient is relatively rare, either deficiencies or toxicities of micronutrients are commonly observed.

Table 13-1 lists some important roles of some of these elements and their typical deficiency symptoms. Such symptoms are usually studied by growing plants in a liquid or sand medium in which each element is added in measured quantities. These media allow scientists to see the exact effects of varying the amount of or even totally eliminating a particular element.

Fig. 13-4. Herbicide (dicamba) injury on soybeans.

Such controlled conditions are not possible in nature, so the symptoms in the field or garden are not always as clear as the descriptions might suggest. The diagnosis of a nutrient deficiency requires, first, a familiarity with the particular plant and its fertility history. For instance, if a potted plant has been growing in the same soil for many years without the addition of fertilizer, some macronutrient deficiencies are to be expected. Likewise, fields that recently received excessive rainfall are likely to be deficient in water-soluble nutrients such as nitrogen.

The Nitrogen Cycle

Because nitrogen is such an important element to living organisms, required for both proteins and nucleic acids, it is commonly the limiting element in agricultural systems. Nitrogen is a mobile nutrient, which means that it is transported from older tissues to growing tissues when adequate supplies of nitrogen are not available, leaving the older tissues deficient in nitrogen and making them yellow prematurely. Nitrogen-deficient plants typically look stunted and yellowed, with the oldest tissue expressing the symptoms most severely.

Tremendous quantities of nitrogen-containing materials are added to agricultural soils each year. An apparent contradiction is that nitrogen in its gaseous form (N_2) is the predominant component of our atmosphere

Table 13-1. Some Important Elements for Plant Growth and Their Deficiency Symptoms in Plants

Element	Deficiency Symptoms
Nitrogen (N)	Stunted growth, light green color; yellowing of leaves beginning with oldest foliage
Phosphorus (P)	Stunted growth; dark green, sometimes bluish to purplish coloration of foliage; thin and weak shoots
Potassium (K)	Dieback; poor shoot growth; yellow and necrotic spots on leaves; browning of tips and edges of leaves
Magnesium (Mg)	Symptoms appear on edges of younger tissue first and include yellowing and sometimes reddish color; leaves may appear cupped
Calcium (Ca)	Youngest foliage distorted and irregular; terminal buds may die; root growth poor
Sulfur (S)	Similar to nitrogen deficiency except that young leaves may be pale green first
Boron (B)	Poor growth of buds, new growth, and whole plant; surface and internal cracking on stems, fruits, and vegetables
Iron (Fe)	Young leaves light green to yellow, especially in interveinal areas; in severe cases, leaves become dry and may be shed
Zinc (Zn)	Interveinal yellowing; may become necrotic or purplish; leaves may be small; defoliation in some cases beginning with lower leaves; reduced fruit production
Manganese (Mn)	Interveinal yellowing; necrotic spots on leaves common; whole leaves may become brown in severe cases

Fig. 13-5. Nitrogen deficiency symptoms typically involve stunting of growth and yellowing of lower leaves.

Fig. 13-6. Blossom end rot on peppers. This problem is believed to involve calcium availability and is most prominent on tomatoes and peppers that receive an irregular water supply.

(over 78%), and yet nitrogen is such an important limiting factor in agriculture. Nitrogen fertilizers added to agricultural soils also pollute surface waters and groundwater, through runoff and percolation, respectively. Surface water pollution contributes to algal blooms that often result in eutrophication, or premature aging, of lakes and ponds. Excessive nitrates in groundwater are a health hazard that makes the water unpotable unless it is first treated. Because nitrogen is a limiting factor in agricultural production as well as an environmental pollutant, an understanding of the nitrogen cycle can lead to improved agricultural efficiency and improved environmental protection.

Nitrogen is an important component of many complex organic compounds in living organisms. When plants are eaten by animals, amino acids from the plant proteins are used to create animal proteins. The "essential" amino acids are those that we cannot synthesize through our own metabolic processes and that must be obtained by eating amino acids synthesized by other organisms. An efficient way to get the full array of amino acids necessary for building human proteins is to eat meat and other animal products, which contain high concentrations of protein. There is currently considerable interest in breeding plants, such as high-lysine corn, that contain higher levels of the essential amino acids so that we can more easily obtain the necessary compounds directly from plants. This would be a much more efficient way to feed the growing world population a diet with adequate

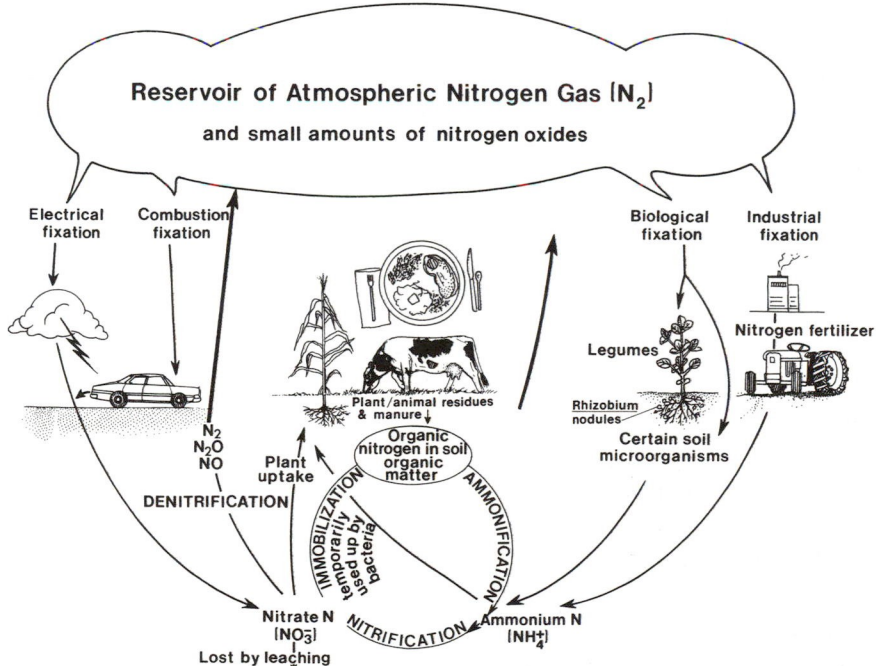

Fig. 13-7. The nitrogen cycle.

protein, since considerable energy efficiency is lost when humans eat animals rather than plants to obtain these proteins.

Nitrogen is lost from agricultural lands when crops are harvested or animals graze on forage crops. It must be continually added if subsequent crops are to be grown. Nitrogen is not returned to soils until plants and animals die and decay or the waste products of animals and humans are put into the soil. Farmers have long known that manures and plant debris are a good source of nitrogen and other important plant nutrients and have used them as soil amendments for use by the next crop. In agriculture, we are most interested in forms of nitrogen available for plant uptake and how to maintain sufficient quantities for optimal plant growth. Because nitrogen is so important in biological processes, its chemical form constantly changes due to the activity of the many kinds of living organisms. As in the carbon cycle, a balance in the nitrogen cycle is critical to the sustainability of life on earth.

In natural ecosystems, organisms die and their tissues decay. Some of the nitrogen is converted into organic compounds used for the growth of the decay organisms, primarily bacteria and fungi (this is the **immobilization** part of the cycle). Some nitrogen is released from the decaying tissues in an inorganic form, ammonium (NH_4^+) (**ammonification**). Some ammonium is lost to the atmosphere after its conversion to ammonia, NH_3. This loss is noticeable in the odor of fresh manure, in which active microbial decay of organic matter is occurring. Urine of animals also releases significant quantities of NH_3 to the atmosphere.

Ammonium in soil is available for plant uptake for use in plant growth and has the added advantage of being bound tightly by soil particles and not subject to leaching into groundwater. Unfortunately, ammonium N does not remain in the soil for long periods because bacteria in the soil continually convert ammonium to another inorganic form, nitrate (NO_3^-) (**nitrification**). Nitrate N is more easily absorbed by plants than ammonium but has the major disadvantage of having a negative charge, which reduces its binding to soil particles. This makes nitrate N highly vulnerable to runoff and leaching from agricultural soils. Nitrogen as NO_3^- is an important water pollutant.

Under anaerobic (lacking oxygen) conditions, which are common after heavy rains, in low spots where water collects, and in "microenvironments," or small pockets of soil in which oxygen is absent, other bacteria convert nitrate N back to the gaseous atmospheric form, N_2 (**denitrification**). Significant amounts of nitrogen can be quickly lost from poorly drained soils due to denitrification. Nitrogen is thus lost from agricultural soils due to harvesting, runoff, leaching, and denitrification.

Sources of Nitrogen

Nitrogen can be added to agricultural soils in several ways besides the addition of manure or plant debris. Lightning, with its powerful electrical charge through the atmosphere, contributes a significant amount of nitrate to soils each year. Lightning converts atmospheric N_2 to nitrate N, which is washed from the air during rain. A small amount of nitrogen is also added to both soils and the atmosphere in the form of nitrogen oxides from the exhaust of automobiles, trucks, buses, and airplanes.

Certain bacteria are capable of "fixing" atmospheric N_2 to ammonium, a form useful to plants. Some of these bacteria are free-living in soil and water. Other nitrogen-fixing bacteria are found in the genus *Rhizobium* and are parasitic on leguminous plants such as beans, peas, soybeans, alfalfa, and clover. They are closely related to *Agrobacterium tumefaciens*, the crown gall bacterium, and exist in small galls on the root system of their host plant, absorbing nutrients from the plant. Their nutritional requirements are more than offset by the benefit to the plant of their nitrogen-fixing activity. Particularly efficient nitrogen-fixing strains of *Rhizobium* species have been selected over the years and are deliberately added to the seeds of legumes at planting time to encourage infection by these parasitic bacteria. A current focus of molecular biology is a better understanding of the genes responsible for nitrogen-fixation. The goal is to increase the efficiency of microbial nitrogen fixation in agricultural soils and perhaps to enable plants to fix their own nitrogen through genetic engineering.

Despite all the natural nitrogen inputs, most crops still require additional nitrogen. Most of this is supplied through synthetic fertilizers produced through industrial fixation processes. Industrial fixation converts atmospheric nitrogen to the ammonium form, which is very energy-intensive and costly. Improvements in our ability to provide optimal amounts of nitrogen to improved crop cultivars will increase not only total food supply but also protein levels, which are insufficient in the diets of more than a billion people.

Recent concern over the problems of runoff and leaching of nitrate from soils has led to increased interest in "slow-release" fertilizers. Manures and other organic materials are natural slow-releasing fertilizers because the nitrogen becomes available to plants gradually as the organic materials decay and nitrogen is converted to inorganic forms. Organic materials such as manure and compost also provide other macro- and micronutrients to plants in the decay process and improve soil structure. Nitrogen from

Fig. 13-8. *Rhizobium* root galls on peanut, a leguminous plant.

synthetic fertilizers is identical to that from natural sources except that the amount may be more concentrated. Thus, it is possible to cause "nitrogen burn" or damage to plants if too much synthetic fertilizer is added. However, plant roots absorb nitrogen in the same way whether the source is synthetic or natural. Slow-release forms of synthetic fertilizers are also widely available. One example is sulfur-coated pellets of urea. Soil bacteria metabolize the sulfur and the nitrogen is released gradually. The goal is to provide a continuous supply of nitrogen to the plant without contributing to pollution problems. Another method of application of synthetic fertilizers that is less likely to contribute to environmental problems is the frequent addition of low concentrations of liquid fertilizers in irrigation systems.

Micronutrients

Most commercial fertilizers supply the three macronutrients required in greatest quantity for plant growth: nitrogen, phosphorus, and potassium. The chemical forms and relative amounts of these elements vary with the time of application, the soil type, and the crop. Micronutrients are not commonly added to agricultural soils except in special circumstances. For instance, cole crops, such as cabbage and broccoli, are particularly susceptible to boron deficiency in some areas when the soil is sandy.

Most micronutrients are available in the majority of soils at sufficient concentrations for good plant growth. When deficiency symptoms are observed, the micronutrient is frequently present but unavailable in very acid or very basic soils. The pH factor is used to measure how acid a soil is; pH affects how soluble many chemicals are in water. As the pH of the soil water changes, some compounds become more soluble, while others become insoluble and unavailable for plant uptake. Figure 13-9 demonstrates some of these changes in solubility. Certain elements such as iron and manganese become particularly soluble at low pH levels, to the point of causing plant toxicity.

Because soil pH has such a significant effect on the availability of many nutrients, soil pH should be adjusted to the recommended range of 6.0–7.0 for most plants by the use of lime (to increase pH) or sulfur (to reduce pH) before micronutrients are added to soils where deficiencies are suspected. If micronutrient amendments are still necessary after adjustment of the soil pH, care must be taken to avoid toxicity from adding excessive amounts.

Diagnosis of Environmental Problems

It is important to know whether plant symptoms are due to an infectious disease caused by a parasite or to some environmental imbalance or pollutant. Diseases caused by infectious pathogens may spread and increase in severity without proper management. They often respond to various chemical treatments or may be reduced through the use of genetically resistant cultivars in future plantings. Environmental problems, on the other hand, may be caused by exceptional weather conditions that occur only rarely. A hard freeze in Florida citrus groves or a hail storm in an apple orchard cannot be prevented or cured. However, many problems caused by environmental factors are chronic problems that can be corrected with the adjustment of soil pH, addition of plant nutrients, care in the application of pesticides, or planting at a more appropriate time, depth, and spacing.

In diagnosing a plant problem, several clues help to separate environmental problems from diseases caused by infectious pathogens. Plants suffering from pathogen infection generally show symptoms at various stages of development, depending on when the infection occurred. There is often evidence of the pathogen, such as bacterial ooze, fungal spores and mycelium, or nematodes. A transitional zone of dying cells often occurs between healthy

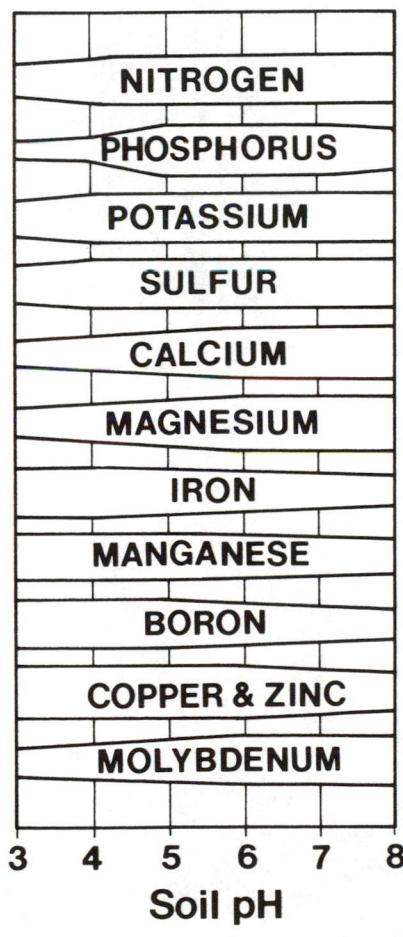

Fig. 13-9. Effects of pH on availability of elements in soil. At low pH, iron and manganese may reach concentrations that cause toxicity symptoms in plants. A number of important elements become less available for plant use in acid (low-pH) soils.

Fig. 13-10. Microelements are necessary for plant growth, but may cause toxic effects when present at higher concentrations. **Left,** *Ficus* plant with boron deficiency symptoms; **right,** *Ficus* plant leaf with boron toxicity symptoms.

Fig. 13-11. Sugar maple trees (*Acer saccharum*). The tree on right is healthy and vigorous. The tree on left shows signs of environmental stress due to parking lot paving. The stressed tree is also infected by *Armillaria mellea*, a root-rotting fungus that commonly attacks stressed eastern hardwoods.

and dead tissue in a disease lesion. If plant tissue has been chemically or physically injured, the line between healthy and damaged tissue is generally very distinct and lacks the zone of dying cells typical of parasitic infections. Generally, only certain species are infected because parasites tend to be highly specific. In contrast, environmental problems often affect many kinds of plants in an area to some degree.

Air Pollution

One of the most difficult kinds of environmental problems to diagnose and quantify is the chronic damage caused by air pollutants. Industrial activities, automobiles, and the burning of fossil fuels for energy have increased the concentration of certain chemicals in the atmosphere. Some air pollutants are released into the atmosphere and damage plants directly. These are considered **primary pollutants**. For example, **hydrogen fluoride** (HF) causes marginal necrosis in sensitive plants. It is a common pollutant from ore refineries, especially those that produce aluminum, as well as ceramic factories. The incomplete combustion of fuels is a source of other important pollutants. They are present in exhaust from internal combustion engines and are also released from factories. Both **nitrogen oxides** (NO_x) and **sulfur dioxide** (SO_2) are common pollutants that cause toxic effects in plants at relatively low levels (less than 3 ppm).

Both sulfur dioxide and nitrogen oxides cause interveinal yellowing of affected tissues and general growth suppression at lower concentrations, while tissues may show bleaching and bronzing at higher concentrations. An important source of sulfur dioxide is the burning of high-sulfur coal for energy production. Sulfur dioxide and nitrogen oxides combine with water in the atmosphere to produce sulfuric acid and nitric acid, respectively,

Table 13-2. Major Air Pollutants and Their Effects on Plants

Pollutant	Source	Symptoms
Ozone (O_3)	Exhaust of internal combustion engines; strastosphere; lightning formed in sunlight; major component of "smog"	Upper surface stippling, mottling, bleaching; tissue collapse; stunting; flowering and bud formation depressed
Peroxyacyl nitrates (PAN)	Exhaust of internal combustion engines; formed in sunlight; important component of "smog"	"Silver leaf"; necrotic spots, especially on lower leaf surfaces; stunting; young leaves most susceptible to damage
Sulfur dioxide (SO_2)	Combustion of high sulfur coal; factory stacks; engine exhaust	Interveinal necrosis of leaves; general yellowing; growth suppression
Nitrogen oxides (NO_x)	Combustion; engine exhaust	Similar to SO_2; SO_2 and NO_x combine with atmospheric water to form "acid rain"
Hydrogen fluoride (HF)	Pottery, cement, ceramics, and brick industries; ore smelters; phosphate fertilizer factories	Marginal necrosis and tip dieback of leaves; "tipburn" of grasses and conifers

which results in **acid rain**.

Two important pollutants also originate from engine exhaust but must be chemically converted in the presence of sunlight before they are chemicals toxic to plants. These **secondary pollutants** are **ozone** (O_3) and **peroxyacetyl**

Fig. 13-12. Ozone (O_3) injury symptoms on tobacco. Bel W3 tobacco is particularly sensitive to ozone injury and is used an indicator plant for the presence of ozone in the atmosphere. The leaf on the right is from a plant grown outdoors in Waltham, MA. A healthy tobacco leaf on the left shows no injury.

Fig. 13-13. Many urban areas suffer high levels of airborne ozone. Dark areas had hourly measurements that violated the Federal air quality standards more than once a year for three years.

nitrate (PAN), which are the toxic components of **smog**. Ozone enters leaves through stomata and causes necrotic specks to form on upper leaf surfaces. High concentrations of ozone can cause premature defoliation of sensitive plants, while chronic damage from lower concentrations of ozone inhibits photosynthesis. Ozone is the most destructive air pollutant and is the predominant air pollutant in the eastern and midwestern United States. Acute ozone damage has been documented in some crops in some areas, such as spinach grown in New Jersey. Yield reductions due to chronic ozone effects have been demonstrated in a number of other crops such as cotton, soybeans, and citrus. Symptoms of PAN damage are sometimes called "silver leaf" because the lower leaf surfaces are damaged. At higher concentrations, damage may be similar to that caused by ozone. PAN is an important component of smog in highly populated areas of the western United States, especially in the region around Los Angeles.

Forest Decline

A worldwide problem that has attracted much attention in recent years is **forest decline**. In some areas such as the San Bernadino National Forest east of Los Angeles, visible injury, growth reduction, and loss of pollution-sensitive trees in hundreds of thousands of acres of forest are certainly due to smog from the city. In many other areas, however, more subtle evidence of a problem has been noticed. In Germany, many forest areas, including the famous Black Forest, contain dead and dying trees. Other trees show needle loss, yellowing, reduced feeder roots, and decreased annual trunk growth. This phenomenon has been called *Waldsterben* or forest decline. Older trees show the greatest damage, including less mycorrhizal development and more disease by fungal root pathogens such as *Armillaria* and *Phytophthora*. Many species of plants have been affected. This suggests a widespread environmental problem, and the direct or indirect effects of air pollutants are suspect.

Chronic effects of pollutants are difficult to assess. These pollutants damage plants at high concentrations, as many experiments have demonstrated. In nature, a subtle decline due to lower concentrations of these same pollutants is much more difficult to prove. One method of study is to filter the air or enclose plants in rain-exclusion containers, leaving the plant in its natural environment but eliminating exposure to air pollutants. The growth of plants in filtered air or protected from acid rain can be compared to that of neighboring plants that remain exposed to the pollutants. Other studies compare plant growth and health in forests where pollutant concentrations are high with plant growth and health from areas where the air is cleaner.

Acid Rain

Many theories describe how air pollutants may cause forest decline. Some involve direct damage to plants, while others consider the indirect effects on soil and water. The effects of acid rain have been particularly difficult to assess. There is evidence that acid rain contributes to lower soil pH, which in turn may result in toxic levels of aluminum. A magnesium deficiency hypothesis suggests that acid deposition leaches Mg and Ca from leaves and soil. Some scientists have suggested that lower soil pH may affect

Fig. 13-14. Marginal necrosis of maple leaves caused by exposure to hydrogen flouride.

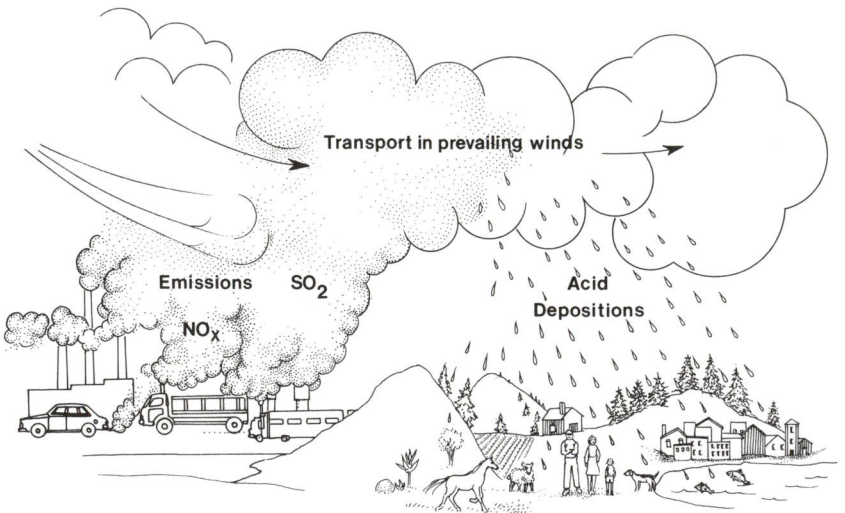

Fig. 13-15. The origin of acid rain. SO_2 = sulfur dioxide; NO_x = nitrogen oxides.

plant roots, their uptake of water and nutrients, and the availability of certain nutrients and may have direct effects on the microorganisms in the soil, including pathogens.

The effects of environmental changes on forest health are difficult to determine because of the numerous other stresses that affect the forest ecosystem. Ozone levels, drought, and forest management practices are also blamed for the apparent decline in many eastern forests. Direct correlations require long-term studies of plant growth, continuous monitoring of pollutant concentrations, and environmental monitoring of other factors to determine the causes of forest decline. Air pollutants and acid rain certainly affect the growth of some plants to a degree, but direct measurement of the chronic effects has not yet been accomplished.

Sugar Maple Decline

In eastern North America, the apparent decline of the sugar maple has heightened concerns about this resource and the possible effects on syrup production. Sugar maple trees frequently exhibit dieback and foliage that is yellow and smaller than normal or shows premature fall coloring. Acid rain has been suggested as a contributing factor in this decline, but other factors must also be considered. For instance, recent summer droughts and unusual winters with little snow cover certainly contribute to stress. A recent infestation of pear thrips has taken its toll. Certain pests and diseases are more common on stressed trees and they, too, have increased. Examples include the sugar maple borer, certain canker diseases, Armillaria root rot, and tent caterpillars. Overtapping of trees may also contribute to their decline, especially during periods of environmental stress. Although it is tempting to blame acid rain for this decline, demonstrating any direct effects on forest trees has been very difficult because of the complex nature of the problem.

Fig. 13-16. Red spruce dieback in New Hampshire. Injury has been linked to acid rain depositions in soil and foliage.

Air Quality Standards

Air pollution affects the health of all members of an ecosystem, including humans. The severity of the effects is difficult to assess, but federal air quality standards have been established in an attempt to protect air quality. Each year these standards are exceeded in areas of concentrated vehicle traffic and near certain industrial areas. Industrial smokestacks, known as point-source pollution, can be monitored, and polluters can be fined. Burning of high-sulfur coal for energy production in the Midwest is considered an important source of the acid rain that precipitates over the eastern United States and Canada. Although it is technically possible to prevent the emission of sulfur dioxide by installing "scrubbers" in smoke stacks, the cost is high. Should midwestern consumers bear the entire cost of preventing acid rain in eastern states? Cleaner automobile exhaust is produced by cars with catalytic converters, but gas mileage is reduced, increasing transportation costs. Despite more stringent auto exhaust regulations, traffic continues to increase, causing higher ozone and PAN atmospheric pollution. Should drivers be restricted to alternate-day driving to reduce air pollution? Should the private automobile be banned from cities and only public transportation be used?

Unfortunately, air pollutants do not honor political boundaries. The pollution suspected of causing forest decline in West Germany has its source in eastern European countries, where air quality standards are more lax. Pollutants are detected at ever-higher concentrations even in areas far from their source, such as Antarctica. On some days, views of the Grand Canyon in Arizona are blurred by air pollution produced hundreds of miles away. These toxic compounds not only detract from the natural beauty of our environment but affect the health of all organisms.

Pollution levels have become a part of regular weather reports on television news in many areas, and warnings to reduce outdoor activities during periods of heavy pollution are common in cities in the summer. Citizens must balance their own health and that of the world ecosystem against the inconvenience and expense of improving air quality. Strictly enforced air quality standards are the only real solution to the problem of air pollution.

Selected Readings

Houston, D. R. 1987. Forest tree declines of past and present: Current understanding. Canadian Journal of Plant Pathology 9:349-360.

Kandler, O. 1990. Epidemiological evaluation of the development of *waldsterben* in Germany. Plant Disease 74:4-12.

McLaughlin, S. B. 1985. Effects of air pollution on forests, a critical review. Journal of the Air Pollution Control Association 35(5):512-534.

Mohnen, V. A. 1988. The challenge of acid rain. Scientific American 259(2):30-38.

Schulze, E.-D. 1989. Air pollution and forest decline in a spruce (*Picea abies*) forest. Science 244:776-783.

Plant Diseases in a Hungry World

We began Chapter 1 with a consideration of the Irish potato famine of the 1840s. Some of the factors that made this historical event particularly disastrous were directly related to the biology of *Phytophthora infestans* and the potato disease, late blight. Others were more sociological or economic. In 1845 in Ireland, a rapidly growing human population was dependent almost entirely on a single plant species, the potato. Although the famine was widespread, it was worst in Ireland because of political conditions that isolated the population and prevented transport of food from areas where it was more abundant. The Irish peasants had very restricted areas in which to grow their food. The potatoes were vegetatively propagated and, therefore, genetically uniform and susceptible to devastation when exposed to a virulent fungal pathogen from which they had been isolated.

Much has changed in the world since the Hungry Forties, but famines still occur. Many of the same factors that led to the death and suffering of the Irish in the 1840s deny good health and even survival to millions of people in the world today. The statistics are disturbing. More people died from hunger between 1983 and 1985 than were killed in World War I and World War II combined. The most widely accepted standard of measurement of the existence of hunger is the infant mortality rate, and in some countries, more than 10% of the infants die before they are a year old (Table 14-1). The Hunger Project, a nonprofit charitable organization dedicated to the elimination of world hunger, estimates that approximately 15 million children die each year from hunger. Each day 35,000 human beings, mostly women and children, die as a consequence of hunger and starvation. The effects of hunger and malnutrition are well known but are frequently ignored by the well-fed people of the world. Impaired physical and mental development and general lethargy and susceptibility to disease deprive hungry people of the energy necessary for enjoyment of even the most basic qualities of life.

Certainly no problem is more complex than that of how to feed the earth's human population. To address its complexities with simplistic solutions is to fail to provide an adequate solution. Many academic disciplines can contribute to the solution of the problem of a quality existence for all people on earth. Historians, anthropologists, sociologists, and political scientists examine the political and social interactions of humans that confound our ability to cooperate sufficiently for political and economic stability and self-determination. Economists contribute theories on the distribution of wealth and economic development within the constraints of available resources. Ecologists evaluate the **carrying capacity** of the earth's ecosystem; they attempt to determine the human population that can be adequately maintained without permanent disruption and, potentially, destruction of our fragile environment. Agricultural scientists also contribute to our ability to provide a reliable and sustainable food supply. Norman Borlaug, a plant pathologist, received the Nobel Peace Prize in 1970 for his contributions to the development of "miracle" high-yielding wheat, which led to the media phrase **Green Revolution**. The Nobel committee does not present a specific prize for contributions to agricultural science, but they recognized that inadequate food supplies are directly related to political unrest and wars.

Table 14-1. Infant Mortality Rate (IMR)[a]

Country	Population[b]	IMR[c]
Bangladesh	107.1	140
Brazil	141.5	63
India	800.3	101
Indonesia	174.9	88
Nigeria	108.6	124
Pakistan	104.6	125

[a]Source: The Ending Hunger Briefing Workbook. 1984. The Hunger Project, San Francisco, CA; used by permission.
[b]Estimates from mid-1987, Population Data Sheet.
[c]Hunger is considered a chronic, society-wide condition when the IMR of a nation is greater than 50; that is, more than 50 children per 1,000 live births die in the first year of their lives.

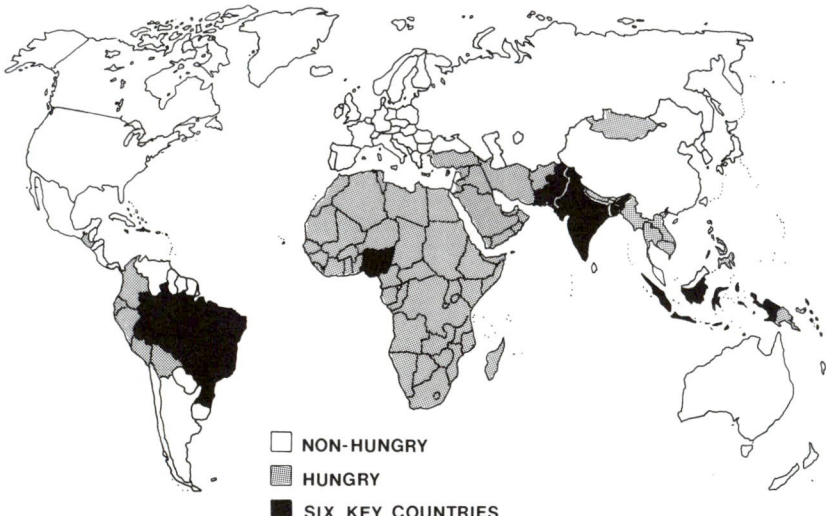

NON-HUNGRY
HUNGRY
SIX KEY COUNTRIES

Fig. 14-1. Hungry and nonhungry countries and the six key countries for ending hunger. If hunger were ended in these six countries, approximately 60% of the world's hunger would be ended.

World Population

The growing world food problem cannot possibly be adequately addressed in a single chapter, but it is possible to consider some important aspects of this challenge that directly relate to the science of plant pathology. The world population of human beings is increasing, so food supplies must also increase. In some areas of the world, particularly in Africa, Mexico, and parts of Asia, growth rates are more than 3.5%. The world population, which now exceeds 5 billion, reached its first billion around the year 1800. The population doubled by 1930, and doubled once more by 1975. At current rates, it is expected that 50% more people will be on earth in the year 2000 than were present in 1975. Even with extensive food shipments and economic aid, it is hard to imagine the technological, political, and sociological problems in providing for a population that increases so quickly. To build schools and housing and provide jobs and medical care in addition to food for a population that doubles in less than 25 years is beyond the capabilities of most governments.

Population experts expect that the world population will not stabilize before reaching 8 billion and perhaps will reach 12 billion by the year

Fig. 14-2. Norman Borlaug, winner of the Nobel Peace Prize.

2045. This increased human population will require approximately 300% more food than is produced today. Unusual weather conditions that would reduce crops by even as little as 2.2% could result in the starvation of 264 million people. What are the limitations to increased food production? Is it possible to produce and distribute sufficient food for so many people?

Arable Land

One way to increase agricultural production is to bring more land into cultivation. This is an area of considerable controversy because of disagreement about how much arable, or cultivable, land exists in the world. Total cultivable land is estimated at 7 billion acres (3.2 billion hectares), of which 3 billion are now under cultivation. Most easily farmed and very productive land is already subject to intensive agriculture. In addition, much good agricultural land is lost every year. Land is converted to building sites or is used for road building and urban sprawl. Other land is lost to agriculture because of severe erosion and **desertification** (desert formation). In the United States, four square miles of prime farmland are shifted to nonagricultural uses such as roads and urban expansion each day. Much of the most valuable farmland in the United States, especially around cities and rivers, has been permanently lost because the land brought a better price as building lots for homes than from agricultural production. To reduce the loss of farmland in areas of high population density, land "banks" are being created by providing tax incentives to those who keep their land in production rather than selling it for development. Florida, California, and many northeastern states are rapidly, and permanently, losing significant areas of potential agricultural production to other uses. In the poorest countries, where population growth rates are high, agricultural land must be taken out of food production to provide living areas for people.

Erosion continues to strip away the topsoil from areas of intensive agricultural production. This topsoil, which accumulated for millions of years beneath prairies and forests, is exposed to wind and water erosion

Table 14-2. World Population Growth: 1800–1983[a]

Year	Number of People (billion)	Time Taken to Reach (years)
1800	1.0	2–5 million
1930	2.0	Approximately 130
1960	3.0	30
1975	4.0	15
1983	5.0	8

[a] Source: Ending Hunger: An Idea Whose Time Has Come. 1985. The Hunger Project, San Francisco, CA; used by permission. Data taken from J. vanderTak, C. Haub, and E. Murphy, 1979, Our Population Predicament: A New Look, Population Bulletin 34(5):4, and from World Population Data Sheet, 1983, Population Reference Bureau, Washington, DC.

when tilled. One inch of topsoil is lost from farmland in Iowa every 14 years, which reduces corn yields by three to six bushels per acre. In 1984, worldwide loss of topsoil exceeded new soil formation by 25 billion tons. Water erosion carries away valuable topsoil to rivers and oceans in tremendous quantities. The Mississippi River carries 300 million tons of soil into the Gulf of Mexico each year; every major river system in the world carries a similar load of runoff.

To increase production, new farmland must be brought into cultivation, but what remains uncultivated at this time is frequently on steeply sloping areas that lose their topsoil rapidly due to erosion unless they are carefully managed. In Asia, where population density has been high for many centuries, complex terrace systems help maintain topsoil for sustained production. Contour tillage that produces the alternating strips familiar in midwestern farmlands also reduces soil erosion and became popular after the Dust Bowl days of the 1930s in the United States.

Water Resources

Agricultural production is limited by water in many areas of the world. Irrigation is necessary for production throughout the world, including in

Fig. 14-3. Rice culture in the highlands of the Philippines.

many states in the United States. Rivers have been diverted, and water rights have been the subject of many court battles in recent years. Recent droughts have led to enforced water restrictions in urban areas, including bans on car washing and on watering of lawns. However, water use by individuals and cities accounts for only a small proportion of total water use. Agriculture accounts for approximately 70% of the water consumption in the United States.

The large green circles surrounded by brown desert, visible from airplanes when one is flying over western states, are evidence of how water can bring agricultural production to otherwise unproductive areas. In many cases, irrigation water is pumped from underground aquifers in which it has been accumulating for millions of years. This **fossil water**, such as the Ogallala Aquifer that stretches from Texas beneath the High Plains as far as South Dakota, is not being replaced by infiltration as quickly as it is being removed. Nearly one fifth of U.S. irrigated land uses similar water sources. As a result, the cost of pumping water from greater and greater depths reaches a point where water costs become excessive, and land must be abandoned. The removal of water in some areas has been so rapid and extensive that sink holes have developed, where land collapses into empty recesses left after the water is pumped out. Improvement in irrigation techniques, such as drip irrigation, which greatly reduces losses to evaporation, and other new techniques in dryland agriculture may allow sustained agricultural production even as irrigation water becomes scarcer.

Huge tracts of arable land, especially in Africa, have become desert in areas where land has been overcultivated, overgrazed, and stripped of nearly

Fig. 14-4. Center pivot irrigation system used on corn in Minnesota. In many states, agriculture would be severely limited if irrigation water were not available.

all its trees and shrubs by people searching for wood for fuel. In 1987, about 18 million acres (8 million hectares) of rainforest of the Amazon basin were cleared, and the process continues to the present. Some areas were cut to harvest valuable trees; others were burned to clear land for farming. Much of this farmland remains productive for only a few years because of the particular problems of soils beneath the rainforest. Erosion and short-term fertility restrict the use of these soils for agriculture. The farmland is then abandoned, and farmers move on to new sites, which requires the cutting of more forest.

There is considerable concern that the loss of significant tracts of forest in the world will disrupt both the carbon cycle (Chapter 8) and the **hydrological cycle**, which affects rainfall rates and patterns throughout the world. The dense vegetation of the world's rainforests significantly affects temperature and precipitation through the absorption of CO_2 and loss of water by evapotranspiration. Unusual droughts and weather patterns also contribute to the loss of arable land due to erosion and **salinization**, or salt accumulation, where water is insufficient to elute accumulating salts from the soil. A 1984 report by the United Nations Environment Programme

Fig. 14-5. When too much water is pumped from below the surface, subsidence may occur. These pictures are from Arizona.

estimated that 35% of the world's land surface is threatened by desertification. Cereal breeding programs are developing corn and sorghums that require less water and can provide a crop in drought-prone areas such as the Sahel of Africa. Development of cultivars with salt tolerance is another goal, which would allow the use of water sources with high salt levels that damage currently used cultivars.

Soil Fertility

The fastest growing areas of population are concentrated in the subtropical and tropical regions of the world. Many of their soils are shallow and particularly vulnerable to destruction and can support production for only a few years. Many of them are quite acid and contain high concentrations of iron and aluminum detrimental to the growth of many food crops. More research must be done on how to increase and maintain the productivity of such soils.

Fertilizers are a mainstay of agricultural production in developed countries. In 1982, 115 million metric tons of inorganic fertilizer were produced and applied to agricultural lands. As discussed in Chapter 13, the energy costs in producing inorganic fertilizer are high, but yield increases have also been substantial. Since the cost of production of inorganic fertilizers is directly related to the price of oil, a nonrenewable resource, much attention has been focused on reducing the need for high inputs of inorganic fertilizer. New tillage methods have reduced erosion, increased water use efficiency, lowered the need for fertilizer, and reduced polluting runoff and leaching from agricultural lands. Organic matter, such as manures and composts, can be returned to the land to lower the need for inorganic fertilizers. Nitrogen, which accounts for more than 80% of the energy used

Fig. 14-6. Desertification (Sahel region of the Sahara).

in manufacturing fertilizers, can be made available to plants through biological fixation (Chapter 13), using improved strains of *Rhizobium* species that parasitize various leguminous plants, free living nitrogen-fixing bacteria, and genetic engineering of the nitrogen-fixing process.

Agricultural production is directly related to available land, quality of the soil, and availability of water and fertility. The previously given statistics demonstrate that there is reason for great concern about each of these major factors. The judicious and sustained use of land for agriculture and the critical inputs of water and fertility have scientific, social, and political aspects that must be approached both locally and from a global perspective.

Food Crops

An important lesson from the Irish potato famine is the danger of a population relying on a single plant species. Is the population of the world in the 1990s any less vulnerable than the Irish were in the 1840s? Approximately 800,000 species of plants exist in the world, of which at least 3,000 are edible. Of these perhaps 150 are commonly eaten, but only about 12 species contribute significantly to the world's food supply. These include wheat, rice, corn (maize), potato, sweet potato, cassava (manioc), banana, coconut, soybean, common bean, sugarcane, and beet. Most of the calories consumed by people on our planet are provided by only three species of plants: wheat, rice, and corn.

This heavy reliance on only three plant species by over 5 billion people reflects a vulnerability similar to that of the Irish population in 1845. A significant decrease in the annual yield of even one of these crops would significantly reduce the total worldwide calories available that year. These crops are also the major stored food of the world. In 1981, enough stored

Table 14-3. Population Growth Rates of the World's Nations[a]

Nation	Annual Growth Rate (%)	1989 Population (million)
World's fastest-growing nations[b]		
Kenya	4.1	24.1
Syria	3.8	12.1
Iraq	3.8	18.1
Zimbabwe	3.6	10.1
Tanzania	3.6	26.3
World's slowest-growing nations[c]		
Hungary	−0.2	10.6
West Germany	−0.1	61.5
East Germany	0.1	16.6
Italy	0.0	57.6
Greece	0.1	10.0

[a]Source: World Population Data Sheet, 1989, Population Reference Bureau, Washington, DC; used by permission.
[b]With populations of 8 million or more.
[c]With populations of 10 million or more.

grain existed for only 40 days of world consumption, down from approximately 102 days in 1961. Improved food storage is of critical importance in areas of high population growth, particularly in tropical areas where warm temperatures increase the rate of food degradation by pathogens and decay organisms. In addition, insect and rodent pests cause significant losses every year. Hungry months between harvests are a normal way of life for one to two billion people every year.

The Green Revolution

How can yields of major crop species be increased? Breeding programs have produced significant yield increases through repeated crosses and selection of promising parental lines. High-yielding cultivars of wheat and rice were first released in the 1960s. High-yielding rice cultivars were released from the International Rice Research Institute (IRRI) in Los Baños, The Philippines. They contain a dwarfing gene that produces plants of low stature. Such cultivars can more efficiently utilize high fertilizer inputs that greatly increase yield without causing lodging (falling over) of grain stalks during rain and wind near harvesttime. The introduction of these rice cultivars contributed to increased food production and even food export from India in the 1970s. Yield increased from 0.79 tons per hectare in 1950 to 2.6 tons in 1975. Improved wheat cultivars that also possessed dwarfing genes led to similar yield increases.

While there is no doubt that total food production increased when these new cultivars were made available, there were also accompanying problems. Critics point to the social upheaval in which small farmers were sometimes placed at a disadvantage because they were less able to obtain credit to

Fig. 14-7. Pile of wheat at a grain terminal on the Snake River, Washington, 1983.

purchase the fertilizers necessary for the higher yields. In many areas, the "land races" of wheat and rice that had been selected by farmers over many centuries for their adaptation to the local requirements were lost, causing a significant decrease in the genetic diversity of the crops. Besides genetic vulnerability, the new cultivars were susceptible to both traditional and novel pest and pathogen problems that led to significant losses in some areas and increased pesticide use in others.

Fig. 14-8. The "miracle" rice cultivar, IR-8, an early release compared to more recent cultivars with multiple resistance to insect pests and diseases. They were developed by the International Rice Research Institute in The Philippines.

Fig. 14-9. Bananas in La Lima, Hondurus, showing genetic diversity.

The media coined the phrase "Green Revolution," but the scientists who created the "miracle" grains never considered the early releases to be the final answer to world food production. On the contrary, breeding programs are ongoing, resulting in the continuing release of new cultivars with improved resistance to many of the pests and pathogens that attacked the earlier releases. No plant geneticist or pathologist would say that use of one or a few cultivars of any important crop is a reliable or desirable situation. The tremendous yield increases brought about through the use of the dwarfing genes were but a first step in a program that must continue as long as the human population continues to grow and to disrupt the environment.

What is the future of the Green Revolution? Both its successes and the problems that accompany them illustrate very well that improvements in the agricultural sciences can be successfully implemented only after the appropriate economic, sociological, and political problems have been solved. However, agricultural scientists must develop the most productive and sustainable production methods to provide the foundation on which other disciplines can build policies to ensure that the new cultivars and production methods will be successfully implemented.

Genetic Diversity in Agriculture

Genetic resistance and overall genetic diversity are critical for the long-term sustainability of the world's food supply. The more diverse the genetic background of our crop species, the less likely it is that a single new pest or pathogen will cause devastating losses. The intense collection of both wild and cultivated genetic resources for the cereals aids in their preservation for future use and for immediate use in the ongoing breeding programs. In addition, overall genetic diversity is necessary to ensure the adaptability of important species that may be faced with long-term climatic changes

Fig. 14-10. Guava, avocado, and mango cultivar collection, Seychelles.

if global warming continues or if major shifts in weather patterns take place.

Where do we find sources of this important genetic diversity? The **centers of origin** of the major food species remain the critical areas that must be preserved. Unfortunately, the genetic wealth that holds the future of world food production is primarily found in areas of the world where population growth and economic constraints are the greatest. Maintenance of the gene pool of wild crop species and their close relatives requires their preservation in the centers of origin, so that they remain available for scientific improvement of plant species. Many areas rich in genetic diversity are rapidly being lost as all available land is used for cities, homes, and agriculture.

Besides increasing genetic diversity in major crop species, maintenance

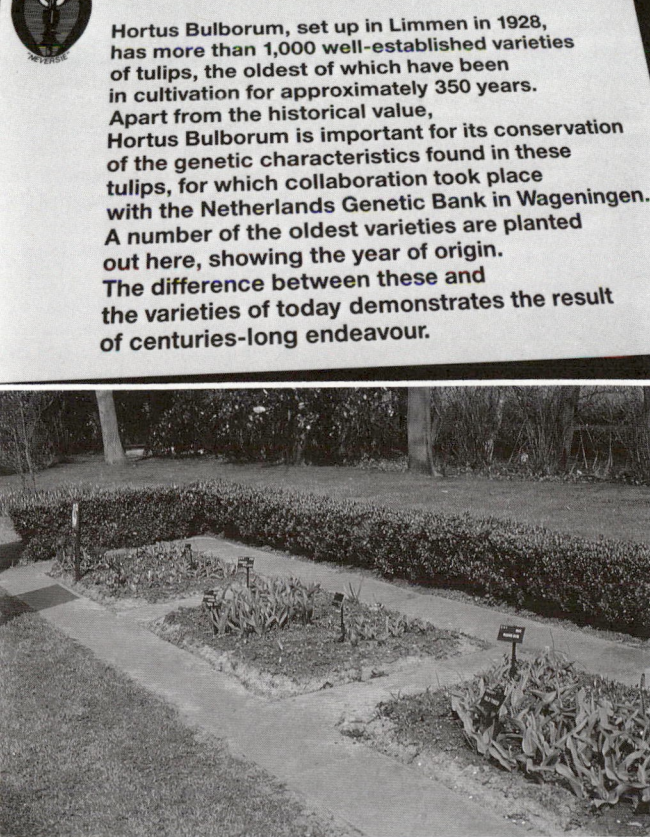

Fig. 14-11. Genetic resources for tulip breeding in the Netherlands.

of a reliable food supply demands that we increase and expand the plant species on which we depend. This involves improvement of minor crops that are already well exploited and increasing production of less familiar edible plants. Some currently minor cereals are important food crops in Africa and produce well even under stressful conditions and in poor soils. Millet, grown in North America primarily for bird seed, and grain sorghums deserve much more attention. Agricultural scientists of temperate climates are sometimes accused of having a "grain mentality," which suggests that more attention should go to root and tuber crops, particularly those also adapted for growth in tropical climates. Yams, taro, and sweet potatoes are crops being improved for better yields and higher quality. In fact, crop research and development centers have been established throughout the world with just these goals in mind.

Growing out of previous Rockefeller and Ford Foundation projects, the **international agricultural research centers**, supported through the **Consultative Group on International Agricultural Research** (CGIAR), were established in 1971. Some centers were established in the centers of origin of important food crops. The CGIAR sponsors these research centers to foster the improvement of food crops of world importance. They not only select and develop new cultivars but are active in the collection, preservation, and exploitation of the diverse species and cultivars of those species. Many of these research centers are in the tropics. Most important food crops of the world originated in the tropics, which is not surprising since the greatest concentration of genetic diversity is present there. Research centers that concentrate on a wide variety of tropical food crops include the organizations listed in Table 14-4.

Fig. 14-12. U.S. clonal collection of apples, grapes and other fruits in Geneva, NY.

In recognition of the importance of expanding the number of plant species that feed the world, the National Academy of Sciences published the book *Underexploited Tropical Plants with Promising Economic Value* in 1975. Some of the species grow only in tropical areas and have received little research attention because most agricultural scientists in the world live and work on crops of importance in temperate climates. There has been particular interest in a number of long-neglected crops from the Andes including *arracacha, ulluco*, and *oca* (root and tuber crops), *nunas* (high-protein beans), and *naranjilla, cherimoya*, and *pepino dulce* (nutrient-rich fruits). *Quinoa* is an important crop that was cultivated as a sacred grain for many years in South America by the Incas. Another neglected high-protein seed producer was cultivated in Mexico before the arrival of the Spanish and belongs to the genus *Amaranthus*, which includes the common garden weed known as pigweed. **Amaranths** are well adapted to the arid, stressful climates of many areas not particularly productive for traditional cereals. The cultivation of amaranths was banned by Spanish church officials because the seeds were used by the natives in religious ceremonies, but now there is new effort to reestablish the crop, particularly for use in marginal

Fig. 14-13. University of Illinois scientists examine stored genetic stock of corn.

croplands.

Another crop that is not very familiar to people who live in temperate climates is cassava, or manioc, (*Manihot esculenta*) (Chapter 9). This is an important food crop for over 300 million people in tropical areas of Africa and South America. The starchy root is harvested and processed into many kinds of food, including a product used in tapioca pudding. Despite the cyanogenic glucosides in the plant tissue, which must be removed before it can be eaten, the plant is an efficient calorie producer, and the starchy roots can be left in the ground up to 18 months until needed. One goal of breeding programs is to reduce the level of cyanide in the

Fig. 14-14. The international agricultural research centers. See Table 4 for names.

Table 14-4. International Agricultural Research Centers Supported by the Consultative Group on International Agricultural Research

Abbreviation	Full Name and Location
CIAT	Centro Internacional de Agricultura Tropical; Cali, Colombia
CIMMYT	Centro Internacional de Mejoramiento de Maiz y Trigo; Mexico City, Mexico
CIP	Centro Internacional de la Papa; Lima, Peru
IBPGR	International Board for Plant Genetic Resources; Rome, Italy
ICARDA	International Center for Agricultural Research in the Dry Areas; Aleppo, Syria
ICRISAT	International Crops Research Institute for the Semi-Arid Tropics; Hyderabad, India
IFPRI	International Food Policy Research Institute; Washington, DC, USA
IITA	International Institute of Tropical Agriculture; Ibadan, Nigeria
ILCA	International Livestock Centre for Africa; Addis Ababa, Ethiopia
ILRAD	International Laboratory for Research on Animal Diseases; Nairobi, Kenya
IRRI	International Rice Research Institute; Manila, Philippines
ISNAR	International Service for National Agricultural Research; The Hague, Netherlands
WARDA	West Africa Rice Development Association; Monrovia, Liberia

tissues to enhance its edibility for humans and animals, but there is some concern that the plant may be subject to more pest and disease problems if this happens. Many insects and pathogens that attack cassava have developed cyanide detoxification mechanisms, so apparently the cyanide serves a protective purpose.

Threatening Diseases

While the major diseases of temperate food crops are familiar, they have not been eliminated despite many years of study. Genetic resources in the centers of origin of these crops hold the promise of improved protection and reduced need for pesticides through genetic resistance. Genetic variations that become widely distributed due to their popularity always portend potential danger, as seen in the increased susceptibility of Victoria oats and Texas-male-sterile corn to new races of previously minor pathogens.

Fig. 14-15. Amaranthus species produce a high-protein seed and grow well in poor soils with little water. **Top,** flowering plants; **bottom,** harvested seed from different cultivars.

Some important diseases remain restricted to only certain areas of the world and require constant vigilance to prevent their movement into currently uninfested areas. Examples include the citrus canker (*Xanthomonas citri*), recently reintroduced into Florida, where 99% of the citrus crop is susceptible to infection. So far, the bacterium has not been detected in citrus-growing areas of the Southwest and California. Soybean rust (*Phakopsora pachyrhizi*) is an important pathogen in Asia, but quarantines have so far prevented its introduction into North America. Downy mildews are a major limiting factor in the production of corn (maize) and related crops such as sorghum in tropical Asia. Eight downy mildew species attack maize, but only one species, *Peronosclerospora sorghi*, has been introduced to the Western Hemisphere. This occurred in Texas in 1961. Since that introduction, the fungus has spread to other southern states as well as to areas of Central and South America.

Two fatal diseases of coconut palm have become important in recent years. The cadang-cadang viroid is restricted to the Eastern Hemisphere, while the lethal yellowing mycoplasma is a problem in the Western Hemisphere. African cassava mosaic virus, transmitted by white flies, can reduce yields by as much as 40%. Modern tissue culture methods are making pathogen-free propagative material available to help reduce infection by this otherwise uncontrollable virus. The virus has not yet infected cassava in South and Central America. The Moko disease (*Pseudomonas solanacearum*) of banana that destroyed millions of trees in Central America in the 1960s has not been transmitted to Asia or Africa, the major area of banana and plantain production in the world. South American leaf blight of rubber remains a devastating disease only in the Western Hemisphere.

Fig. 14-16. Cassava, or manioc (*Manihot esculenta*), is an important root crop in tropical countries.

The important cash crop, cacao, has important disease problems in both hemispheres but, so far, Monilia pod rot is found only in South America, while the swollen shoot virus disease is important in Africa. Witchweeds (*Striga* species), the parasitic weeds that threaten many important monocot crops, are important primarily in Africa and Asia.

Plant pathologists are working to reduce our vulnerability to plant diseases on a world scale. Increased availability of technological improvements in computers has made a world crop disease data base a possibility. New immunoassay techniques have increased the speed and accuracy of detecting microscopic pathogens at quarantine inspection stations. Current cultivars, breeding lines, and germ plasm can be taken to areas where diseases are endemic to aid in the selection of resistant cultivars in preparation for the eventual spread of pathogens to new areas. Resistant cultivars can be

Fig. 14-17. Soybean rust (*Phakopsora pachyrhizi*). **Top,** rust pustules on soybean leaf; **bottom,** microscopic section of a uredium filled with uredospores.

maintained for emergency use in case pathogens are accidentally introduced. Global data and modern communication technologies can improve emergency responses for rapid eradication of newly infected crops, in the same way that emergency responses are made to introduced animal

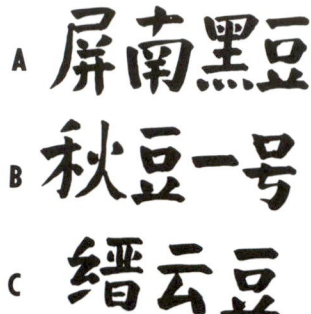

Fig. 14-18. Chinese symbols for three cultivars of soybean reported to be rust-resistant.

Fig. 14-19. Downy mildew (*Peronosclerospora sorghi*) on sorghum. **Left,** sporangial stage; **right,** infected leaves shred and release oospores.

pathogens.

While monoculture production of major food crops will not disappear, mixed farming systems are being studied, particularly for food production in poorer countries. Small plots of land planted to a number of crop species can be highly productive and reliable because poor yields in certain crops due to weather, diseases, or pests can be compensated for by good yields in other species. Farming mixed species probably results in fertility and pest management benefits that are not yet well understood. Sustainable year-round production from small plots of land is crucial in tropical areas, where population pressures continue to increase.

World Biological Diversity

The continued environmental disruption caused by bringing more land into cultivation and deforestation in many areas for fuel, lumber, and living space has implications beyond the loss of genetic diversity for agricultural purposes. The greatest concentration of genetic diversity lies in a band around the Equator between the Tropic of Cancer and the Tropic of Capricorn known as "the tropics." The tropics, which contain more biological and environmental diversity than any other region of the world, include a range from the hot and humid lowlands to temperate-like climates at higher elevations.

Perhaps 5–10 million species of organisms exist on earth; some scientists estimate even 30 million. Most have not yet been evaluated for their usefulness. These species hold great potential for food production as well as for medicinal, fiber, dye, lumber, and ornamental uses. We may now be losing as many as 10,000 species per year. These estimates come from studies that suggest that the loss of a single plant species results in the elimination of up to 15 other species unable to survive without that plant.

Fig. 14-20. Symptoms of cassava mosaic virus.

Genes that have successfully evolved over millions of years are associated with the species alive today on earth. The total collection of genes that exists has been called the **gene pool**. One analogy used to explain the importance of the gene pool is to compare it to a library in which the various species represent individual volumes of evolutionary history. The value of the library is directly related to the number of volumes it contains because each volume contains information that can contribute to our evolutionary future. The loss of each species on earth reduces the value of the library and the evolutionary flexibility of life on earth. This same gene pool contains the genetic resources from which we will create improved crop cultivars for the future through traditional breeding techniques and the newer genetic engineering methods.

Many scientists are demanding that we increase the conservation of these genetic resources in **gene banks** of living and stored plant materials collected from their native lands. This has led to political conflicts, as poorer countries have had reduced access to their genetic resources because of collectors from technologically advanced countries. In the United States, collections of "germ plasm" of important plants are maintained in different areas for preservation, study, and selection of breeding materials. For instance, the clonal repository in Geneva, New York, contains samples of apples collected from wild species and trees in abandoned orchards that originally grew from seed. In Fort Collins, Colorado, the **national seed storage facility** maintains seed collected from throughout the world, using modern storage techniques, such as **cryostorage** (low temperature freezing) to help prolong the life of seed. The cost of planting and reharvesting the seed periodically as well as describing and maintaining data of the entire collection is high.

Although no doubt exists about the value of germ plasm collections of individual species, the limitations to collecting and maintaining a small amount of plant material require that sections of natural ecosystems be preserved. The **Minimal Critical Size of Ecosystems Project**, sponsored by

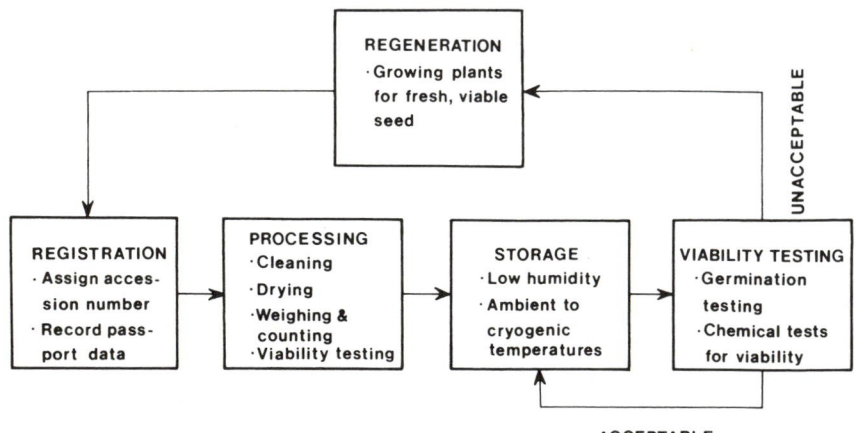

Fig. 14-21. Maintenance process of a plant seed bank.

the World Wildlife Fund and Brazil's National Institute for Amazon Research, is a long-term study to determine how large preserves must be for maintenance of stable ecosystems. As the Amazon rainforest is cleared for pastures, uncut areas of varying sizes are being evaluated to determine the stability of species maintenance over time. Ecosystem preservation helps maintain not only specific species of obvious importance at this time but also maintains the pests, pathogens, protective microbes, related species, and other organisms important for the continued evolution of the ecosystem in a changing environment.

Although the economic benefits of species conservation are obvious, a more critical question remains. What level of biological diversity is necessary to maintain a stable world ecosystem? Humans are causing significant ecological disruptions affecting global temperature, atmospheric CO_2 levels, fresh water distribution and availability, and total genetic diversity on earth. Now, when the evolutionary potential of a diverse and flexible gene pool is probably of critical importance, we are destroying tremendous numbers of species each year. It has been estimated that, at current rates, 30–70% of all plant species will be extinct by the year 2000. Since many ecologists relate ecosystem stability to genetic diversity, biological diversity becomes significant for far more than the direct benefits that individual species play

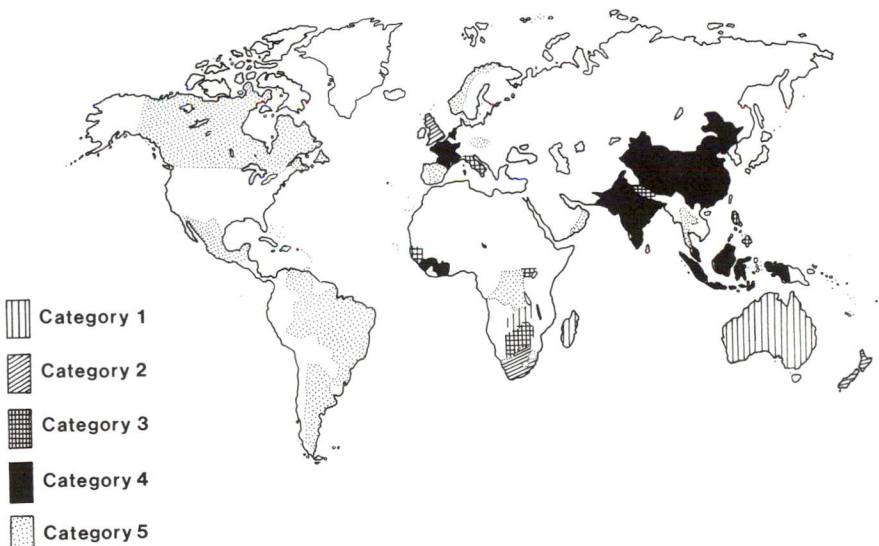

	Category 1
	Category 2
	Category 3
	Category 4
	Category 5

Fig. 14-22. National conservation strategy (NCS) development around the world, July, 1985. Category 1: substantial consensus document produced and endorsed by the government. Category 2: substantial consensus document produced but not yet endorsed. Category 3: process of preparing such a document definitely happening. Category 4: course of action initiated that looks likely to lead to an NCS. Category 5: other involvement (an NCS at exploratory stage; strategic planning for resource management at subnational level).

Fig. 14-23. Exploring for wild potato species in the Andes.

in our lives for such practical uses as food and fiber.

The human ecological impact on the world is challenging not only the sustainability of our human environment but the very survival of life on earth. The political, social, economic, and ecological choices that humans make in the next decade will have far-reaching effects on the earth's environment. These decisions will determine whether life becomes more uncertain in an increasingly inhospitable environment or whether we can find an ecological niche that will provide a reliable and sustainable quality of life for the human population as well as for other members of our ecosystem. The science of plant pathology will continue to play a critical role in this quest.

Selected Readings

Brown, L. R., Durning, A., Flavin, C., French, H., Jacobsen, J., Lowe, M., Postel, S., Renner, M., Starke, L., and Young, J. 1990. State of the World (Worldwatch Institute). W. W. Norton, New York.

DeBruijn, G. H., and Fresco, L. O. 1989. The importance of cassava in world food production. Netherlands Journal of Agricultural Science 37:21-34.

Hawkes. J. G. 1983. The Diversity of Crop Plants. Harvard University Press, Cambridge, MA.

Hunger Project, The. 1985. Ending Hunger. Praeger, New York.

Jabs, C. 1984. The Heirloom Gardener. Sierra Club Books, San Francisco.

Janzen, D. H. 1973. Tropical agroecosystems. Science 182:1212-1219.

Kahn, E. J., Jr. 1984. The Staffs of Life. Little, Brown, and Co., New York.

Lewin, R. 1989. How to get plants into the conservationists' ark. Science 244:32-33.

Mellor, J. W., and Gavian, S. 1987. Famine: Causes, prevention, and relief. Science 235:539-545.

Plucknett, D. L., Smith, N. J. H., Williams, J. T., and Anishetty, N. M. 1987. Gene Banks and the World's Food. Princeton University Press, Princeton, NJ.

Sattaur, O. 1988. Native is beautiful. New Scientist 118(1615):54-57.

Shell, E. R. 1989. Seeds in the bank could stave off disaster on the farm. Smithsonian 20(10):94-105.

Thurston, H. D. 1973. Threatening plant diseases. Annual Review of Phytopathology 11:27-52.

U.S. Congress, Office of Technology Assessment. 1987. Technologies to Maintain Biological Diversity. OTA-F-330. U.S. Government Printing Office, Washington, DC.

Glossary

acid rain—precipitation with a low pH, due to the presence of nitric and sulfuric acid formed by the combination of air pollutants (NO_x and SO_2) with water

acute toxicity—ability of a single dose of a compound to poison

ADI—acceptable daily intake; a regulatory term used by the Food and Drug Administration to determine limits such as pesticide residue tolerances

adventitious—describing structures growing from an unusual place, such as roots arising from the base of a monocot stem

aeciospore—dikaryotic spore produced by rust fungi that infects the alternate host species in heteroecious rusts; produced in a structure called an aecium (pl. aecia)

aerobic—requiring oxygen

aflatoxin—a fungal metabolite produced by *Aspergillus flavus* and closely related fungi; causes toxic effects in humans and other animals

agar—a non-nitrogenous gelatinlike mixture of polysaccharides extracted from certain red algae (seaweeds) and used for preparation of culture media on which microbes are grown

agglutination—the clumping together of antibodies and antigens that allows detection of an antigen such as a virus in plant sap

alkaloids—nitrogen-containing ring compounds produced by plants that cause physiological effects in animals

allele—an alternative form of a gene

alternate host—one of two or more species of plants that can be parasitized by the same organism; often used to describe the hosts of heteroecious rusts

alternation of generations—a reproductive cycle in which a haploid phase alternates with a diploid phase

amino acid—the units or "building blocks" of proteins

anaerobic—without oxygen

anastomosis—hyphal fusion

angiosperm—a flowering plant; a plant having its seeds enclosed in an ovary

annual—a plant that completes its life cycle in one year or less

annual ring—a single year's growth layer in wood

anther—the pollen-bearing portion of a stamen

antheridium (pl. antheridia)—a male reproductive organ

antibiotic—a compound produced by a microorganism that is toxic to other microorganisms

antibody—a protein produced by the immune system of warm-blooded animals in reaction to a foreign antigen, such as a pathogen

antigen—a protein that stimulates the production of antibodies by the immune systems of warm-blooded animals

apical dominance—suppression of lateral buds by the hormonal influence of the terminal (apical) bud

apical meristem—a meristem at the tip of a plant, such as the end of a shoot or root

appressorium (pl. appressoria)—a swelling on a germ tube or hyphal tip of parasitic fungi that adheres to the host surface before a penetration peg penetrates the host cells

aquifer—area in which groundwater accumulates

arable—able to be cultivated for agriculture

ascogonium (pl. ascogonia)—a specialized cell that gives rise to the hyphae that produce asci

Ascomycetes—fungi that typically produce haploid, septate hyphae, ascospores (sexual spores) in an ascus, and conidia (asexual spores)

ascospore—the sexual spore of an Ascomycete, produced in a saclike ascus

ascus (pl. asci)—sac typically containing four or eight ascospores

asexual—not sexual; imperfect

autoecious—describing a rust fungus that completes its entire life cycle on a single host plant

autotroph—an organism that can synthesize its nutritive substances from inorganic sources; e.g., plants capable of photosynthesis (*see* heterotroph)

auxin—a plant growth regulator that controls cell elongation and stimulates root production

bacteriophage—a virus that infects bacteria

bacterium (pl. bacteria)—a prokaryotic, microscopic, single-celled organism that increases by binary fission

bark—all tissues outside the vascular cambium of a woody plant, generally including the cork layers, cork cambium, and phloem

Basidiomycetes—fungi that typically produce dikaryotic mycelium and basidiospores (sexual spores); this group includes rusts, smuts, and most mushrooms

basidiospore—the haploid sexual spore of Basidiomycetes

berry—a fleshy fruit containing seeds

binomial, Latin—a two-word Latin name, part of a system that names all living organisms and originated with C. Linnaeus

biocide—a compound that is toxic to all forms of life

biological control—the use of living organisms to reduce disease, usually through competition or antagonism

biotechnology—the use of tissue culture and/or genetic engineering to produce new kinds of organisms

bud—an embryonic shoot of a plant, composed of a meristem and leaf primordia and often protected by scales

callus—a mass of undifferentiated plant cells produced in tissue culture or on wounds

canker—a necrotic lesion on a stem or branch, often sunken into the tissue

carbon cycle—the continuous circulation of carbon atoms from inorganic carbon dioxide (CO_2) to organic molecules and back to CO_2

carcinogen—something that causes cancer

carrying capacity—a term used by ecologists for the number of members of a species that an ecosystem can maintain

Casparian strips—impervious thickenings between the cells of endodermis in roots

cell membrane—the membrane that bounds a cell and helps control the movement of substances into and out of the cell

cell wall—a protective, resistant, but permeable polysaccharide structure external to the cell membrane; typically found in plants, fungi, some bacteria, and certain other organisms

cellulose—a carbohydrate that is an important component of cell walls

center of origin—the area where a plant originated

chlamydospore—a thick-walled asexual resting spore produced from hyphal or conidial cells

chloroplast—an organelle in eukaryotic cells in which photosynthesis takes place

chromatid—one of two daughter strands of a duplicated chromosome formed during mitosis or meiosis

chromosome—the organelles that contain genes (DNA); chromosomes are found in the nucleus of eukaryotic cells

chronic toxicity—poisoning due to low levels of exposure to a compound over a period of time

cleistothecium (pl. cleistothecia)—a closed fruiting body produced by powdery mildews in which asci and ascospores are produced

coat protein—the protective layer of protein surrounding the nucleic acid core of viruses

collenchyma—supporting tissue in leaves and stems

compartmentalization—the process by which some woody plants produce barrier tissues to prevent decay

conidium (pl. conidia)—an asexual spore

conjugation—the temporary fusion of bacterial cells during which genetic material is transferred

cork—external protective tissue impermeable to water and gasses and the primary component of bark

cork cambium—a lateral meristem in woody plants that produces cork

corm—an upright, thickened underground stem

cortex—parenchyma cells in stems and roots, usually bounded by the epidermis and the vascular tissue

cotyledon—first leaf or leaves of a germinating plant

crop rotation—the alternation of different crop species in the same soil to improve fertility and reduce disease and pest problems

cross-resistance—resistance in fungi to chemically related compounds; the use of mild strains of viruses to inoculate plants for protection against infection by more severe strains

cryostorage—storage at very low temperatures for long-term maintainence

cultivar—a cultivated variety; group of closely related plants of the same species that differ in certain minor details such as color or fruit

cultural practices—various means of disease management: application of nutrients, irrigation practices, type of cultivation, etc.

cuticle—the protective layer composed primarily of wax and cutin on epidermal cells of plants; also the outer layer of a nematode

cutin—a fatty substance that is an important component of the cuticle of epidermal cells

cytokinins—growth regulators that control cell division and are important for shoot stimulation of callus in tissue culture

cytoplasm—all living substances of a cell except the nucleus and the cell wall

cytoplasmic inheritance—genes not located in the nucleus, i.e., those in mitochondria and chloroplasts (*see* maternal inheritance and extrachromosomal inheritance)

damping off—death of seedlings before or shortly after emergence, often by toppling over at the soil surface

days to harvest—the minimum number of days required between a final application of a pesticide and the time of crop harvest

deciduous—describing plants that shed leaves at the end of the growing season

Delaney Clause—part of Section 409 of the Federal Food, Drug, and Cosmetic Act that allows no food additives (including pesticide residues) that have been shown to cause cancer in animals at any dosage

deoxyribonucleic acid (DNA)—the double-stranded, helical molecule that contains genetic code information; each repeating unit, or nucleotide, is composed of deoxyribose (a sugar), a phosphate group, and a purine (adenine or guanine) or a pyrimidine (thymine or cytosine)

desertification—formation of deserts in areas that were previously productive

Deuteromycetes—fungi that produce no sexual stage (synonym: Imperfect Fungi)

dicotyledon—a large group of angiosperms in which the seedling plant has two cotyledons, a tap root branches to form lateral roots, and leaves generally have a netlike vein pattern

dikaryotic—composed of cells in which each cell contains two different haploid nuclei

dioecious—describing plants that have separate male and female individuals

diploid—describing the nuclear state in which chromosomes are present in pairs (*see* haploid)

disease—any disturbance of a plant that interferes with its normal growth and development, economic value, or aesthetic quality; a continuously, often progressively affected condition in contrast to injury, which results from momentary damage

disease pyramid—a memory aid similar to the disease triangle but including, in addition, the factor of time in the development of an epidemic

disease triangle—a memory aid that lists the three important components necessary for disease: susceptible plant, virulent pathogen, and favorable environment

disinfect—to remove a pathogen from infected plant tissues, e.g., the use of systemic fungicides or hot water treatments to kill seed-infecting pathogens

disinfest—to remove a pathogen from the surface of a plant or inanimate object such as tools or tables

dominant allele—an allele of a gene that causes the same phenotype in the homozygous or heterozygous state (*see* recessive allele and incomplete dominance)

downy mildew—a host-specific, obligate parasite that infects above-ground plant parts and belongs to the Oomycetes

EBDC—ethylenebisdithiocarbamate, an important group of broad-spectrum protective fungicides

ecology—the study of the interrelationships of organisms in their environment

economic damage threshold—a level of damage that justifies the cost of a control measure to prevent further damage

ecosystem—a community of living organisms and their environment

ectoparasite—a parasite that feeds from the outside of the host

electroporation—a means of gene transfer in which an electrical current is applied to protoplasts, creating temporary pores in the cell membrane that allow entry of foreign DNA

embryo sac—the female gametophyte in angiosperms

endodermis—a layer of cells in roots that separates the cortex from the vascular tissue and pericycle

endoparasite—a parasite that enters host tissues

endophyte—strictly defined, a plant inside another organism; used to describe parasitic fungi in grasses that produce compounds toxic to mammals and some insects

endospore—an asexual spore developed within a bacterial or fungal cell that is highly resistant and thick-walled

enzyme—a biological catalyst composed of protein

enzyme-linked immunosorbent assay (ELISA)—a serological method in which one antibody is linked to an enzyme so that a colored compound is formed after addition of a substrate to indicate detection of an antigen

EPA—Environmental Protection Agency

epidemic—an increase of disease in a population with time

epidemic rate—a measure of the rate of increase of disease in a population, often designated as r in mathematical descriptions of epidemics

epidemiology—the study of epidemics

epidermis—the outermost layer of nonwoody stems, leaves, and roots

epiphytotic—an epidemic in a plant population

epizootic—an epidemic in an animal population

eradication—a disease management method in which a pathogen is removed from a growing site or plant part, e.g., by crop rotation, sanitary practices, or seed treatments

ergot—a sclerotium, or resting structure, produced by *Claviceps purpurea* and other closely related fungi in infected flowers of parasitized grain plants, especially rye

ergotism—a disease caused by ingestion of grain contaminated with alkaloids of ergot fungi, often called the holy fire or St. Anthony's fire in the Middle Ages

etiolated—describing a plant that is weak and spindly due to insufficient light

eukaryote—an organism containing a membrane-bound nucleus and other organelles, i.e., all higher plants, animals, fungi, and protists (*see* prokaryote)

exclusion—a disease management method in which a pathogen is prevented from coming in contact with a plant, e.g., by quarantines or pathogen-free propagation of seeds or planting parts

exogenous—having its origin from outside

extrachromosomal inheritance—non-nuclear genes, i.e., those in mitochondria and chloroplasts (*see* cytoplasmic inheritance and maternal inheritance)

facultative—able to exist under variable conditions

FFDCA—the Federal Food, Drug, and Cosmetic Act, which sets legal tolerances of pesticides on food products at harvest

fiber—a cell in wood that strengthens the tissue

FIFRA—the Federal Insecticide, Fungicide, and Rodenticide Act, which legislates the conditions for use of pesticides

fission—asexual reproduction, as in cell division in bacteria

flagellum (pl. flagella)—the whiplike organ of locomotion found on bacteria, zoospores, and other organisms

forma specialis (pl. formae speciales)—a subspecies that has no apparent morphological differences from other similar subspecies but differs in host range

fossil water—water in aquifers that has accumulated over very long periods of time and is not quickly replaced through infiltration from recent rainfall

fruiting body—a fungus structure that contains or bears spores

fumigant—a pesticide that must be contained in a space or in soil, where it volatilizes and kills pests

fungicide—a compound that is toxic to fungi

fungistasis—inhibition of growth of fungi; often imposed by soil microbes and overcome by the nutrient-rich rhizosphere of a host plant, allowing soilborne fungal pathogens to infect the plant

fungus (pl. fungi)—a eukaryotic organism that requires a source of organic nutrients, is commonly composed of threadlike hyphae or mycelium, and reproduces by asexual and/or sexual spores

gall—tumorlike overgrowth on a plant, composed of unorganized cells

gallery—tunnels formed by bark beetles and larvae feeding beneath the bark

gamete—sex cell, e.g., egg or sperm

gametophyte—the haploid stage of a plant (*see* sporophyte)

gene—a unit of heredity; a segment of DNA that codes for a protein

gene pool—the genes of organisms considered collectively

gene-for-gene theory—a theory that describes the dynamic genetic interaction between hosts and parasites

genetic code—the system of triplet codons composed of nucleotides of DNA or RNA that determine the amino acid sequence of a protein

genetic engineering—the transfer of specific genes between organisms using enzymes and laboratory techniques rather than biological hybridization

genetic recombination—the variation in genetic makeup of offspring compared to parents following sexual reproduction

genetics—the study of heredity

genome—the entire genetic information of an organism

genotype—the genetic composition of an organism

genus (pl. genera)—a taxonomic category that includes closely related species

germ plasm—the genetic resources of an organism

germ theory—the theory that infectious or contagious disease is caused by microbes (germs)

germ tube—hypha resulting from the outgrowth of the spore wall and/or cytoplasm during germination

gibberellins—growth regulators that affect stem elongation

Gram stain—a dye used for identification of bacteria

greenhouse effect—a theory that atmospheric temperatures will rise with the increasing concentration of certain gasses such as CO_2, due to the trapping of heat from solar radiation as occurs inside a greenhouse on a sunny day

groundwater—water contained in aquifers in the soil, sometimes in "underground rivers" but more often in smaller accumulations mixed with sand

growth regulator—a compound active at very low concentrations that controls growth processes in plants (synonym: hormone)

guard cells—specialized epidermal cells comprising a stoma that expand and contract to control transpiration

haploid—nuclear state in which only one of each kind of chromosome is present (*see* diploid)

Hartig net—the mantle of hyphae on root surfaces and between cortical cells of roots supporting ectomycorrhizal fungi

haustorium (pl. haustoria)—specialized outgrowths that increase the surface area of parasites to aid in nutrient and water absorption from hosts, found in fungi and parasitic plants

heartwood—the often dark-colored wood in the center of a tree that no longer functions in conduction

hemicellulose—a carbohydrate similar to cellulose that is an important component of cell walls

herbicide—a compound that is toxic to plants, generally used for weed control

herbivore—an organism that eats plants

hermaphrodite—an organism with both male and female sexual organs

heteroecious—a rust fungus that requires two different host species to complete its life cycle

heterotroph—an organism that cannot manufacture its own organic compounds and must obtain nutrients from other organisms (*see* autotroph)

heterozygous—having alternate forms of an allele of a gene on homologous chromosomes

homozygous—having the same form of an allele of a gene on homologous chromosomes

honeydew—a secretion attractive to insects produced by certain fungi such as *Claviceps* species and the spermogonia of some rusts; a sweet sticky secretion given off by aphids and certain other related organisms

hormone—*see* growth regulator

host range—the group of plants that can be infected by a parasite

hybrid vigor—increased vigor of hybrid offspring over either parent (synonym: heterosis)

hybridoma—a cell produced by the fusion of an antibody-producing cell and a lymphoma (cancer) cell for production of monoclonal antibodies

hydathode—a channel that opens along the sides of leaf for exudation of excess fluid

hydrological cycle—the cycling of water molecules between gaseous form (water vapor), liquid, and ice

hyperparasite—a parasite of a parasite

hypersensitive reaction—rapid death of host cells following infection

hypha (pl. hyphae)—the threadlike structures that comprise fungi; filaments bound by a cell membrane and cell wall, containing nuclei (see *mycelium*)

hypovirulence—greatly reduced ability to cause disease

immunoassay—a detection method based on antibodies

Imperfect Fungi—fungi that produce no sexual stage (synonym: Deuteromycetes)

incomplete dominance—the interaction of alleles of a gene that produces an intermediate phenotype, as in the production of pink flowers when red and white alleles are present in a heterozygous individual

indicator plant—a plant used to detect the presence of a pathogen because of the distinct symptoms it develops upon infection

infected—describing a host in which a parasite has established a parasitic relationship

infectious—capable of causing infection and spreading disease from plant to plant

infested—superficially contaminated by pathogens; refers to plant surfaces, soil, or tools

inflorescence—a cluster of flowers on a plant

initial inoculum—the inoculum that initiates an epidemic, often the surviving population of a pathogen after winter or a dry season (synonym: primary inoculum)

inoculum—a pathogen or its structures capable of initiating an infection

insecticide—a compound that is toxic to insects

integrated pest management—a combination of strategies to reduce losses to pests and pathogens based on environmental and economic considerations

intercellular—between cells

inversion—mutation in which a piece of a chromosome breaks off and reattaches in a reverse direction

karyogamy—nuclear fusion of two haploid nuclei to produce one diploid nucleus (*see* plasmogamy)

Koch's postulates—the procedure used to prove the pathogenicity of an organism, i.e., its role as the causal agent of a disease

latent period—the time between infection and the appearance of symptoms and/or the production of new inoculum; the time after a vector has acquired a pathogen and before it can be transmitted

LC$_{50}$—the lethal inhaled concentration of a compound for 50% of a test population

LD$_{50}$—the lethal oral dose of a compound for 50% of a test population

leaf—organ of a plant whose primary function is photosynthesis, usually composed of a flat blade and a petiole that connects to the stem

legume—a plant in the pea family, such as beans, alfalfa, clover, and soybeans

lenticel—a lens-shaped opening in bark composed of loosely packed cells that functions in gas exchange

lesion—localized area of diseased tissue

ligase—the enzyme that causes pieces of DNA or RNA to fuse to each other

local lesion—a small localized spot produced on a leaf upon mechanical inoculation with a virus

macroconidium (pl. macroconidia)—a large asexual spore

macrocyclic—describing a rust fungus that produces all five spore stages in its life cycle

macronutrient—an element needed in relatively large quantities for plant growth, e.g., nitrogen (N), phosphorus (P), and potassium (K)

maternal inheritance—non-nuclear genes, i.e., those in mitochondria and chloroplasts (*see* cytoplasmic inheritance and extrachromosomal inheritance)

mating types—compatible strains necessary for sexual reproduction in some fungi

meiosis—reduction division of one diploid nucleus to four haploid nuclei

meristem—a mass of undifferentiated cells that divide; apical meristems are found at shoot and root tips

mesophyll—tissue composed of cells with chloroplasts between the epidermal layers of a leaf

microconidium (pl. microconidia)—a small asexual spore

microcyclic—describing a rust fungus that produces only teliospores and basidiospores

micron—one millionth of a meter (synonym: micrometer)

micronutrient—an element needed in relatively small quantities for plant growth and likely to be toxic at higher concentrations, e.g., iron (Fe) and manganese (Mn)

microorganism—a microscopic organism (synonym: microbe)

middle lamella—the layer between plant cells composed primarily of pectin

mitochondrion (pl. mitochondria)—a membrane-bound organelle found in eukaryotic organisms that functions in respiration

mitosis—nuclear division in which the chromosome number remains the same

mode of action—how a compound functions

mollicute—a prokaryote that lacks a cell wall

monoclonal antibody—an antibody showing specificity to a single antigenic fitting site; produced from hybridomas formed by fusing specific antibody-producing cells with lymphoma (cancer) cells

monocotyledon—a large group of angiosperms whose seedlings possess one cotyledon, with roots that usually arise adventitiously from the base of the stem, and leaves with parallel veins

monoculture—the production of the same plant species in the same soil in successive years or in large, uninterrupted areas

monocyclic pathogen—a pathogen that completes only one disease cycle during a growing season

morphology—study of the form and structure of an organism

multiline—a physical mixture of cultivars that vary primarily in their resistance genes

mutagen—a chemical that causes mutations

mutation—a sudden, random genetic change

mutualistic—describing a relationship that benefits both organisms

mycelium (pl. mycelia)—the threadlike filaments constituting the vegetative body of a fungus (*see* hypha)

mycoherbicide—a parasitic fungus used as a biological control of an unwanted plant

mycology—the study of fungi

mycoplasmalike organism (MLO)—a prokaryote that lacks a cell wall and causes diseases in plants with the typical symptoms of yellowing, stunting, and witches' brooms

mycorrhiza (pl. mycorrhizae)—a fungus that has a mutualistic relationship with plant roots

mycotoxin—a toxic compound produced by a fungus, common in the genera *Aspergillus, Penicillium, Alternaria,* and *Fusarium*

necrotic—dead or dying

nectary—a nectar-secreting gland in a flower

nematicide—a compound toxic to nematodes

nematode—a roundworm that may be parasitic on animals or plants or be free-living

nitrogen cycle—the cycling of the element nitrogen from gaseous forms to various inorganic forms such as NH_4 and NH_3 and organic forms such as nucleic acids and proteins

nitrogen fixation—the conversion of gaseous N_2 to a form available for plant uptake

NOEL—no observable effect level

nucleotide—a unit of a nucleic acid, either DNA or RNA

nucleus—the membrane-bound body in eukaryotic cells that contains the chromosomes

obligate parasite—a parasite that requires living host tissue for growth

omnivore—an organism that uses all types of food, both plant and animal, as nutrient sources

oncogen—a compound that induces benign or malignant tumors

oogonium (pl. oogonia)—the one-celled female reproductive structure of Oomycetes

Oomycetes—fungi that typically produce a nonseptate diploid mycelium, oospores (sexual spores), and sporangia that may contain zoospores (asexual spores)

oospore—sexual spore of an Oomycete

ooze—mass of bacterial cells mixed with host fluids

organelle—a membrane-bound structure within a cell having a specialized function, e.g., mitochondria and chloroplasts

organic—describing a molecule containing carbon atoms; pertaining to living organisms

osmosis—the diffusion of water from an area of lower concentration to one of higher concentration

ovule—the enclosed structure that, after fertilization, becomes a seed

ozone (O_3)—a component of smog formed from oxygen in the presence of sunlight

ozone layer—a protective later of ozone in the upper atmosphere that reduces ultraviolet radiation

PAN—peroxyacyl nitrate; a component of smog formed from the exhaust of automobiles and other internal combustion engines in the presence of sunlight

parasite—a organism living in or on another living organism and obtaining nutrients from it

parenchyma—plant tissue composed of living cells that usually function in photosynthesis or storage; **palisade parenchyma** consists of one or more layers of columnar cells often located just below the upper epidermis; **spongy parenchyma** is more loosely packed tissue often located below the palisade parenchyma

parthenogenic—self-fertile, requiring no male for reproduction

pasteurization—removal of pathogenic microorganisms from soil by the use of heat without total sterilization so that beneficial microorganisms remain

pathogen—an entity that causes disease

pathology, plant—the study of plant diseases (synonym: phytopathology)

pathovar—synonym for "subspecies" when referring to bacterial plant pathogens; a strain of a bacterial species differentiated by pathogenicity on different hosts

pectin—an important carbohydrate component of cell walls and the middle lamella

penetration peg—the specialized, narrow, hyphal strand on the underside of an appressorium that penetrates host cells

perennial—a plant that persists for more than two years or growing seasons

pericycle—cells within the endodermis of a root, from which lateral roots arise

perithecium (pl. perithecia)—flask-shaped fruiting body with an opening, at least at maturity, and containing asci

pesticide—a general term for a compound that is toxic to a pest

petal—delicate leaflike part of a flower, usually conspicuously colored

petiole—the leaf stem that attaches to the main stem of a plant

phenotype—the appearance of an organism

phloem—tissue composed of living cells adapted for transport of products of photosynthesis

photosynthesis—the process by which energy from the sun is "fixed" in chemical bonds of organic sugars formed from water and carbon dioxide in water, with the release of oxygen; in eukaryotic cells photosynthesis takes place in chloroplasts

phytopathology—the study of plant diseases (synonym: plant pathology)

phytotoxicity—damage to plants, often caused by a chemical

pistil—the female or seed-bearing organ of a flower, typically composed of the ovary, style, and stigma

plant pathology—the study of plant diseases (synonym: phytopathology)

plasmid—a small circular piece of DNA that exists independently of the chromosome(s) in bacteria and a few eukaryotes, such as yeast

plasmodesma (pl. plasmodesmata)—cytoplasmic strands that connect living cells

plasmogamy—fusion of two sex cells (*see* karyogamy)

pollen tube—a tube formed after germination of a pollen grain that carries gametes to the ovule

polycyclic pathogen—a pathogen that completes several to many disease cycles in a growing season

polygenic—involving many genes

polyploid—containing several to many sets of chromosomes

powdery mildew—a white, powdery, superficial Ascomycete fungus that is an obligate parasite and generally is found only on one or a few closely related species of plants

principle of independent assortment—Mendel's principle that genetic traits are inherited independently of each other

principle of segregation—Mendel's principle that genetic traits are inherited randomly from each parent

prokaryote—a cell lacking a membrane-bound nucleus or organelles, e.g., a bacterium

proof of pathogenicity—steps taken to prove that an organism is the pathogen that causes a disease (*see* Koch's postulates)

propagule—any structure of an organism capable of renewing the organism's growth

protein—a nitrogen-containing organic compound composed of units called amino acids

protoplast—a plant cell bounded only by a cell membrane after its cell wall has been degraded by enzymes

pseudothecium (pl. pseudothecia)—a fruiting body containing asci with an opening at the top through which ascospores are discharged; similar to a perithecium

Puccinia pathway—the region through which rust uredospores move from southern areas through all grain-producing areas of the United States to Canada each season

pyramiding genes—accumulating a number of resistance genes in a single cultivar through breeding

quarantine—legal restriction in the movement of plants and animals to prevent the spread of diseases

ray cell—tissue that extends radially in the secondary xylem and phloem of a woody plant

receptive hypha—the part of a rust spermogonium that receives the nucleus of a spermatium

recessive allele—an allele whose presence is masked by a dominant allele in the heterozygous condition

recombinant DNA—genetic combinations produced through genetic engineering

replication—repetition; the method by which new virus particles are produced

resistance—the inherent ability of an organism to overcome or retard the activity of a pathogen; in pests, the ability to withstand exposure to certain pesticides

respiration—the breakdown of organic molecules to release energy for cell functions

restriction enzymes—enzymes that cleave DNA at a particular base sequence

ribosome—a subcellular organelle involved in protein synthesis

rhizomorph—specialized fusions of many strands of hyphae that appear rootlike

rhizosphere—the narrow region of soil modified by root influence

rhizosphere competence—the ability to survive in the rhizosphere

ribonucleic acid (RNA)—molecules involved in the transcription of DNA and its translation to protein

root cap—a group of cells on a root that protects the growing tip

root exudates—the various compounds that leak from growing and expanding sections of roots as well as from broken cells at exit points of lateral roots

root hair—threadlike, single-celled outgrowths from a root epidermal cell

rust—a disease caused by a specialized group of Basidiomycetes that often produces spores of a rusty color

salinization—the accumulation of salt in soils

sanitation—destruction of infected and infested plants or plant parts; elimination of disease inoculum and insect vectors

saprophyte—an organism that obtains nutrients from dead organic matter

sapwood—the generally lighter colored wood that functions in conduction

sclerenchyma—cells with thick secondary walls such as fibers and sclerids (stone cells)

sclerotium (pl. sclerotia)—a resting structure produced by many species of fungi, composed of masses of mycelium protected by a distinct dark-colored rind that resists desiccation

secondary growth—tissues produced from lateral meristems, including cork tissue from cork cambium and secondary xylem and phloem from the vascular cambium, that increase the girth of a woody plant

secondary inoculum—inoculum produced in secondary disease cycles during an epidemic, often asexual spores in fungi

seive tube—a series of living cells in the phloem arranged end to end for food transport in plants

selective medium—a nutrient medium that supports the growth of only certain organisms

septate—having septa

septum (pl. septa)—cross wall, as in a hypha or spore

serology—the use of antibodies to detect antigens

serum—blood fluid

sexual reproduction—reproduction involving fusion of two haploid nuclei (karyogamy) to form a diploid nucleus and reduction division back to haploid nuclei (meiosis), resulting in genetic recombination

smuts—a group of fungi in the Basidiomycetes that typically releases masses of black dusty spores at maturity

soil inhabitant—an organism that maintains its population in soil over a period of time

soil invader—an organism that is generally eliminated or greatly reduced in soil after a period of time

solarization—heating of soils with sunlight, usually through plastic ground covers, to reduce pathogen and pest populations

species—a kind of organism, one that is genetically isolated from other organisms

specific epithet—the second word in a Latin binomial

spermatium (pl. spermatia)—a spore produced in a spermogonium of a rust that fertilizes a receptive hypha, which leads to the formation of a dikaryotic mycelium (synonym: pycniospore)

spermogonium (pl. spermogonia)—the structure produced after infection by rust basidiospores, composed of spermatia and receptive hyphae (synonym: pycnium, pl. pycnia)

spiroplasma—a prokaryotic organism that lacks a cell wall and exists in a helical shape in the phloem of diseased plants or on plant surfaces

spontaneous generation, theory of—the theory that the microorganisms found in dead and decaying matter arose from the process of decay

sporangium (pl. sporangia)—a spore container

spore—a reproductive body of fungi and lower plants

sporophyte—the diploid stage of a plant (*see* gametophyte)

stamen—the male reproductive organ of a flower, composed of an anther and a stalk or filament

stele—the central cylinder of a plant root composed of the pericycle, xylem, and phloem

sterilization—the total destruction of living organisms by the use of heat or chemicals

stolon—an underground stem that grows horizontally and may produce roots along it

stoma (pl. stomata or stomates)—air exchange pores commonly found on the underside of leaves

stylet—a hollow structure used to pierce cells for feeding, found in many insects and nematodes

super race—a race of a pathogen capable of overcoming numerous resistance genes

susceptible—not immune; prone to infection

synnema (pl. synnemata)—fused or compacted spore-bearing hyphae with conidia at the apex (synonym: coremium; pl. coremia)

systematics—the study of the kinds of organisms and the relationships between them

taxonomy—the scientific classification of organisms

teliospore—the thick-walled spore of rusts and smuts in which karyogamy and meiosis occur

teratogen—something that causes birth defects

tolerance—in pesticide regulation, the legal pesticide residue allowed at harvest

totipotency—the concept that even specialized cells contain all of the genetic information for an organism and, therefore, any cell should be able to regenerate into any tissue or into an entire plant

toxin—a poisonous substance

tracheid—a nonliving tapered cell in xylem that transports water and minerals

transcription—the production of a complementary strand of RNA from a segment of DNA

transduction—the transfer of genes from one organism to another by viruses, especially in bacteria

transformation—the transfer of genetic material to a new organism

transgenic—possessing a gene from another species; used to describe the organisms that have been the subject of genetic engineering

translation—the assembly of a chain of amino acids for a protein from a segment of messenger RNA in the ribosomes

translocation—the movement of a segment of a chromosome and its reattachment to another chromosome; the movement of food through a plant in the phloem

transpiration—the loss of water by evaporation from plant surfaces, primarily through stomata

transposon—a segment of a chromosome capable of changing its location

trichome—a leaf hair or scale

triplet codon—a set of three nucleotide bases in DNA or RNA that code for an amino acid

triploid—having three sets of chromosomes

tuber—an underground stem adapted for storage, typically produced at the end of a stolon

tylose—an overgrowth of a parenchyma cell into a xylem vessel through a pit in the secondary vessel wall

uredospore—the asexual, dikaryotic, often rusty-colored spore of a rust fungus, produced in a structure called a uredium (pl. uredia); the "repeating stage" of a heteroecious rust fungus, capable of infecting the host plant on which it is produced

vascular cambium—the lateral meristem that produces cells for the secondary xylem and phloem

vascular tissue—tissue that transports food and water; composed of phloem and xylem

vascular wilt disease—a xylem disease that disrupts normal uptake of water and minerals, resulting in wilting and yellowing of foliage

vector—a living organism that transmits a pathogen

vegetative propagation—asexual reproduction; in plants, the use of cuttings, bulbs, tubers, and other vegetative plant parts to grow new plants

vessel—a structure composed of nonliving cells connected end to end and perforated at the ends for transport of water and minerals

virion—a virus particle

viroid—a small plant-pathogenic RNA with no protein coat

virulent—strongly pathogenic

virus—a submicroscopic agent that causes disease and multiplies only in living cells, composed of nucleic acid and usually a protein coat

winter burn—damage to tissues during winter due to water deficiency

witches' broom—the excessive proliferation of weak shoots or roots from a point; caused by mites, some fungi, MLOs, etc.

wood—secondary xylem

xylem—vascular tissue that functions in water and mineral transport; secondary xylem is wood

zoospore—an asexual spore capable of locomotion by means of flagella

Figure Credits

Chapter 1

Figs. 1-1, 1-7, 1-10, 1-18. Drawings by Nancy Haver.

Fig. 1-2. Courtesy of W. C. Sparks, University of Idaho, Aberdeen.

Figs. 1-3, 1-13, 1-17, and 1-19. Courtesy of H. D. Thurston, Cornell University, Ithaca, NY.

Fig. 1-4. Courtesy of the International Potato Center (CIP), Lima, Peru.

Fig. 1-5. Used, by permission, from the 1978 Annual Report of the International Potato Center (CIP), Lima, Peru. (Redrawn by Nancy Haver)

Fig. 1-6. Courtesy of H. Murphy, University of Maine, Orono.

Figs. 1-8 and 1-9. From: Woodham-Smith, C. 1962. The Great Hunger. Harper and Row, New York.

Figs. 1-11 and 1-12. Reprinted, with permission, from: Hooker, W. J., ed. 1981. Compendium of Potato Diseases. American Phytopathological Society, St. Paul, MN. Plate 29 and Fig. 45D and C, respectively.

Fig. 1-14. Used by permission of the Bettman Archive, New York.

Fig. 1-15. From: Woodham-Smith, C. 1962. The Great Hunger. Harper and Row, New York.

Fig. 1-16. Courtesy of The American Phytopathological Society.

Chapter 2

Figs. 2-1, 2-3 through 2-9. Drawings by Nancy Haver.

Fig. 2-2. Courtesy of H. D. Thurston, Cornell University, Ithaca, NY.

Fig. 2-10. Reprinted, with permission, from: Zentmyer, G. A. 1980. *Phytophthora cinnamomi* and the Diseases it Causes. American Phytopathological Society, St. Paul, MN. Fig. 27.

Fig. 2-11. Courtesy of T. A. Zitter, Cornell University, Ithaca, NY.

Fig. 2-12. Courtesy of P. B. Hamm, Oregon State University, Corvallis.

Fig. 2-13. Courtesy of C. C. Powell, APS Slide Set 22, slide 12 (1979).

Fig. 2-14. Courtesy of M. Daughtrey, Cornell University, Ithaca, NY.

Fig. 2-15. Courtesy of W. F. Wilcox, New York State Experiment Station, Geneva.

Fig. 2-16. Courtesy of G. A. Payne, North Carolina State University, Raleigh.

Chapter 3

Figs. 3-1 and 3-10. Taken by Gail Schumann.

Fig. 3-2. Reprinted, by permission, from: Schieber, E., and Zentmyer, G. A. 1984. Coffee rust in the Western Hemisphere. Plant Disease 68:89-93. Fig. 1.

Figs. 3-3, 3-5, and 3-6. Courtesy of H. D. Thurston, Cornell University, Ithaca, NY.

Fig. 3-4. Reprinted, by permission, from: Oliver, F. W. 1913. Makers of British Botany. Cambridge University Press, New York.

Fig. 3-7. Courtesy of Sandoz Ltd., Basle, Switzerland.

Figs. 3-8, 3-9, 3-12 (redrawn), 3-15, 3-19. Drawings by Nancy Haver.

Fig. 3-11. Reprinted, with permission, from: Thurston, H. D. 1984. Tropical Plant Diseases. American Phytopathological Society, St. Paul, MN.

Fig. 3-12. Courtesy of J. Kloppenberg, University of Wisconsin, Madison. Redrawn from the original.

Figs. 3-13 and 3-14. Used by permission of C. Krebs, Issaquah, WA.

Figs. 3-16 and 3-17. Courtesy of B. H. Waite, APS Slide Set 15, Slides 32 and 33 (1980).

Fig. 3-18a. Reprinted, by permission, from: Beckman, C. 1987. The Nature of Wilt Diseases of Plants. American Phytopathological Society, St. Paul, MN.

Figs. 18b and c. Reprinted, by permission, from: VanderMolen, G. E., Beckman, C. H., and Rodehorst, E. 1987. The ultrastructure of tylose formation in resistant banana following inoculation with *Fusarium oxysporum* f. sp. *cubense*. Physiological and Molecular Plant Pathology 31:185-200. Figs. 5 and 6.

Chapter 4

Fig. 4-1. Courtesy of B. H. Waite, APS Slide Set 15, slide 39 (1980).

Fig. 4-2. Adapted, by permission, from Agrios, G. N. 1988. Plant Pathology, 3rd ed. Academic Press, New York. Fig. 1-2.

Figs. 4-3, 4-6 through 4-10, 4-12. Drawings by Nancy Haver.

Fig. 4-4. Courtesy of S. V. Thomson, Utah State University, Logan.

Fig. 4-5. Courtesy of S. V. Beer, Cornell University, Ithaca, NY.

Fig. 4-11. Reprinted, by permission, from Wynn, W. K. 1976. Appressorium formation over stomates by the bean rust fungus: Response to a surface contact stimulus. Phytopathology 66:136-146. Figs. 9, 13, and 16.

Figs. 4-13 and 4-14 (left). Courtesy of A. L. Jones, Michigan State University, East Lansing.

Fig. 4-14 (right). Courtesy of the Plant Pathology Herbarium, Cornell University, Ithaca, NY.

Fig. 4-15. Courtesy of the National Archives, Washington, DC. Photo 83-FB-10068.

Fig. 4-16. Adapted, by permission, from Campbell, R. N. 1979. Fire blight. Natural History (May), p. 67.

Figs. 4-17 and 4-18. Courtesy of the Florida Department of Agriculture and Consumer Services, Gainsville.

Fig. 4-19. Taken by Gail Schumann.

Chapter 5

Figs. 5-1 through 5-5. Drawings by Nancy Haver.

Fig. 5-6. Used, with permission, from Priestley, R. H. 1978. Plant Disease Epidemiology. P. R. Scott, and A. Bainbridge, eds. Blackwell Scientific Publications, Oxford. Fig. 1. Adapted from the original.

Fig. 5-7. Courtesy of H. D. Thurston, Cornell University, Ithaca, NY.

Figs. 5-8, 5-14, and 5-17. Courtesy of Monsanto Company, St. Louis, MO.

Fig. 5-9. Courtesy of M. S. Mount, University of Massachusetts, Amherst.

Figs. 5-10 and 5-13. Used, by permission, from Genetic Engineering: A National Science. Monsanto Company, St, Louis, MO. Redrawn from the originals.

Fig. 5-11. Courtesy of K. Wood, Promega, Madison, WI.

Fig. 5-12. Courtesy of S. Lindow, University of California, Berkeley.

Fig. 5-15. Courtesy of University of California, Berkeley.

Fig. 5-16. Courtesy of Crop Genetics International, Hanover, MD.

Fig. 5-18. Courtesy of R. J. Cook, USDA-ARS, Washington State University.

Chapter 6

Fig. 6-1. Courtesy of R. Kirkby, Centro Internacional de Agricultura Tropical, Debre Zeit, Ethiopia.

Fig. 6-2. Reprinted from Spears, J. F. 1968. Golden Nematode Handbook. Agricultural Handbook 353. U.S. Department of Agriculture, Washington, DC. Fig. 8.

Fig. 6-3. Courtesy of J. Galindo, Centro Agronomico Tropical Investigacion y Ensenanza, Turrialba, Costa Rica.

Fig. 6-4. Drawing by Nancy Haver.

Fig. 6-5. Courtesy of B. M. Zuckerman, University of Massachusetts, Amherst.

Fig. 6-6. Courtesy of North Dakota State University, Fargo.

Fig. 6-7. Courtesy of K. F. Baker, Corvallis, OR. Redrawn from the original.

Fig. 6-8. Courtesy of University of Massachusetts, Amherst.

Fig. 6-9. Courtesy of W. Atlee Burpee & Co., Warminster, PA.

Fig. 6-10. Used (redrawn), by permission, from VanderPlank, J. E. 1984. Disease Resistance in Plants, 2nd ed. Academic Press, Orlando, FL. Figs. 13.2 and 13.7.

Fig. 6-11. Reproduced with the permission of the Minister of Supply and Services Canada, 1990, from James, C. 1971. A Manual of Assessment Keys for Plant Diseases. Agriculture Canada Publication 1458. Keys 3.1.1 and 1.6.1.

Fig. 6-12. Courtesy of L. Farrar, Auburn University, Auburn, AL.

Figs. 6-13, 6-14, and 6-16. Courtesy of W. F. Moore, Extension Plant Pathology, Mississippi State University, Mississippi State.

Fig. 6-15. Used, by permission, from Moore, W. F. 1970. Origin and spread of southern corn leaf blight in 1970. Plant Disease Reporter 54:1104. Figs. 1–6. Redrawn from the original.

Chapter 7

Figs. 7-1, 7-11, and 7-15. Drawings by Nancy Haver.

Fig. 7-2. Courtesy of C. W. Roane. APS Slide Set: Ascomycetes, slide 18 (1988).

Fig. 7-3. Courtesy of R. T. Hanlin. APS Slide Set: Ascomycetes, slide 20 (1988).

Fig. 7-4. From: F. J. Schneiderhan, trans. 1933. The Discovery of Bordeau Mixture. American Phytopathological Society, St. Paul, MN.

Fig. 7-5. Courtesy of Cornell University, Ithaca, NY.

Fig. 7-6. Courtesy of Monsanto Company, St. Louis, MO.

Fig. 7-7. Used, by permission, from Delp, C. J., ed. 1988. Fungicide Resistance in North America. American Phytopathological Society, St. Paul, MN. Fig. 2. Redrawn from the original.

Fig. 7-8. Drawn, by permission, from Gianessi, L. P. and Puffer, C. A. 1989. Use of Selected Pesticides in Agricultural Crop Production—National Summary. Quality of the Environment Division, Resources for the Future, Washington, DC.

Fig. 7-9. Courtesy of Cornell University Cooperative Extension, Ithaca, NY.

Fig. 7-10. Courtesy of G. Johnson and M. J. Jackson, Montana State University Cooperative Extension, Bozeman.

Fig. 7-12. Courtesy of W. J. Manning, University of Massachusetts, Amherst.

Fig. 7-13. Courtesy of G. N. Agrios, University of Florida, Gainesville.

Fig. 7-14. Courtesy of G. J. Weidemann. APS Slide Set: Ascomycetes, slide 15 (1988).

Fig. 7-16. Reprinted, by permission, from Smith, D. H. 1984. Compendium of Peanut Diseases. American Phytopathological Society, St. Paul, MN. 1984. Plates 1 and 3.

Fig. 7-17. Courtesy of Neogen Corporation, East Lansing, MI.

Chapter 8

Figs. 8-1 through 8-6, 8-8 through 8-10, 8-20. Drawings by Nancy Haver.

Fig. 8-7. Courtesy of D. N. Schumann, Kalamazoo, MI.

Figs. 8-11, 8-12, and 8-16. Courtesy of Cornell University, Ithaca, NY.

Fig. 8-13. Courtesy of R. A. Rohde, University of Massachusetts, Amherst.

Figs. 8-14 and 8-15. Courtesy of R. S. Hussey, University of Georgia, Athens. (Society of Nematologists slide set, slides 17 and 18)

Fig. 8-17. Courtesy of B. M. Zuckerman, University of Massachusetts, Amherst.

Fig. 8-18. Courtesy of G. L. Barron, University of Guelph, Guelph, Ontario. (Society of Nematologists slide set, slide 94)

Fig. 8-19. Courtesy of M. A. McClure, Egerton University, Njoro, Kenya. (Society of Nematologists slide set, slide 87)

Fig. 8-21. From: *The Community Journal*, Wading River, NY, January 22, 1986.

Fig. 8-22. Courtesy of New York's Food and Life Sciences Quarterly 18 (1–2), 1988, and J.-Y. Parlange, Cornell University, Ithaca, NY.

Fig. 8-23. Courtesy of ENSR Corporation, Acton, MA.

Chapter 9

Fig. 9-1. Courtesy of Staatliche Graphische Sammlung München, Munich, Fed. Republic of Germany. Redrawn from the original.

Fig. 9-2. Courtesy of the Department of Plant Pathology, North Dakota State University, Fargo.

Figs. 9-3, 9-12, 9-16, 9-18, 9-20. Drawings by Nancy Haver.

Fig. 9-4. Courtesy of Malcolm Shurtleff, University of Illinois, Urbana.

Fig. 9-5. Courtesy of the Department of Veterinary Science, North Dakota State University, Fargo.

Figs. 9-6 and 9-7. Courtesy of C. R. Funk, Rutgers University, New Brunswick, NJ.

Fig. 9-8. Courtesy of M. F. Brown and H. G. Brotzman, University of Missouri, Columbia.

Fig. 9-9. Courtesy of J. Stack, Ecoscience Laboratories, Amherst, MA.

Fig. 9-10a. Reprinted, by permission, from: Meronuck, R. A. 1987. The significance of fungi in cereal grains. Plant Disease 71:187-291. Fig. 1B.

Fig. 9-10 (top). Reprinted, by permission, from: Meronuck, R. A. 1987. The significance of fungi in cereal grains. Plant Disease 71:187-291. Fig. 1B.

Fig. 9-10 (bottom). Courtesy of R. A. Meronuck, University of Minnesota, St. Paul.

Fig. 9-13. Courtesy of E. M. Dutky, University of Maryland, College Park.

Fig. 9-14. Reprinted, by permission, from: J. L. Maas, ed. 1984. Compendium of Strawberry Diseases. The American Phytopathological Society, St. Paul, MN. Plate 77.

Fig. 9-15. Courtesy of S. Colleen O'Keefe-Safir, Chateau Grand Traverse, Ltd., Traverse City. MI.

Fig. 9-17. Reprinted, by permission, from: M. C. Shurtleff, ed. 1980. Compendium of Corn Diseases, 2nd ed. The American Phytopathological Society, St. Paul, MN. Fig. 40.

Fig. 9-19. Courtesy of R. Duran, Washington State University, Pullman, and R. Cruz, Broomfield, CO.

Chapter 10

Fig. 10-1. Reprinted, by permission, from: Zadoks, J. 1982. Cereal rusts, dogs, and stars in antiquity. Garcia de Orta, Serie de Estudos Agronomicos, Lisboa 9(1-2):13-20.

Fig. 10-2. Courtesy of the Food and Agriculture Organization of the United Nations (FAO), Rome.

Fig. 10-3. Courtesy of the U.S. Department of Agriculture Soil Conservation Service.

Fig. 10-4. Reprinted, by permission, from Rust Scoring Guide, Research Institute for Plant Protection, Wageningen, The Netherlands.

Fig. 10-5. Courtesy of M. F. Brown and H. G. Brotzman, University of Missouri, Columbia.

Fig. 10-6. Courtesy W. Q. Loegering, U.S. Department of Agriculture.

Fig. 10-7a. Courtesy of G. D. Statler, North Dakota State University, Fargo.

Fig. 10-7b. Courtesy D. L. Long, U.S. Department of Agriculture.

Figs. 10-8, 10-10, 10-15. Drawings by Nancy Haver.

Figs. 10-9 and 10-12. Courtesy of Cornell University, Ithaca, NY.

Fig. 10-11. Courtesy of J. A. Browning, APS Slide Set 1, slide 8 (1958).

Fig. 10-13A. Courtesy of E. Herrling, University of Wisconsin, Madison.

Fig. 10-13B. Reprinted, by permission, from Brown, M. F., and Brotzman, H. G. 1979. Phytopathogenic Fungi—A Scanning Electron Steroscopic Survey. University of Missouri Press. Fig. 1, p. 315.

Fig. 10-14. Courtesy J. C. Carter, APS Slide Set 13, slide 15 (1963).

Fig. 10-16. Courtesy of E. G. Kuhlman, U.S. Department of Agriculture—Forest Service.

Fig. 10-17. Courtesy of Cooperative Extension, University of Massachusetts, Amherst.

Fig. 10-18. Courtesy of H. R. Powers, Forestry Sciences Lab, Athens, GA.

Chapter 11

Figs. 11-1, 11-3, 11-12, 11-18, 11-19, 11-21, 11-22, 11-25. Drawings by Nancy Haver.

Figs. 11-2, 11-4, 11-16, 11-17, 11-23. Taken by Gail Schumann.

Figs. 11-5, 11-6 (left), and 11-7. Courtesy of Shade Tree Laboratories, University of Massachusetts, Amherst.

Fig. 11-6 (right). Courtesy of L. F. Grand, North Carolina State University, Raleigh.

Fig. 11-8. Courtesy of D. W. French, University of Minnesota, St. Paul.

Fig. 11-9 (left). Courtesy of the USDA Forest Service and the American Chestnut Foundation.

Fig. 11-9 (right). Reprinted from "Scenes in Fairmount Park," The Art Journal, 1878, Appleton and Co.; received courtesy of the American Chestnut Foundation.

Fig. 11-10 (top). Courtesy of the American Chestnut Foundation.

Figs. 11-10 (bottom) and 11-11. Courtesy of the Pennsylvania Blight Commission and the American Chestnut Foundation.

Figs. 11-13 and 11-14. Courtesy of D. W. Fulbright, Michigan State University, East Lansing.

Fig. 11-15a. Courtesy of P. M. Wargo, USDA Forest Service, Hamden CT.

Fig. 11-15b. Courtesy of D. R. Houston, USDA Forest Service, Hamden CT.

Fig. 11-15c. Courtesy of Cornell University, Ithaca, NY.

Fig. 11-20. Courtesy of G. Hudler, Cornell University, Ithaca, NY.

Fig. 11-24. Courtesy of University of Massachusetts, Amherst.

Fig. 11-26. Courtesy of U.S. Department of Agriculture, Animal and Plant Health Inspection Service, Science and Technology.

Fig. 11-27. Adapted, by permission, from Eplee, R. E. 1981. Striga's status as a plant parasite in the United States. Plant Disease 65:951-954. Fig. 2.

Chapter 12

Fig. 12-1. Used by permission of The Metropolitan Museum of Art, Purchase, 1968, Rogers Fund. (68.66).

Fig. 12-2. Courtesy of Cornell University, Ithaca, NY.

Fig. 12-3. Courtesy of L. Nault, APS Slide Set 21, slides 20 and 39 (redrawn) (1978).

Fig. 12-4. Courtesy of H. W. Israel, Cornell University, Ithaca, NY.

Fig. 12-5 and 12-6. Adapted, by permission, from Agrios, G. N. 1988. Plant Pathology, 3rd ed. Academic Press, New York. Figs. 14-3 and 14-8.

Fig. 12-7. Taken by Gail Schumann.

Fig. 12-8. Courtesy of H. D. Thurston, Cornell University, Ithaca, NY.

Fig. 12-9. Redrawn, by permission, from Clark, M. F., and Adams, A. N. 1977. Characteristics of the microplate method of enzyme-linked immunosorbent assay for the detection of plant viruses. Journal of General Virology 34:475-483.

Fig. 12-10. Courtesy of Agridiagnostics Associates, Cinnamonson, NJ.

Fig. 12-11. Redrawn, by permission, from a drawing by Bunji Tagawa, in Milstein, C. 1980. Monoclonal antibodies. Scientific American. Oct. p. 68.

Fig. 12-12 (top). Drawing by Nancy Haver.

Figs. 12-12 (bottom), 12-13, and 12-14. Courtesy of T. M. Nourse, Nourse Farms Inc., Deerfield, MA.

Fig. 12-15. Reprinted, by permission, from Potatoes for the Developing World. International Potato Center, Lima, Peru. 1984. p. 116.

Fig. 12-16. Courtesy of K. Maramorosch, Rutgers University, New Brunswick.

Fig. 12-17. Courtesy of T. O. Diener, University of Maryland, College Park.

Figs. 12-18 and 12-19. Courtesy of R. McCoy, Champlain Island Agro-Associates. Isle La Motte, VT.

Fig. 12-20. Courtesy of R. E. Davis, USDA-ARS, Beltsville, MD.

Chapter 13

Figs. 13-1, 13-7, 13-9, 13-15. Drawings by Nancy Haver.

Fig. 13-2. Courtesy of Pennsylvania State Extension Service, APS Slide Set 13, slide 1 (1963).

Fig. 13-3. Courtesy of J. M. F. Yuen, University of Florida, Opopka.

Fig. 13-4. Courtesy of E. L. Knake, University of Illinois, Urbana.

Fig. 13-5. Courtesy of A. V. Barker, University of Massachusetts, Amherst.

Fig. 13-6. Courtesy of A. Sherf, APS Slide Set 7, slide 39 (1958).

Fig. 13-8. Courtesy of J. C. Wynne, North Carolina State University, Raleigh.

Fig. 13-10. Courtesy of R. B. Marlatt, APS Slide set 22, slides 43 and 44 (1979).

Fig. 13-11. Courtesy of P. M. Wargo, U.S. Department of Agriculture—Forest Service, Hamden CT.

Fig. 13-12. Courtesy of W. J. Manning, University of Massachuesetts, Amherst.

Fig. 13-13. Source: U.S. Environmental Protection Agency.

Fig. 13-14. Courtesy of R. Kostka-Rick, University of Massachusetts, Amherst.

Fig. 13-16. Courtesy of G. C. Smith, University of Massachusetts, Amherst.

Chapter 14

Fig. 14-1. Redrawn, by permission, from The Ending Hunger Briefing Workbook. 1984. The Hunger Project, San Francisco, CA.

Fig. 14-2. Courtesy of University of Minnesota, St. Paul, and N. Borlaug, CIMMYT, Mexico.

Figs. 14-3, 14-8, 14-9, 14-16, and 14-20. Courtesy of H. D. Thurston, Cornell University, Ithaca, NY.

Figs. 14-4, 14-7, and 14-13. Source: U.S. Department of Agriculture.

Fig. 14-5. Used by permission of T. Moore, Tucson, AZ.

Figs. 14-6 and 14-10. Reprinted from the IBPGR Annual Report of 1984; used by permission of the International Board for Plant Genetic Resources, Rome, Italy.

Fig. 14-11. Taken by Gail Schumann.

Fig. 14-12. Courtesy of J. Drozdowski, Athol, MA.

Fig. 14-14. Drawing by Nancy Haver.

Fig. 14-15. Courtesy of N. Marban-Mendoza, Centro Agronomico Tropical de Investigacion y Ensenanza (CATIE), Turrialba, Costa Rica.

Figs. 14-17 and 14-18. Source: Foreign Disease–Weed Science Research Unit, USDA-ARS, Fort Detrick, MD.

Fig. 14-19. Courtesy of G. Odvody, Texas Agricultural Experiment Station, Corpus Christi.

Fig. 14-21. Source: U.S. Congress, Office of Technology Assessment. 1987. Technologies to Maintain Biological Diversity. OTA-F-330. U.S. Government Printing Office, Washington, DC. Redrawn from the original.

Fig. 14-22. Source: IUCN Bulletin Supplement (International Union for the Conservation of Nature and Natural Resources, Gland, Switzerland; 1985) and U.S. Office of Technology Assessment, 1987. Redrawn from the original.

Fig. 14-23. Reprinted, by permission, from Potatoes for the Developing World. 1984. International Potato Center, Lima, Peru, p. 56.

Index